AF576082

Construction and Demolition Waste: Challenges and Opportunities

Construction and Demolition Waste: Challenges and Opportunities

Editors

Anibal C. Maury-Ramirez
Jaime A Mesa

MDPI • Basel • Beijing • Wuhan • Barcelona • Belgrade • Manchester • Tokyo • Cluj • Tianjin

Editors

Anibal C. Maury-Ramirez
Engineering Faculty
Universidad El Bosque
Bogota
Colombia

Jaime A Mesa
Mechanical Engineering
Department
Universidad del Norte
Barranquilla
Colombia

Editorial Office
MDPI
St. Alban-Anlage 66
4052 Basel, Switzerland

This is a reprint of articles from the Special Issue published online in the open access journal *Sustainability* (ISSN 2071-1050) (available at: www.mdpi.com/journal/sustainability/special_issues/DemolitionWaste).

For citation purposes, cite each article independently as indicated on the article page online and as indicated below:

LastName, A.A.; LastName, B.B.; LastName, C.C. Article Title. *Journal Name* **Year**, *Volume Number*, Page Range.

ISBN 978-3-0365-5050-3 (Hbk)
ISBN 978-3-0365-5049-7 (PDF)

Contents

About the Editors

Anibal C. Maury-Ramirez

Aníbal Maury-Ramírez (Prof.dr.ir) is a part-time professor of a Master's course in Sustainable Businesses Management and a researcher at the Clean Production Research Group (Choc Izone) of the Engineering Faculty of Universidad El Bosque (Colombia). Additionally, he is the current President of the Alumni Chapter of Ghent University in Colombia and a visiting researcher at Tampere University of Technology, now Tampere University (Finland), and Ghent University (Belgium). He is a Civil Engineer (Universidad del Norte (Colombia)) and Doctor of Civil Engineering (Ghent University (Belgium)). He earned his post-doctorate at the Hong Kong Polytechnique University (China). Dr. Maury Ramírez is an associate researcher (Minciencias), author of more than 20 high-impact articles, and 50 articles in the proceedings of international conferences as well as several books. His main scientific contributions lie in the development of sustainable building materials, particularly green walls and roofs, photocatalytic materials, nanomaterials, and sustainable concrete (i.e., supplementary cementitious materials, recycled aggregates, and permeable concrete). Recently, he has also been involved in the design of sustainable buildings and communities, particularly formulating circular economy models and studying urban metabolism. Among others, he was the Director of Post-Graduate Studies on Civil Engineering at Pontificia Universidad Javeriana Cali (Colombia) and the Dean of the Engineering Faculties from Universidad de La Salle and Universidad El Bosque (Colombia).

Jaime A Mesa

Jaime A. Mesa (Dr.) works as a full-time professor in the Department of Mechanical Engineering at Universidad del Norte (Colombia). Dr. Mesa has degrees in Mechanical Engineering from Universidad Pontificia Bolivariana (Colombia) and an M.Sc. and Ph.D. (Summa Cum Laude) in Mechanical Engineering from Universidad del Norte (Colombia). Dr. Mesa is a junior researcher (Minciencias) and an author of several articles in top peer-reviewed journals and proceedings of international conferences. His research interests are product design, modular design, design thinking, eco-design, lifecycle analysis, sustainability, circular economy, and engineering education around sustainability topics. He has been employed as a professor at several Colombian universities, including Pontificia Universidad Javeriana Bogota, Universidad Pontificia Bolivariana, Universidad Tecnológica de Bolívar.

Preface to "Construction and Demolition Waste: Challenges and Opportunities"

Due to the significant construction material demands for the rising number of buildings and urban infrastructure, the use of construction and demolition waste (C&DW) in building materials becomes a crucial issue for the sustainable development of our planet, especially in developing countries, where construction industry is growing very rapidly. The extraction of natural aggregates and minerals from mountains and rivers has destroyed vital ecosystems, which had also generated social conflicts for accessing to clean water, food and others natural resources. Moreover, the manufacture of building materials releases a significant amount of greenhouse emissions, particulate matter and solid waste. For example, only for producing 1 ton of Portland cement, approx. 1 ton of CO_2 is released to the atmosphere. Based exclusively on Portland cement, with around 9% of carbon dioxide emissions, the construction industry is currently an important producer of global warming gases. Similarly, causing a negative impact on landscape, environment and human health, the construction and demolition processes are responsible for 30 to 40% of the different world economic sector's solid waste.

Although many efforts have been done made by civil society, governments, industries, and academia to valorize C&DW in building materials, highly articulated efforts are required at global, national, and local scales to develop truly sustainable construction sectors that facilitate the massive use of sustainable building materials. Therefore, this e-book is based on the "Special Issue on Construction and Demolition Waste: Challenges and Opportunities", which presents at first, several novel and innovative building materials using C&DW and other residues. Second, management systems to develop circular economy models (CE) in the construction sector, particularly referred to applied cases in France, Western Balkans and Colombia. Finally, the e-book includes two comprehensive reviews on conceptual challenges of C&DW valorization.

First, regarding the novel and innovative building materials using C&DW and other residues, the following articles are included: (a) Mechanical Properties of Concrete Using Recycled Aggregates Obtained from Old Paving Stones, (b) Water-Washed Fine and Coarse Recycled Aggregates for Real Scale Concretes Production in Barcelona, (c) Properties of Concrete with Recycled Aggregates Giving a Second Life to Municipal Solid Waste Incineration Bottom Ash Concrete, (d) Construction and Demolition Waste (CDW) Recycling –As Both Binder and Aggregates –In Alkali-Activated Materials: A Novel Re-Use Concept, and (e) Ecotoxicity of Concrete Containing Fine-Recycled Aggregate: Effect on Photosynthetic Pigments, Soil Enzymatic Activity and Carbonation Process.

Second, regarding the management systems towards CE, the e-book includes the following articles: (a) Circular Economy in the Construction Sector: A Case Study of Santiago de Cali (Colombia), (b) Challenges and Opportunities for Circular Economy Promotion in the Building Sector, and (c) Circular Economy in Construction and Demolition Waste Management in the Western Balkans: A Sustainability Assessment Framework.

Finally, the e-book contains two comprehensive reviews on conceptual challenges: (a) Life Cycle Assessment on Construction and Demolition Waste: A Systematic Literature Review and (b) Some Remarks towards a Better Understanding of the Use of Concrete Recycled Aggregate: A Review.

From this e-book, it can be concluded that even though a wide variety of novel and innovative building materials using C&DW has been developed worldwide, more incentives are required (e.g. through public policies) to really convert the local and national construction sectors in sustainable businesses which appropriate the circular economy as production and consumption systems that promote, at least, the efficiency in the use of materials, water and energy. This has not only the potential to develop new sustainable business models based on research, this also might transform existing companies into more sustainable businesses, which results very important for the current economy post-pandemic scenario.

Anibal C. Maury-Ramirez and Jaime A Mesa
Editors

MDPI

Article

Mechanical Properties of Concrete Using Recycled Aggregates Obtained from Old Paving Stones

Ana María Bravo-German [1], Iván Daniel Bravo-Gómez [1], Jaime A. Mesa [2] and Aníbal Maury-Ramírez [3,*]

[1] Department of Civil and Industrial Engineering, Engineering and Sciences Faculty, Pontificia Universidad Javeriana Cali, Cali 760031, Colombia; anabravog25@hotmail.com (A.M.B.-G.); idbravo_2@hotmail.com (I.D.B.-G.)

[2] Department of Industrial Engineering, Engineering Faculty, Pontificia Universidad Javeriana, Bogotá 110231, Colombia; mesajaime@javeriana.edu.co

[3] Engineering Faculty, Universidad El Bosque, Bogotá 111711, Colombia

* Correspondence: amaury@unbosque.edu.co; Tel.: +57-1-6489-000 (ext. 1285)

Abstract: Nowadays, construction, maintenance, reparation, rehabilitation, retrofitting, and demolition from infrastructure and buildings generate large amounts of urban waste, which usually are inadequately disposed due to high costs and technical limitations. On the other hand, the increasing demand for natural aggregates for concrete production seriously affects mountains and rivers as they are the source of these nonrenewable goods. Consequently, the recycling of aggregates for concrete is gaining attention worldwide as an alternative to reduce the environmental impacts caused by the extraction of nonrenewable goods and disposal of construction and demolition waste (C&DW). Therefore, this article describes the effect on the mechanical properties of new concrete using recycled aggregates obtained from old paving stones. Results show that replacing 50% by weight of the fine and coarse aggregate fractions in concrete with recycled aggregate does not meaningfully affect its mechanical behavior, making the use of recycled aggregates in new precast paving stones possible. Therefore, the latter can reduce environmental impacts and costs for developing infrastructure and building projects.

Keywords: paving stones; aggregates; C&DW; sustainability; mechanical properties; concrete

Citation: Bravo-German, A.M.; Bravo-Gómez, I.D.; Mesa, J.A.; Maury-Ramírez, A. Mechanical Properties of Concrete Using Recycled Aggregates Obtained from Old Paving Stones. *Sustainability* **2021**, *13*, 3044. https://doi.org/10.3390/su13063044

Academic Editor: José Alvarez

Received: 16 February 2021
Accepted: 5 March 2021
Published: 10 March 2021

Publisher's Note: MDPI stays neutral with regard to jurisdictional claims in published maps and institutional affiliations.

1. Introduction

The consumer society, together with the technological revolution, has led to the most massive production of waste in humanity's entire history. This problem has led to most countries seeking solutions to decrease pollution rates on the planet [1]. According to John [2], the construction industry produces 40% of the different world economic sectors' waste. On the other hand, with the pivotal role that the construction industry has in developing countries, it is convenient to adopt urgent measures to achieve sustainable development [3].

The construction industry is still one of the largest waste generators today [4–6]. However, historically it has been a necessary pillar for the development of our communities. In general, the construction sector's pollution occurs in most stages: from the extraction of raw materials, the manufacture of materials, to the different activities carried out during the construction, operation, and end of the life cycle of buildings and infrastructure. This causes the depletion of various nonrenewable resources, as well as water and air pollution, in addition to excessive energy consumption [7].

1.1. Global Context

In other industrial sectors, recycled materials are typically competitive when there is difficulty in obtaining virgin raw materials and suitable locations for storage. Therefore, the proposal for sustainable concrete is based on the substitution of stone aggregates of natural origin for construction and demolition waste (C&DW), and it proves to be very interesting for infrastructure and buildings, enabling the implementation of circular economic models [8–10]. The government requirement to mitigate the environmental impact of the different construction projects through the proper management of the C&DW potentiates their reincorporation into the construction production chain through recycling. However, to use this waste in new projects, it is necessary to evaluate the physical, chemical, mechanical, and durability characteristics of the C&DW [11–13]. The use of C&DW as aggregates of concrete mixtures without knowing their properties can generate projects with concrete properties in a fresh and hardened state that are undesirable and unsafe for its users.

In the construction sector, a large amount of waste of different types is generated, but only a part of it can be reincorporated in the same sector, either by reusing or recycling it. C&DW must be inert and uncontaminated. Although there are significant fluctuations in Colombia, it can be estimated that the usable waste is 80% of the CDW, which is made up of materials such as bricks or blocks, concrete, rock, excavation material, steel, wood, and others [14]. The remaining 20% that is not usable in the construction sector (wood, plastics, packaging, and inert materials with organic matter) should be sent to specific recycling plants for industrial symbiosis or disposed in landfills [15].

To recycle concrete waste into new building products, intensive research worldwide, demonstrating that in general, the use of recycled aggregates from concrete in new fabrication of concrete has resulted in enhanced improvement of mechanical properties [16–19]. Perez-Benedicto et al. [20] studied the mechanical behavior of concrete made with recycled aggregate coming from discarded concrete prefabricated units, demonstrating that including coarse aggregate replacement can offer excellent quality for structural applications providing similar compressive strength respect conventional concrete. Cakir and Dilbas [21] demonstrated that durability of concrete can be increased up to 60% using recycled aggregate and an optimized ball mill method. The study included comparison of several aggregates such as natural aggregate, recycled aggregate, silica fume, basalt fiber and recycled aggregate and optimized ball mill method.

Vedrtnam et al. [22]. studied the response of cement-based composites under direct flame conditions, demonstrating improvement in the residual compressive strength compared to conventional concrete after the thermal exposure. In addition, it was found minimum damage in the microstructure of the material when residues of PET bottles are included in the mixture. Zareei et al. [23]. analyzed the combination of recycled waste ceramic aggregates and waste carpet fibers to produce high strength concrete. They used combinations from 20% up to 60% of recycled waste ceramic to replace natural coarse aggregate. As result it was observed an increase of compressive, splitting tensile, flexural and tensile strengths by 13%, 15%, 3% and 21% respectively. Zahid-Hossain et al. [24]. also studied the inclusion of recycled material as rubber and polypropylene fibers into concrete mixtures to evaluate their mechanical behavior. As result it was demonstrated that compressive strength, splitting tensile strength and flexural strength decreases as the crumb rubber content increases, and increase with the fiber content. In addition, it was observed a reduction in the propagation speed of failures in the material, providing a more gradual failure propagation.

On the other hand, several studies have demonstrated decrease in mechanical properties when replacing virgin aggregate with recycled aggregate. Alam et al. [25] found that replacements of 25% using recycled aggregate can reduce compressive strength of concrete in 15%. Similarly, Meherier [26] studied the replacement of aggregates using crumb rubber and obtained reductions of 20% in concrete compressive strength. Limbachiya et al. [27] found that replacements with recycled concrete higher than 30% affect drastically compressive strength. However, all previous studies did not follow a standard procedure and therefore results can vary from region to region and according to environmental and methodological tasks.

1.2. Colombian Context

In Colombia, large amounts of construction and demolition waste produced by the construction industry are inadequately disposed (Figure 1a,b). Similarly, large amounts of nonrenewable resources are converted illegally, by industrial processing, into construction materials (Figure 2a,b).

Figure 1. Inadequate disposal of construction and demolition waste (C&DW) in Colombia. (**a**) "Estación de Transferencia (EDT) de la Carrera 50" located in Cali [28]; (**b**) Cauca River [29].

Figure 2. Illegal extraction of raw materials for construction in Colombia. (**a**) Rock extraction from Pance River [30]; (**b**) sand extraction from Cauca River [31].

Based on those above-mentioned environmental and social problems, a circular economy strategy from the national government has been launched. Particularly for C&DW, more than 22 million tons generated yearly in major cities should be waste managed [32]. On the other hand, the most widely used materials in the construction industry have historically been: aggregates, wood, concrete, steel, and glass. Except for aggregates and wood, the rest are composite materials that are made from nonrenewable raw materials. They are also the predominant materials in the last hundred years in large, intermediate cities and, unfortunately, even in the most remote rural areas [33].

So far, the construction of infrastructure and buildings has become one of the country's main economic activities. Despite the social benefits that the previous process has brought, cities and rural environments face significant environmental challenges due to the demand for nonrenewable resources generated by urban centers. For example, the problem of consuming nonrenewable natural resources for concrete production in Colombia is growing, with increases in concrete production of around 6% per year [34]. This generates critical environmental problems derived from the manufacture of concrete, such as the requirement of large quantities of stone material used in the natural stage as fine and coarse aggregates, and for producing Portland cement [35]. Commonly sand from the riverbed is used as a fine aggregate, while extraction and crushing of rock from quarries, usually rocky mountains, is carried out for the coarse aggregate. This situation poses significant technological challenges such as reduction, reuse, and recycling applied to all the country's productive sectors for the genuinely sustainable development of Colombian cities and municipalities.

In the case of Colombia for example, Diosa [36] performed an experimental analysis of mechanical properties of high-strength concrete from recycled sources, obtaining significant improvements in terms of mechanical resistance and durability. Hurtado [37] studied the effect of partial replacement of Portland cement by ash from the paper industry in the manufacture of mortar samples; results demonstrated that it is possible to obtain better mortar workability but less mechanical resistance. Moreover, Londoño [38] developed an analysis to determine the technical and financial suitability of using in situ recycled aggregates to fabricate prefabricated elements for construction.

In summary, the high demand for concrete, the natural resources for its production, and the generation of construction and demolition waste motivate the development of studies in which some waste is used and incorporated into the production of materials such as concrete [39,40]. Therefore, in this article, the mechanical performance of new concrete paving stones made with recycled aggregates from old paving stones in the municipality of Almaguer (Cauca, Colombia) were evaluated to be used for the same application in the municipality. The novelty of this research is focused in the use of aggregate to recirculate material within the same application and location, without incur in additional costs and using conventional crushing and laboratory equipment. The proposed method provides useful insights to recycle C&DW in the Colombian context and similar countries to fabricate added value products and lower environmental impact. The subsequent sections of this paper are organized as follows: Section 2 summarizes the materials and methods employed to conduct the experimentation and analyze the mechanical performance of concrete mixtures. Results are presented in Section 3, while Sections 4 and 5 correspond to discussion and conclusions.

2. Materials and Methods

This section summarizes the materials used to manufacture the paving stones and the methodology to analyze and compare mechanical properties. Materials mainly consist of used pavers, cement, coarse and fine aggregates (from natural sources and recycled). The proposed methodology started performing an aggregate sampling to classify materials, followed by a characterization of components, mixture design, evaluation of mechanical properties, and selecting the most suitable mixture concentrations to manufacture new paving stones.

2.1. Materials

Table 1 shows the materials employed to manufacture and analyze concrete samples. It included used paving stones converted into recycled aggregates, natural aggregates, and Portland cement. Such materials are processed and converted into new concrete samples, which are later analyzed and tested to measure and compare conventional values of mechanical properties (flexural and compressive strength).

2.2. Method

The proposed methodology for this study consisted of five phases, which are oriented to generate and establish suitable concrete mixtures. Figure 3 shows the overall methodology followed in this article.

Table 1. Description of components involved in the manufacturing of paving stones.

Component	Sample Picture	Description
Used Pavers: Initial form—before grinding and transformation into recycled aggregate		Used pavers present cracks, voids, flaking, damage of edges, and require replacement. The study employed 200 used pavers for analyzing and generate new ones (converting used pavers into recycled aggregates) and compare pavers manufactured using natural aggregates.
Aggregates: From natural source (quarry) and used paving stones (recycled)		Detailed characterization can be found in further sections.
Portland Cement: Obtained from a local supplier		Ordinary Portland cement (OPC) Density of 3.1 g/cm^3 Consistency of 26.4% (Portland type I) Detailed characterization can be found in further sections.

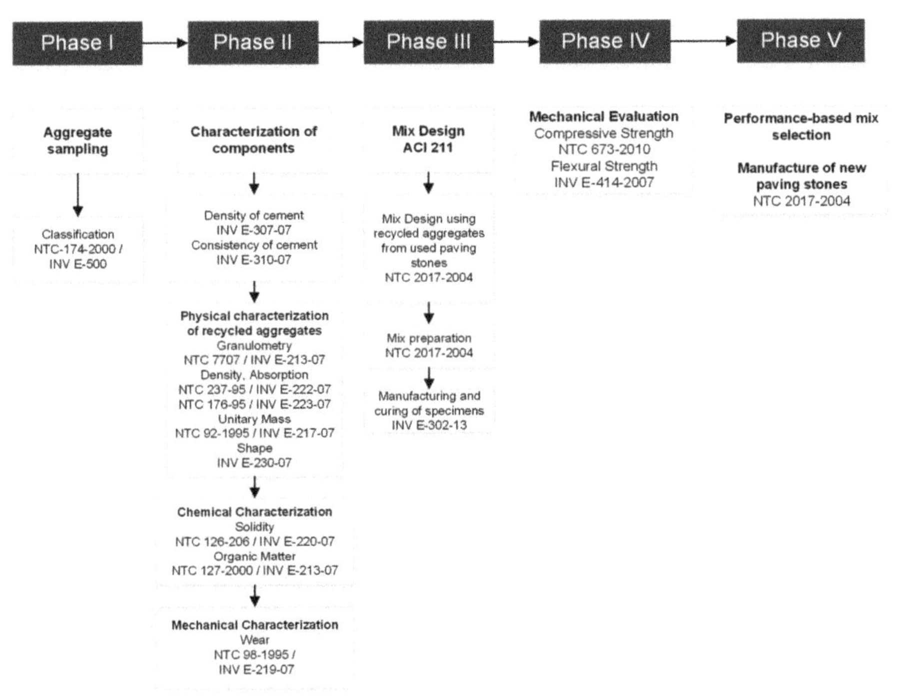

Figure 3. Methodology followed during this research project.

As a first stage, the process started with collecting used paving stones, which corresponded later to recycled aggregates (coarse and fine). Such components were triturated until obtaining the desired granulometry according to the NTC 174 standard [41]. In a parallel process, natural aggregates were also obtained from a quarry to compare and create different concrete mixtures, varying the percentage of recycled material. The second stage was denominated characterization of components, and it aimed to determine the most relevant properties of all materials used during the experimentation process; such materials comprise cement and aggregates. Here, physical characterization was performed analyzing granulometry, specific weight and absorption, the unitary mass of CDW and natural materials, shape, elongation, compressive indexes, and density and consistency of Portland cement. Besides, a chemical characterization that included sulfide resistance and organic material analysis. Later, mechanical characterization covered the response under wear conditions. The third state covers the mixture design, the manufacturing of concrete specimens for mechanical tests after curing processes. Evaluation of mechanical properties was performed as the fourth stage, and it included experimental tests for compression and flexural strength of concrete specimens. Lastly, step five concludes the proposed methodology by manufacturing the paving stones using the most suitable mixture obtained from tests and analysis of previous phases. Table 2 summarizes the overall composition of the nine concrete samples selected for analyzing and performing mechanical tests.

Table 2. Composition of concrete samples for evaluating mechanical properties of concrete.

Mixture	Type of Aggregate		Number of Samples (28 Days of Age)	
	Coarse	**Fine**	**Flexural**	**Compressive**
M1	100% NAT + 0% RC	100% NAT + 0% RF	3	3
M2	100% NAT + 0% RC	50% NAT + 50% RF	3	3
M3	100% NAT + 0% RC	0% NAT + 100% RF	3	3
M4	50% RC + 50% NAT	100% NAT + 0% RF	3	3
M5	50% RC + 50% NAT	50% RF + 50% NAT	3	3
M6	50% RC + 50% NAT	0% NAT + 100% RF	3	3
M7	0% NAT + 100% RC	100% NAT + 0% RF	3	3
M8	0% NAT + 100% RC	50% RF + 50% NAT	3	3
M9	0% NAT + 100% RC	0% NAT + 100% RF	3	3
Total samples			27	27

M: mix, NAT: natural aggregate, RC: coarse recycled aggregate, RF: fine recycled aggregate.

To develop the five phases previously shown in Figure 3 it was necessary to perform several task and analysis activities to obtain a proper mixture of concrete including recycled aggregates. Such phases were performed using as reference the Colombian normative (NTC standards), and national road standards (INVIAS) which are adaptations of American Society for Testing and Materials (ASTM) standards (Appendix A). A brief description of each phase is presented as follows:

Phase 1 consisted of obtaining the samples of recycled and natural aggregates and classifying them according to the standard NTC-174 /INV E-500. Later, phase II included physical, chemical and mechanical characterization of aggregates and cement. Density and consistency of cement were determined using the INV-E-307/310 standards; physical characterization of aggregates comprised granulometry (NTC 7707, INV-E 213), density (NTC 237, INV E-222), water absorption (NTC 176, INV E-223), unitary mass and shape (NTC 92, INV E-217 and INV E-230); chemical characterization covered solid analysis (NTC 126, INV E-220) and organic matter (NTC 127, INV E-213); and, wear behavior (NTC 98, INV E-219) as mechanical characterization.

Phase 3 included mixture design, which is performed using the American Concrete Institute ACI 211 method (also equivalent to NTC 2017 standard). This method consists of selecting amounts of cement, aggregates, water and additives to produce cost-effective concretes able to obtain a desired mechanical strength, durability, stability, unitary mass, and appearance. In this case nine mixtures were proposed varying the percentage of natural and recycled aggregates. Phase 3 also comprised the curing process of concrete specimens following the standard INV E-402. Then, when specimens were fabricated, it is necessary to perform mechanical tests to validate and compare flexural and compressive strength (NTC 673 / INV E-414) respect to conventional concrete mixtures in Phase 4. Lastly, Phase 5 consisted of selecting the best performance mixtures in terms of mechanical properties to fabricate new paving stones. Such fabrication was developed using the standard NTC 2017.

3. Results

This section comprises the results of five phases previously mentioned in the methodology. Section 4 is later presented to elucidate findings and interesting topics identified from the obtained results.

3.1. Aggregate Sampling

Natural aggregate samples were obtained from the company Canteras de Ingeocc S.A. (Yumbo, Colombia). On the other hand, used paving stones for recycled aggregates were obtained from the Almaguer's municipality (Cauca, Colombia). One hundred used paving stones were processed three times using a jaw crusher until acceptable sizes were obtained for an aggregate. The natural and recycled aggregates were classified into coarse and fine according to standard NTC 174 related to prefabricated paving stones. Approximately 530 kg of both natural and recycled aggregates were used during this study.

3.2. Characterization of Components

Physical and chemical attributes of interest were analyzed to determine the suitability of components to manufacture new concrete mixes. This subsection includes physical, chemical, and mechanical characterization. Physical characterization included several measurements and tests in determining key parameters in aggregates and cement employed: granulometric analysis, specific weight, water absorption, unitary mass, flattening, and elongation indexes for aggregates. In the case of cement, we determined density and consistency. Secondly, chemical characterization was dedicated to analyzing the chemical response of aggregates to sulfides (solidness) and analyzing the organic matter. Lastly, mechanical characterization was the first stage and consisted of a wear analysis for natural and recycled aggregates. Each characterization result is described in detail, as follows in Table 3. Additional properties of materials are summarized in Appendix B.

Table 3. Physical, chemical, and mechanical characterization of aggregates and cement.

Type of Characterization	Parameter	Component	Results
Physical	Granulometric analysis	Aggregate	Natural: Maximum size of 12.50 mm (coarse) and 9.50 mm (fine). Fineness module of 2.96 mm (coarse) and 2.97 mm (fine). Recycled: Maximum size of 12.50 mm (coarse) and 4.76 mm (fine). Fineness module of 5.44 mm (coarse) and 3.06 mm (fine).
	Specific weight	Aggregate	Natural: 2.8 g/cm^3 for fine and 2.93 g/cm^3 for coarse Recycled: 1.98 g/cm^3 for fine and 1.89 g/cm^3 for coarse
	% absorption	Aggregate	Natural: 1.6 for fine and 1.4 for coarse Recycled: 14 for fine and 15 for coarse
	Unitary mass	Aggregate	Natural: 1460 g/cm^3 for fine (l) and 1530 g/cm^3 for fine (c) 1600 g/cm^3 for coarse (l) and 1710 g/cm^3 for coarse (c) Recycled: 1250 g/cm^3 for fine (l) and 1422 g/cm^3 for fine (c) 1123 g/cm^3 for coarse (l) and 1232 g/cm^3 for coarse (c)
	Flattening and elongation Indexes	Aggregate	Natural: % flattening index: 26 and % elongation index: 20 Recycled: % flattening index: 9.51 and % elongation index: 3.27
	Density	Cement	3.1 ± 0.1 g/cm^3
	Consistency	Cement	26.4%

Table 3. *Cont.*

Type of Characterization	Parameter	Component	Results
Chemical	Solidness: resistance to sulfide	Aggregate	Natural % material loss: 1.4 (fine) and 3 (coarse) Recycled: % material loss: 86 (fine) and 66 (coarse)
	Organic Matter	Aggregate	Natural: 2 (number of organic reference) Recycled: 1 (number of organic reference) Both values below 3, which is the limit value for use in concrete
Mechanical	Wear	Aggregate	Natural % wear: 23 Recycled % wear: 64

l: loose; c: compact.

3.3. Mixture Design

This phase consisted of nine steps, which are summarized as follows in Table 4. According to Sánchez de Guzmán [42], calculations and analysis during this phase were performed meeting the requirements of the standard NTC 2017 [43], which implies a compressive strength of 50 MPa and flextraction of 5 MPa. The mixture design did not consider severe conditions (e.g., freeze-thaw cycles).

Table 4. Steps and their outputs during the mixture design phase.

Task	Results and Outputs
(i) Settling selection	An average value of 10 mm was selected. A very dry consistency is recommended for paving stones and implies the use of extreme vibration and possible pressure for achieving the desired compaction (recommended settlement: 0–20 mm).
(ii) Selection of maximum aggregate size	According to granulometry analysis, the maximum and maximum nominal size of aggregates corresponded to 12.5 mm (1/2") and 9.51 mm (3/8"), respectively.
(iii) Air content estimation	Since paving stones are not exposed to extreme conditions (freeze-thaw cycles), the concrete's air content is zero.
(iv) Estimation of mixing water content	This was performed using a linear regression between values provided, and using a value of settlement of 10 mm, which establishes that for settlement of 0 mm and 25 mm are required 201 kg and 208 kg of mixing water content respectively per 1 m^3 of concrete. The resulting value of mixing water content was 203.8 kg per 1 m^3.
(v) Determination of design resistance	According to standard NTC 2017, a paving stone unit must provide a minimum modulus of rupture of 4.2 MPa after 28 days. Such modulus of rupture commonly varies between 10% and 20% of the compressive resistance. Therefore, as an indirect measurement, minimum compressive resistance of the mixture of 42 MPa is required. To compensate for possible fluctuations, it is desirable to include a safety factor. Following the standard NTC 2017, 100 kg/cm^2 were added to the compressive resistance, thus an approximate compressive resistance of 50 MPa (520 kg/cm^2).
(vi) Selection of water/cement ratio	This was established according to the water/cement ratio values provided by. This establishes that for a value of 520 kg/cm^2 the corresponding water/cement ratio is equal to 0.36.
(vii) Calculation of cement content	This was calculated using the value of mixing water content and the water/cement ratio. Thus, it was required 565 kg per m^3 of concrete.
(viii) Estimation of aggregate proportions	They were determined using a graphical method. The recommended value is 52% for fine aggregate and 48% for coarse aggregate.
(ix) Adjustment of water content	Due to aggregate moisture, this was performed assuming a volume of the concrete mixture of 0.033 m^3 to perform a test for one slump and two beams. Moisture was determined for both coarse and fine aggregates.

3.4. Evaluation of Mechanical Properties

Flexural and compression resistance tests were carried out at 28 days of the age of beams and cylinders with different contents of recycled aggregates to evaluate the concrete's mechanical properties (Tables 5 and 6). In terms of flexural strength, it was observed that compared to the reference sample (M1), made with aggregates of natural origin and considering the minimum strength required for paving stones (5 MPa), the mixtures M2, M3 and M4 met with established mechanical requirements. However, given the high deviation found for the M3 mixture and the low average compressive strength, it was not considered suitable in this project. Considering the above, M2 and M4 mixtures, which replace 50% of the fine natural aggregates and 50% of the natural coarse aggregates, respectively, were the only viable options in terms of flexural strength. On the other hand, in terms of compressive strength, mixtures M2 and M4 met with the established design resistance (50MPa) and are comparable to the reference mixture (M1) resistance that used fine and coarse aggregates of natural origin.

Table 5. Summary of results for flexural and compressive strength (after 28 days). Three samples for each mixture.

Mixture	Samples Flexural Strength MPa (28 Days)					Samples Compressive Strength MPa (28 Days)				
	S1	S2	S3	Average	Std Deviation	S1	S2	S3	Average	Std Deviation
M1	5.76	5.95	5.33	**5.68**	0.32	63.51	60.42	58.73	**60.89**	2.42
M2	4.00	4.60	5.11	**4.57**	0.56	57.43	53.68	53.34	**54.82**	2.27
M3	2.64	4.53	4.69	**3.95**	1.14	48.34	48.39	46.35	**47.69**	1.16
M4	5.53	4.70	4.95	**5.06**	0.43	55.33	56.16	56.83	**56.10**	0.75
M5	4.28	3.95	3.71	**3.98**	0.29	42.60	42.00	43.24	**42.61**	0.62
M6	3.16	3.84	4.41	**3.80**	0.63	39.69	39.25	38.77	**39.24**	0.46
M7	4.74	3.96	3.82	**4.17**	0.50	42.26	44.30	42.57	**43.04**	1.10
M8	4.04	4.47	3.99	**4.17**	0.26	37.41	42.51	42.01	**40.64**	2.81
M9	2.79	3.56	3.29	**3.21**	0.39	39.08	37.71	35.96	**37.58**	1.56

Table 6. Summary of results by coarse and fine aggregate combination.

		Avg Flexural Strength MPa				Avg Compressive Strength MPa		
		Fine Aggregate Replacement				Fine Aggregate Replacement		
		0%	50%	100%		0%	50%	100%
Coarse aggregate replacement	0%	5.68	4.57	3.95	0%	60.89	54.82	47.69
	50%	5.06	3.98	3.80	50%	56.10	42.61	39.24
	100%	4.17	4.17	3.21	100%	43.04	40.64	37.58

Figure 4 summarizes the results obtained for flexural and compressive tests, while Figure 5 shows the relationship between flexural and compressive strengths obtained from mechanical tests for all nine specimens.

Finally, when analyzing the correlation between compressive and flexural strength of the results (Figure 5), it was observed that flexural strength represented between 8 and 10% of the compressive strength, such output is similar to the relationship that conventional concrete offers (using natural aggregates).

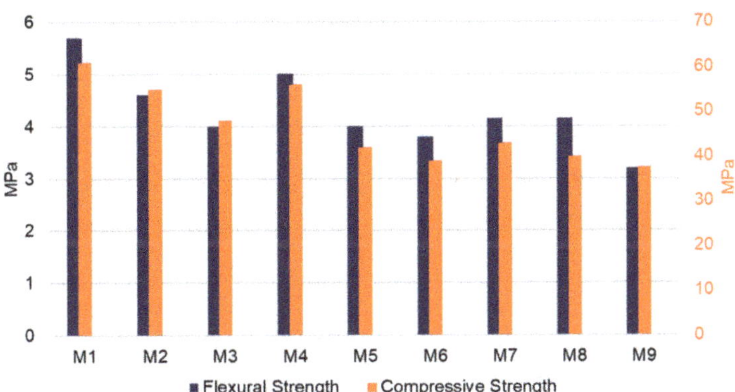

Figure 4. Mechanical properties of concrete mixtures: flexural strength, compressive strength.

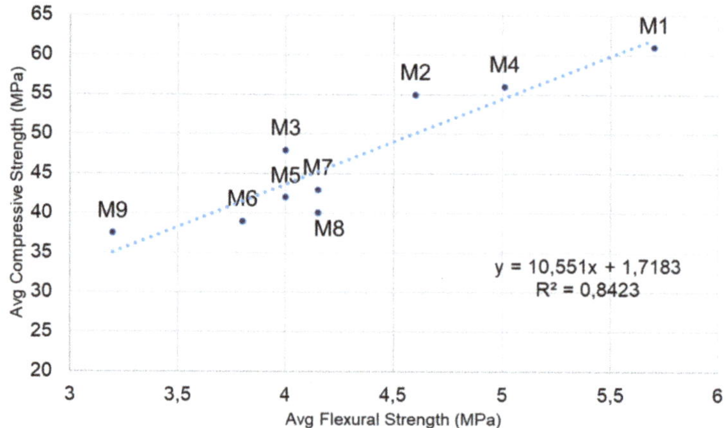

Figure 5. Compressive and flexural strength of concrete mixtures.

3.5. Performance-Based Selection of Mixture

The M2 (50% replacement of fine natural aggregate) and M4 (50% replacement of natural coarse aggregate by recycling) mixtures were selected based on results for flexural (modulus of rupture) and compressive strengths of the concrete mixtures concerning the behavior of the reference mixture and the established design strengths. Figure 6 shows the new paving stones from the M2 mixture, and Table 7 summarizes the most relevant parameters for selected mixtures (M2 and M4) compared to the reference sample M1.

Figure 6. Paving stones made from M2 concrete mix using recycled aggregates.

Table 7. Parameters for selected mixtures (M1, M2, and M4).

Parameters (Avg Values)		Mix		
		M1	**M2**	**M4**
Dimensions (cm)	Length	20.45	20.35	20.3
	Width	15.35	15.25	15.25
	Height	10.20	10.05	10.05
% of absorption		3.9	8.6	8.6
Std deviation		0.67	0.87	0.73
Wear (mm)		13.9	12.3	19.2
Std deviation		0.28	0.14	0.99
Flexural strength (MPa)		9.4	4.45	4.2

4. Discussion

4.1. Physical, Chemical and Mechanical Characterization

According to granulometric measurements, coarse and fine aggregates did not fully satisfy the reference standard (NTC 174). However, these aggregates were used in this study due to their high quality and commercial acceptance (mechanical properties and durability). For recycled aggregates, these satisfy acceptance ranges according to the reference standard (NTC 174). Except for the finest fraction of coarse aggregates (fringe sieve between #200 and #10). Thus, fine recycled aggregate is higher than the fine aggregate while maintaining the same maximum size regarding fineness modulus.

It is observed from the fineness modules that the recycled fine aggregate is higher than the natural one, while the maximum sizes for both types of aggregates are equal. This indicates that concrete with recycled aggregates may require higher water content to achieve workabilities, similar to conventional concrete. Regarding other interest parameters, it was possible to identify several important differences between recycled and virgin aggregates. In the case of absorption and density, recycled aggregates provided values 9 to 11 times higher and 32 to 35% lower than natural ones. Recycled aggregates have a lower density or specific gravity concerning the densities of natural aggregates. Thus, commonly the water content required for a concrete mix with the same workability is higher when recycled aggregates are used and may also be worsened; this is because although in this case the standards are met, the elongation and flattening index of recycled coarse aggregates is relatively low compared to natural ones [44].

Another interesting difference is related to mechanical properties, demonstrating that recycled aggregates proved three times less resistant to wear than natural aggregates (64% of loss). Lastly, unitary mass for both recycled and natural aggregates presented conventional values for concretes (950 to 1950 kg/m^3). Thus, both aggregate materials can reduce cement consumption, plastic deformation and contribute to concrete density once it is solidified. In terms of chemical properties, it was found that despite organic material not affecting the hydration of Portland cement, virgin fine aggregates (sand) provide a higher content compared to recycled ones. This can be explained by the fact that natural aggregates come directly from the mineral quarry. Regarding wear resistance under exposition to sodium sulfate, values of 68% and 86% were registered for recycled coarse and fine aggregates, respectively.

4.2. Mechanical Evaluation

According to Figure 4, when 50% fine aggregate is replaced, the flexural strength is reduced by approximately 20%. In contrast, with the replacement of the coarse aggregate, a reduction of 10% is presented concerning the reference mixture, noting that it gets better bending behavior of the mix when replacing coarse aggregates. Regarding mixtures with a 100% replacement in both fine and coarse, the resistance decreased by approximately 30% and did not satisfy the NTC 2017 standard (4.2 MPa). On the other hand, it is observed that when replacing 50% of the coarse aggregate, the resistance to compression is reduced by approximately 8%. It is a concrete that met the established mechanical requirements; by replacing 100% of the coarse aggregate, the compressive strength is decreased by approximately 30%. Therefore, this replacement was not considered viable due to its high reduction in mechanical performance.

When combining the recycled aggregate replacements between fine and coarse, a decrease of more than 30% and 25% of the mechanical resistance to compression and flexion, respectively, values that do not meet the initial requirement. Reductions of this concrete's mechanical properties are mainly since recycled materials from paving stones provided high absorptions greater than 14%, indirectly indicating that the aggregates also offered high porosity, which implied a considerable volume of voids in the internal structure of the aggregates. In addition, the pores of recycled material include small cracks that, within the concrete, reduce the mechanical properties of mixtures. Although there are different improvement techniques for the mechanical response (e.g., coatings and mineral fillers), the most economical and straightforward way is balancing the mechanical properties with the partial use of natural aggregates. In addition to the environmental benefits of using recycled aggregates, these have better adherence to the matrix, low content of organic matter, and typically low cost. Finally, by replacing 50% of fine natural aggregate with recycled aggregate, it was possible to observe a reduction in compressive strength by approximately 10%, a similar value to replacing 50% of coarse aggregate in the mix. One hundred percent replacement of the fine aggregate content decreases resistance by approximately 20%, which was an admissible value according to the requirement. Nevertheless, mechanical behavior in terms of flexure was less than 4.2 MPa. Therefore, the M7 mixture was not considered for the manufacture of paving stones.

According to Figure 4, mixes M2, M3, and M4 met the mechanical requirements established regarding flexural strength. However, given the high deviation found for the M3 mixture and the low average compressive strength, this study was not considered. Therefore, the M2 and M4 mixtures, which replace 50% of the fine natural aggregates and 50% natural coarse aggregates, respectively, were the two only viable options for flexural strength fulfillment.

4.3. *Properties of Selected Mixes*

Average values for absorption mixtures M2 and M4 are higher than 7%, while the M1 mixture is within the parameter according to the NTC 176 standard [45], which is to be expected taking into account that M2 and M4 mixtures contain recycled material from old paving stones, which are more porous and less dense. In terms of wear, it must be less than 23 mm, according to the NTC 5147 standard [46]. Tested mixtures showed satisfactory results since all of them satisfy that reference parameter. However, it is noteworthy that new paving stones with M2 concrete presented lower wear than M1, contrary to the M4 sample that presented higher wear. Fifty percent replacement of fine aggregates in M2 did not affect the wear of the new paving stone, which indicates that the manufacture of paving stones with such a mixture will promote durability by friction over time. New paving stones made with recycled material decreased their resistance by approximately 50% compared to M1. This decrease due to the recycled material used is highly porous and has possible microcracks derived from the crushing process.

To summarize, it is observed that new paving stones manufactured with a 50% replacement of natural fine and coarse aggregates by recycled aggregates (M2 and M4) satisfactorily comply with the minimum mechanical resistance (4.2 MPa according to the NTC 2017 standard). However, the results in absorption and wear are above the permitted values for these applications. It is recommended to quantify in future research the practical impact on the lifecycle of paving stones, including the interaction between these high percentages of absorption and wear. The results obtained demonstrated that it is possible to fabricate paving stones with similar properties to those fabricated using 100% virgin material. Possible applications of concrete with recycled aggregates include structural applications and buildings, however, it is necessary to perform specific experiments to validate its suitability in such applications. The study summarized in this article did not consider extreme conditions such as exposure to fire or freezing environment.

4.4. *Limitations of the Study*

The experimentation and examination processes followed in this study involved several characterization, physical and chemical analysis, and laboratory tests. Therefore it is relevant to address several limitations regarding the methodology, materials and results obtained:

- Properties of raw material for aggregates (old paving stones) can vary depending on the age, degree of wear and typical use conditions. Hence, it is possible to find differences in results for mechanical properties considering another paving stone with different age and use regime.

- Results of mechanical properties obtained are valid for the aggregate sizes considered in this study. The use of high-technology jaw crushers can play an important role to analyze smaller sizes of aggregate and their influence on mechanical properties and durability of the new recycled material.
- Environmental conditions (humidity and temperature) were not controlled or measured during the development of experiments. Possible differences in results can be observed in regions with colder or hotter weather. Mechanical test were performed in cold conditions only. Therefore, results can vary under other conditions not considered in this study (i.e., mechanical behavior after exposition to fire).
- Characterization processes (density, consistency, absorption, chemical properties, and mechanical wear) followed NTC (Colombian technical standards). Such standards are conventional based on ASTM standards, although methodological differences are not ruled out compared to another standard or newest versions of the ASTM standards.

5. Conclusions

This article described first the evaluation of the mechanical properties of concrete using recycled aggregates obtained from old paving stones, and second the development of two suitable concrete mixtures for manufacturing new paving stones with important replacements of the fraction of natural coarse and fine aggregate. Based on the technical standards, the two concrete mixes with the replacement of 50% by weight of the fine and coarse fraction of natural origin presented adequate flexural strength when evaluated as new paving stones. These results are entirely satisfactory concerning the replacements found in the literature where recycled aggregates are typically used in fractions less than 50% by weight of the natural coarse or fine aggregates of a concrete mix. This is fundamentally due to the high quality and processing of the recycled aggregates used in this project. It is remarkable that the selected recycled aggregates were fundamentally constituted of concrete and did not present significant contamination.

On the other hand, in terms of the relative high absorption and low resistance to wear of the paving stones developed, future research is proposed to improve recycled aggregates against these variables, particularly wear caused by pedestrian. The correlation of existing standardized wear tests with actual applications still demands better approaches that open the possibility of using recycled aggregates without affecting construction safety. Finally, this study demonstrated that use of recycled aggregates has the potential to reduce the environmental impact and, when applied correctly, it can significantly reduce the costs of a construction project. Therefore, results from this project, together with more research results about recycling technology in building materials are the base of a circular economic model proposed for one of the major cities from Colombia, Santiago de Cali (Figure 7). In particular, the Circular Economy Model, a tool led by the municipal planning department of the mayor's office of Santiago de Cali for the construction sector, is composed of four stages that are: (I) production of construction materials, (II) construction, (III) use and operation and (IV) termination of the life cycle of buildings and infrastructure. It also includes the articulation with other productive sectors that generate waste of interest to the construction chain, a strategy known as industrial symbiosis. Furthermore, in the Circular Economy Model, it is very important to connect at all stages with the Environmental and Technological Park (Parque Ambiental y Tecnológico) of Santiago de Cali, a place conceived as a physical space that has infrastructure and shared human capital that reserves investment for sustainable construction companies. As they are inserted in flexible groups, they find support (in situ and in the park) to transform C&DW and other waste into sustainable materials for construction.

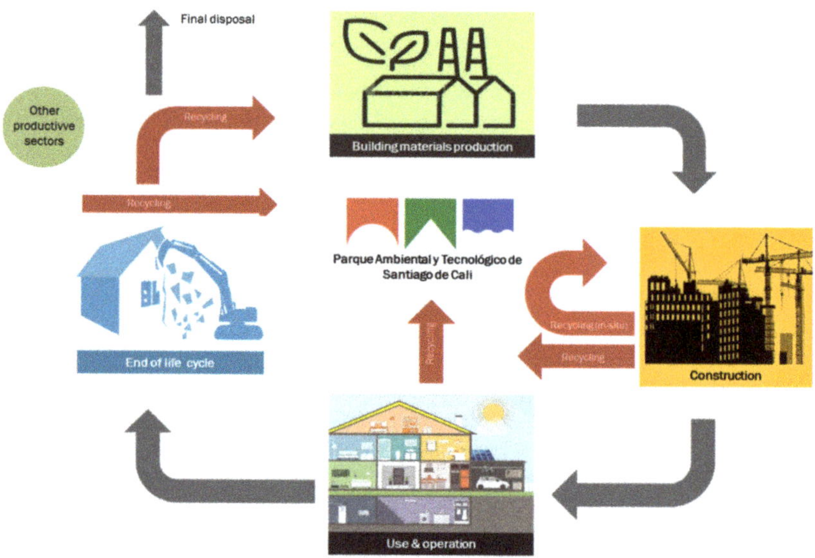

Figure 7. Circular Economy Model for the construction sector of Santiago de Cali which is composed of the phases of production of construction materials, construction, use and operation, and completion of the life cycle of buildings and infrastructure.

Author Contributions: Conceptualization, methodology, and editing, A.M.-R. and J.A.M.; components, classification, and mechanical evaluation, A.M.B.-G. and I.D.B.-G. All authors have read and agreed to the published version of the manuscript.

Funding: This research received no external funding.

Acknowledgments: The authors thank to Pontificia Universidad Javeriana Cali (Colombia) for the technical support given during the experiments reported in this article.

Conflicts of Interest: The funders had no role in the design of the study; in the collection, analyses, or interpretation of data; in the writing of the manuscript, or in the decision to publish the results.

Appendix A. Equivalence between NTC, INV and ASTM Standards

Table A1. Equivalence between NTC, INV and ASTM standards.

NTC/INV Standard (Colombia)	Equivalent International Standard
NTC 174	ASTM C1231
INV E-500	–
INV E-307	–
INV E-310	–
NTC 7707	ASTM C136
INV E-213	–
NTC 237	ASTM C128
INV E-222	–
NTC 176	ASTM C127
INV E-223	–
NTC 92	ASTM C29

Table A1. *Cont.*

NTC/INV Standard (Colombia)	Equivalent International Standard
INV E-217	–
INV E-230	–
NTC 126-206	ASTM C88
INV 3-220	–
NTC 127	ASTM C40
NTC 98	ASTM C131
INV E-219	–
NTC 2017	–
INV E-402	–
NTC 673	ASTM C42
INV E-414	–

Appendix B. Additional Properties of Aggregates and Quantities for Experimental Mixtures

Table A2. Granulometry of recycled aggregate samples (using sieve).

Sizes		Weight and Grade of Samples						
Passing Size	Detained Size	A	B	C	D	E	F	G
3″	2 ½″					2500		
2 ½″	2″					2500		
2″	1 ½″					5000	5000	5000
1 ½″	1″	1250					5000	5000
1″	¾″	1250						
¾″	½″	1250	2500					
½″	3/8″	1250	2500					
3/8″	1/4			2500				
¼″	N°4			2500				
N°4	N°8				5000			
Number of balls		12	11	8	6	12	12	12
Angular speed (RPM)		500	500	500	500	1000	1000	1000

Table A3. Dry weight and absolute volume of components per cubic meter of concrete.

Component	Dry Weight Kg/m^3	Apparent Density g/cm^3	Absolute Volume L/m^3	Proportion
Cement	565	3.1	182	1
Water	203.8	1	203.8	0.36
Air content	–	–	0	0
Coarse aggregate	809.44	2.81	288	1.43
Fine aggregate	876.89	2.68	326.2	1.55
Total	2455		1000	

Table A4. Moisture content for natural aggregates.

Parameter	Fine Aggregate	Coarse Aggregate
Initial weight (g)	381.9	327.7
Dry weight (g)	349.2	315.9
Water content (g)	32.7	11.8
Moisture (%)	8.57	3.60

Table A5. Density and absorption for natural aggregates.

Parameter	Fine Aggregate	Coarse Aggregate
Nominal density (g/cm^3) at 23 °C	2.80	2.93
Apparent density (g/cm^3) at 23 °C	2.68	2.81
Relative density (g/cm^3) at 23 °C	2.72	2.85
Absorption (%)	1.6%	1.4%

References

1. Yang, D.; Liu, M.; Ma, Z. Properties of the foam concrete containing waste brick powder derived from construction and demolition waste. *J. Build. Eng.* **2020**, *32*, 101509. [CrossRef]
2. John, V. Recycling of rubble for the production of building materials. In *Use of Waste as Construction Materials*; EDUFBA: Salvador, Brazil, 2001; Volume 312, pp. 27–45.
3. Terry, M. Waste Minimization in the Construction and Demolition Industry. Bachelor's Thesis, Civil & Environmental Engineering, University of Technology, Sydney, NSW, Australia, 2004.
4. Shen, L.; Tam, V.W. Implementation of environmental management in the Hong Kong construction industry. *Int. J. Proj. Manag.* **2002**, *20*, 535–543. [CrossRef]
5. Francisco, J.T.M.; De Souza, A.E.; Teixeira, S.R. Construction and demolition waste in concrete: Property of pre-molded parts for paving. *Cerâmica* **2019**, *65*, 22–26. [CrossRef]
6. Galán, B.; Viguri, J.; Cifrian, E.; Dosal, E.; Andres, A. Influence of input streams on the construction and demolition waste (CDW) recycling performance of basic and advanced treatment plants. *J. Clean. Prod.* **2019**, *236*, 117523. [CrossRef]
7. Kartam, N.; Al-Mutairi, N.; Al-Ghusain, I.; Al-Humoud, J. Environmental management of construction and demolition waste in Kuwait. *Waste Manag.* **2004**, *24*, 1049–1059. [CrossRef] [PubMed]
8. Lederer, J.; Gassner, A.; Kleemann, F.; Fellner, J. Potentials for a circular economy of mineral construction materials and demolition waste in urban areas: A case study from Viena. *Resour. Conserv. Recycl.* **2020**, *161*, 104942. [CrossRef]
9. Juan-Valdés, A.; García-González, J.; Rodríguez-Robles, D.; Guerra-Romero, M.I.; López Gayarre, F.; De Belie, N.; Morán-del Pozo, J.M. Paving with Precast Concrete Made with Recycled Mixed Ceramic Aggregates: A Viable Technical Option for the Valorization of Construction and Demolition Wastes (CDW). *Materials* **2019**, *12*, 24. [CrossRef]
10. Whittaker, M.J.; Grigoriadis, K.; Soutsos, M.; Sha, W.; Klinge, A.; Paganoni, S.; Casado, M.; Brander, L.; Mousavi, M.; Scullin, M.; et al. Novel construction and demolition waste (CDW) treatment and uses to maximize reuse and recycling. *Adv. Build. Energy Res.* **2019**, 1–18. [CrossRef]
11. Cantero, B.; Bravo, M.; de Brito, J.; del Bosque, I.S.; Medina, C. Mechanical behaviour of structural concrete with ground recycled concrete cement and mixed recycled aggregate. *J. Clean. Prod.* **2020**, *275*, 122913. [CrossRef]
12. Chen, A.; Han, X.; Chen, M.; Wang, X.; Wang, Z.; Guo, T. Mechanical and stress-strain behavior of basalt fiber reinforced rubberized recycled coarse aggregate concrete. *Constr. Build. Mater.* **2020**, *260*, 119888. [CrossRef]
13. De Andrade, G.P.; Polisseni, G.D.C.; Pepe, M.; Filho, R.D.T. Design of structural concrete mixtures containing fine recycled concrete aggregate using packing model. *Constr. Build. Mater.* **2020**, *252*, 119091. [CrossRef]
14. Botero, L. *Sostenibilidad de la Disposición de Escombros de Construcción y Demolición en Bogotá*; Universidad de Los Andres: Bogotá, Colombia, 2003.
15. Pérez, J. *Estudio del Potencial de Reciclaje de Desechos de Materiales de Construcción y Demolición en Santa Fé de Bogotá*; Universidad de Los Andes: Bogotá, Colombia, 1996.
16. Poon, C.S.; Chan, D. Feasible use of recycled concrete aggregates and crushed clay brick as unbound road sub-base. *Constr. Build. Mater.* **2006**, *20*, 578–585. [CrossRef]
17. Bazaz, J.; Khayati, M.; Akrami, N. *Performance of Concrete Produced with Crushed Bricks as the Coarse and Fine Aggregate*; The Geological Society of London: London, UK, 2006; p. 10.
18. Wong, Y.D.; Sun, D.D.; Lai, D. Value-added utilisation of recycled concrete in hot-mix asphalt. *Waste Manag.* **2007**, *27*, 294–301. [CrossRef] [PubMed]
19. Agrela, F.; Barbudo, A.; Ramírez, A.; Ayuso, J.; Carvajal, M.D.; Jiménez, J.R. Construction of road sections using mixed recycled aggregates treated with cement in Malaga, Spain. *Resour. Conserv. Recycl.* **2012**, *58*, 98–106. [CrossRef]
20. Pérez-Benedicto, J.A.; del Río-Merino, M.; Peralta-Canudo, J.L.; de la Rosa-La Mata, M. Mechanical characteristics of concrete with recycled aggregates coming from prefabricated discarded units. *Mater. Construcción* **2012**, *62*, 25–37. [CrossRef]
21. Çakır, Ö.; Dilbas, H. Durability properties of treated recycled aggregate concrete: Effect of optimized ball mill method. *Constr. Build. Mater.* **2021**, *268*, 121776. [CrossRef]
22. Vedrtnam, A.; Bedon, C.; Barluenga, G. Study on the Compressive Behaviour of Sustainable Cement-Based Composites Under One-Hour of Direct Flame Exposure. *Sustainability* **2020**, *12*, 10548. [CrossRef]
23. Zareei, S.A.; Ameri, F.; Bahrami, N.; Shoaei, P.; Musaeei, H.R.; Nurian, F. Green high strength concrete containing recycled waste ceramic aggregates and waste carpet fibers: Mechanical, durability, and microstructural properties. *J. Build. Eng.* **2019**, *26*, 100914. [CrossRef]

24. Zahid-Hossain, F.; Shahjalal, M.; Islam, K.; Alam, M.T.y.M.S. Mechanical properties of recycled aggregate concrete containing crumb rubber and polyproplylene fiber. *Constr. Build. Mater.* **2019**, *225*, 983–996. [CrossRef]
25. Alam, M.S.; Slater, E.; Billah, A.H.M.M. Green Concrete Made with RCA and FRP Scrap Aggregate: Fresh and Hardened Properties. *J. Mater. Civ. Eng.* **2013**, *25*, 1783–1794. [CrossRef]
26. Meherier, M. *Investigation of Mechanical and Durability Properties of Cement Mortar and Concrete with Varying Replacement Levels of Crumb Rubber as fine Aggregate*; University of British Columbia Okanagan: Kelowna, BC, Canada, 2016.
27. Limbachiya, M.; Leelawaty, T.; Dhir, R. Use of recycled concrete aggregate in high-strenght concrete. *Mater. Struct.* **2000**, *33*, 574. [CrossRef]
28. Tiempo, E.l. *Estación de Transferencia EDT Carrera 50 en Santiago de Cali. Problemáticas Asociadas a la Disposición Inadecuada de los RCD*; El Tiempo: Bogotá, Colombia, 2018.
29. Pais, E.l. *Residuos de Construcción Dispuestos Inadecuadamente en el Rio Cauca*; El País: Bogotá, Colombia, 2014.
30. Tiempo, E.l. *Extracción de Roca en el Rio Pance*; El Tiempo: Bogotá, Colombia, 2018.
31. Gobernación del Valle. *Areneros del Rio Cauca Buscan Formalizar su Actividad ante la CVC y el Ministerio de Minas*; Gobernación del Valle: Valle del Cauca, Colombia, 2017.
32. Ministry of Environment and Sustainable Development. *Resolution 472*; Ministry of Environment and Sustainable Development: Bogotá, Colombia, 2017.
33. Bedoya-Montoya, C. *El Concreto Reciclado con Escombros Como Generador de Hábitats Urbanos Sostenibles*; Universidad Nacional de Colombia: Medellín, Colombia, 2003.
34. National Administrative Department of Staticstics—DANE. *First Report on Circular Economy*; Government of Colombia: Bogotá, Colombia, 2020.
35. Oliveira, T.C.; Dezen, B.G.; Possan, E. Use of concrete fine fraction waste as a replacement of Portland cement. *J. Clean. Prod.* **2020**, *273*, 123126. [CrossRef]
36. Diosa, J.S. *Experimental Analysis of the Mechanical Properties of High-Strength Concrete Made with Recycled Concrete Aggregates*; Pontificia Universidad Javeriana Cali: Cali, Colombia, 2020.
37. Hurtado, I. *Evaluation of the Effect of the Partial Replacement of Portland Cement by Ash from the Paper Industry in Mortars and Walls*; Pontificia Universidad Javeriana Cali: Cali, Colombia, 2019.
38. Londoño, J. *Mechanical Behavior of Precast Concrete Elements with Recycled Aggregates within the Source that Generates Them*; Pontificia Universidad Javeriana Cali: Cali, Colombia, 2016.
39. Wagih, A.M.; El-Karmoty, H.Z.; Ebid, M.; Okba, S.H. Recycled construction and demolition concrete waste as aggregate for structural concrete. *HBRC J.* **2013**, *9*, 193–200. [CrossRef]
40. Zhang, C.; Hu, M.; Yang, X.; Miranda-Xicotencatl, B.; Sprecher, B.; Di Maio, F.; Zhong, X.; Tukker, A. Upgrading construction and demolition waste management from downcycling to recycling in the Netherlands. *J. Clean. Prod.* **2020**, *266*, 121718. [CrossRef]
41. ICONTEC. *NTC 174:2018 Concrete. Specifications for Concrete Aggregates*; ICONTEC: Bogotá, Colombia, 2004.
42. Sanchez De Guzmán, D. *Tecnología del Concreto Y del Mortero*; Bhandar Editores: Bogotá, Colombia, 2001.
43. ICONTEC. *NTC 2017:2018 Pavers of Concrete for Pavement*; ICONTEC: Bogotá, Colombia, 2018.
44. Pacheco-Torgal, F.; Jalali, S.; Labrincha, J.; John, V. *Eco-Efficient Concrete*; Woodhed Publishing Limited: Oxford, UK; Cambridge, UK; Philadelphia, PA, USA; New Delhi, India, 2013.
45. ICONTEC. *NTC 176:2019 Method to Determine the Relative Density and Absorption of Coarse Aggregate*; ICONTEC: Bogotá, Colombia, 2019.
46. ICONTEC. *NTC 5147:2002 Test Method for Determining Abrassion Resistance of Floor and Pavement Materials by Sand and Wide Whell*; ICONTEC: Bogotá, Colombia, 2002.

Article

Water-Washed Fine and Coarse Recycled Aggregates for Real Scale Concretes Production in Barcelona

Miren Etxeberria [1,*], Mikhail Konoiko [1], Carles Garcia [1] and Miguel Ángel Perez [2]

[1] Department of Civil and Environmental Engineering, Campus Nord, Universitat Politècnica de Catalunya·BarcelonaTECH, 08034 Barcelona, Spain; Mky95@yandex.ru (M.K.); garciauretacarles@gmail.com (C.G.)
[2] Hercal Diggers S.L., Ctra de Rubí, 286-B, 08228 Terrassa, Spain; maperez@hercal.es
* Correspondence: miren.etxeberria@upc.edu; Tel.: +34-934011788

Abstract: The use of recycled aggregate to reduce the over-exploitation of raw aggregates is necessary. This study analysed and categorised the properties of water-washed, fine and coarse, recycled aggregates following European Normalization (EN) specification. Because of their adequate properties, zero impurities and chemical soluble salts, plain recycled concrete was produced using 100% recycled concrete aggregates. Two experimental phases were conducted. Firstly, a laboratory phase, and secondly, an on-site work consisting of a real-scale pavement-base layer. The workability of the produced concretes was validated using two types of admixtures. In addition, the compressive and flexural strength, physical properties, drying shrinkage and depth of penetration of water under pressure validated the concrete design. The authors concluded that the worksite-produced concrete properties were similar to those obtained in the laboratory. Consequently, the laboratory results could be validated for large-scale production. An extended slump value was achieved using 2.5–3% of a multifunctional admixture plus 1–1.2% of superplasticiser in concrete production. In addition, all the produced concretes obtained the required a strength of 20 MPa. Although the pavement-base was produced using 300 kg of cement, the concrete made with 270 kg of cement per m^3 and water/cement ratio of 0.53 achieved the best properties with the lowest environmental impact.

Keywords: recycled concrete aggregate; recycled aggregate concrete; workability; compressive strength; pavement

Citation: Etxeberria, M.; Konoiko, M.; Garcia, C.; Perez, M.Á. Water-Washed Fine and Coarse Recycled Aggregates for Real Scale Concretes Production in Barcelona. *Sustainability* **2022**, *14*, 708. https://doi.org/10.3390/su14020708

Academic Editor: Jianzhuang Xiao

Received: 15 December 2021
Accepted: 7 January 2022
Published: 9 January 2022

1. Introduction

In 2018, construction and demolition waste was 35.4% of the total waste (2277 million tonnes) generated in the European Union (EU) by all economic activities including households, of which only 54.2% was recovered [1,2]. Consequently, the demolition of concrete structures causes a considerable volume of waste that terminates in landfills. The guidelines of the European Commission [3] are designed to encourage a change in production procedures to embrace a circular, more sustainable and eco-respectful model in which waste is re-introduced into production processes, reducing raw material over-exploitation and maximising material life cycles. Plaza et al. [4] concluded that the benefits of using recycled coarse and fine aggregates to replace natural aggregate partially lie not only in CO_2 emissions reduction in concrete manufacture but also in the significant mitigation of the environmental impacts induced by stockpiling the respective waste. Moreover, about 75–80% of the total concrete components materials are aggregates [5]. Today, the production and use of natural resources such as natural aggregates in concrete production reduce natural resources and increase the volume of atmospheric pollutants [6]. Therefore, recycling waste concrete and concretes from damaged or demolished structures are essential for producing recycled concrete aggregates (RCA), thus mitigating the environmental impacts.

The demand for non-renewable natural resources and industrial products, especially mineral aggregates from quarry extraction, is high for highway construction and maintenance [7]. Therefore, the most widespread practice to achieve pavement sustainability is to lower the quantity of virgin aggregates by partially or fully replacing them with alternative aggregates. In addition, the available literature shows that alternative aggregate, as recycled concrete aggregate, could be effectively used for concrete pavements [8–11]. The mentioned field investigation results indicate that it is possible to produce pavements from recycled aggregates that are equivalent in all aspects to pavements made with conventional aggregates when up to 40% of coarse RCA are used to replace natural aggregates. Gress et al. [12] described that the pavement produced with up to 25% of fine recycled concrete aggregates also achieved adequate properties although with a slight increase of shrinkage value with respect to that in concrete produced only employing coarse RCA. Moreover, recycled concrete made with coarse RCA aggregates have been employed in certain, although few, real structural concrete projects [13]. In Hong Kong [14], from 2022 to 2005, the acceptable behaviour of structural concretes grades C20 and C25 were verified, producing concretes with recycled coarse aggregate replacement levels of 100 and 20%, respectively. Zhang and Zhao [15] also proved 50% coarse RCA in structural concrete production. However, for the structural elements studied by Xiao et al. [16], up to 30% of coarse RCA were used in concrete production. The use of RCA is still limited. However, as mentioned previously, coarse recycled aggregates (C-RCA) have been proven to be suitable for concrete production [17]. Moreover, their use in concrete production as a structural material has been widely analysed and validated in many applications [13,18,19]. However, the use of fine recycled concrete aggregates (F-RCA) is less widespread due to their more negative effect on concrete properties [20].

RCA aggregates have a lower quality than the natural aggregate (NA) because of attached mortar to the stone particles in RCA. In contrast to NA, RCA has the following properties: more water absorption, less bulk density, more abrasion loss and more crushability [21,22]. In particular, fine RCA could also have more dust particles, more organic impurities and also harmful chemicals because of earth mixing with concrete after building demolition [23]. Despite these weaknesses, the un-hydrated cement of the original concrete available in the RCA may play a positive role in its use in structural concrete. In addition, in the case of the use RCA, the specific surface of the aggregates improves the binder/recycled aggregate interface [24].

Coarse and fine RCA in concrete production affects workability and hardened-state (mechanical, physical and durability) properties. Recycled aggregate concrete (RAC) is typically associated with lower workability than natural aggregate concrete (NAC) of the same composition [25]. This is attributed to the poor shaping properties of crushed RCA when compared to NA. If there is proper compensation for water absorption, workability is essentially affected by the shape of the aggregates [26,27]. The reduction of workability is confirmed with the increased replacement of NA with RCA, especially in the finer fraction [28]. Partial absorption of the superplasticiser by the aggregates also occurs, and increased fines in the content are due to a partial loss of the aggregates' mortar during the mixing process [26]. Tobori et al. [29] found that when superplasticiser is added to RAC mix, instead of acting on the cement grains, the absorption of its liquid phase occurs through F-RCA. Evangelista and de Brito [30] found that polymer chains have a larger contact area with fine recycled aggregates than natural ones. Nedeljkovi'c et al. [27], after an exhausted review analysis, concluded that researchers had offered many reasonable explanations on complex flow behaviour of recycled concretes through a combination of experiments and theories. They described that there is no universal approach to obtain and maintain satisfactory workability of mortars/concretes with F-RCA.

It is generally believed that concrete compressive strength decreases as the amount of recycled concrete replacement increases [31], which may be due to the old mortar in fine RCA that makes concrete more porous and less dense [32]. As a result, the concrete produced with 100% RCA (coarse plus fine recycled aggregated) obtained a lower compressive strength [4,27]. However, the tensile strength can improve due to the improvement of the interface transition zone in concretes containing RCA [4,10].

According to Zhang et al. [33], both F-RCA and C-RCA significantly influence the drying shrinkage behaviour of concrete. Total (100%) replacement of natural aggregates with RCA (including both F-RCA and C-RCA) increased the drying shrinkage by more than 100% (102.0–116.9%). In addition, a higher water absorption ratio and a lower density RCA resulted in higher shrinkage strains. However, compared with the influence of C-RCA, the effect of F-RCA is relatively lower. In particular, C-RCA with a 100% replacement ratio increased the drying shrinkage. Sadati and Khayat [9] also determined that the increasing the fine RCA content from 0 to 15% had no significant effects on drying shrinkage of pavement concrete.

Although recycled concrete produced with 100% of recycled aggregate achieves a lower strength and a higher shrinkage than those of conventional concretes, the washing of recycled aggregates could guarantee the quality and consequently the durability of the concrete produced. In order to assure RCA aggregate quality, it is imperative to use innovative recovery plants to manage construction and demolition waste efficiently [34].

The recycled aggregates used in this study were obtained from an innovative recycling plant (see Figure 1) located in Barcelona, Spain. Two types of demolition material are treated separately at the plant: concrete waste (more than 95% is concrete) and mixed waste (with approximately 30% ceramic material). After crushing these to the desired aggregate size (usually 0/20 mm), the cleaning and sieving processes of the mixed recycled aggregates (RMA) and RCA are carried out. Fine (F-) and course (C-) fractions are produced separately, making them suitable for concrete production. In this study, the F-RCA and C-RCA were analysed for use in concrete production.

Figure 1. CDW recovery plant to water-washed recycled aggregate [35].

The production of high-quality plain concrete using 100% water-washed fine and coarse RCA aggregates without employing natural aggregates can be possible if the recycled aggregates fulfil the requirements to be used in concrete production. The objective of this study was to analyse the quality of water-washed C-RCA and F-RCA aggregates produced in the innovative plant and validate them to be used in plain concrete production by building a real pavement-base layer. The production of plain concrete was analysed in two phases: (1) laboratory experiments and (2) on-site pavement-base layer construction. In the first phase, the suitable concrete mix proportions were tested to see if they accomplish the minimum required properties of having an adequate workability (6–8 cm of slump test) 30 min after RAC concrete production and a minimum compressive strength of 20 MPa at 28 days of curing. Moreover, other properties such as the flexural strength, physical properties, drying shrinkage and depth of penetration of water under pressure were also validated in the concrete design. The second phase would verify if the worksite-produced concrete achieved adequate properties similar to those obtained in the laboratory.

2. Materials and Methods

2.1. Materials

2.1.1. Cement and Chemical Admixtures

The cement CEM II A-L 42.5 R (88% clinker, 12% limestone, excluding the set regulator, added in 5%) was used. The composition of the cement is shown in Table 1.

Table 1. Composition of cement as the percentage of total weight.

Cement	SiO_2	CaO	Fe_2O_3	Al_2O_3	MgO	SO_3	Na_2O	K_2O	LOI
CEM II A-L 42.5 R	19.33	62.71	2.65	3.43	1.36	3.52	0.06	0.8	5.28

Two chemical admixtures were employed for concrete production: a multifunctional admixture (P) and a superplasticiser (S). The mix recommended by the manufacturer for the S was 0.3–2.0% and 0.5–1.5% for the P admixtures based on the weight of the cement.

2.1.2. Recycled Concrete Aggregate Production and Its Properties

The production of C-RCA and F-RCA aggregates was conducted in an innovative washing recycling plant. Once all the polluted components, big pieces of steel, wood or plastic, are removed from the demolition waste, the clean concrete waste is crushed to 0/20 mm recycled aggregate fractions. At this point, the water-washed treatment process and different aggregate fraction production starts: (1) the 0/20 mm fractions are transported to the recovery plant by conveyor belt for the next steps in the process. The recycled material is passed through a magnetic separator, and under the water spraying process, the 0/20 mm fractions pass through a 4 mm sieve where the fine and coarse fractions are separated. (2) The recycled aggregate fractions finer than 4 mm go to the hydro-cyclone separators. At this point, the material is divided into three recycled fraction sizes: 0/4 mm, 0/2 mm and a fraction finer than 63 µm (filler and clay). (3) After the cleaning process (with water and brushing), the coarse recycled aggregate fractions (>4 mm) are sieved in three different fractions, 14/20 mm, 8/14 mm and 4/10 mm.

Although the cleaning procedure is conducted via a water system, the entire volume of water employed in the treatment process is reused satisfactorily. In addition, rainwater is also harvested through storage tanks.

Four fractions (0/2 mm, 0/4 mm, 4/10 and 8/20) of RCA, designated FR1, FR2, CR1 and CR2, respectively, were used for concrete production (see Figure 2). The 8/20 mm fraction was produced mixing the fractions 8/14 mm 50% and 14/20 mm 50%. The recycled aggregates were characterised following EN 12620 "Aggregates for concrete" specifications.

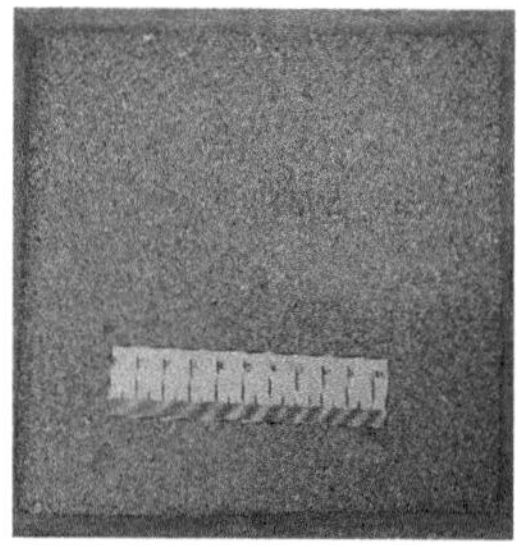

(**a**) FR1: 0/2 mm

(**b**) FR2: 0/4 mm

(**c**) CR1: 4/10 mm

(**d**) CR2: 8/20 mm

Figure 2. The four recycled aggregates (each line in the ruler are 10 mm).

The constituents of C-RCA aggregates (CR1 and CR2) were: Rc (Concrete and mortar) + Ru (unbound aggregate) of 96.27%, Rb (Ceramic) 1.94%, Ra (Asphalt) 1.40% and X (other impurities) 0.4%, determined following the EN 933-11:2009 specifications. According to the EN 12,620 specifications, the RCA aggregates composed of more than 95% concrete are categorised as Type A (Rc90, Rcu95, Rb10, Ra1, FL2 and XRg1).

The grading distribution of the four fractions of recycled aggregates 0/2 (FR1), 0/4 (FR2), 4/10 (CR1) and 8/20 (CR2) are shown in Figure 3. They were determined following EN 933-1 specification. According to the grading distribution, the fine fractions (FR1 and FR2) were categorised as Gf85. The 4/10 gravel fraction was classified as Gc90/15 and the 8/20 fraction as Gc85/20, the categories being highlighted by the Spanish concrete Structural Code [36].

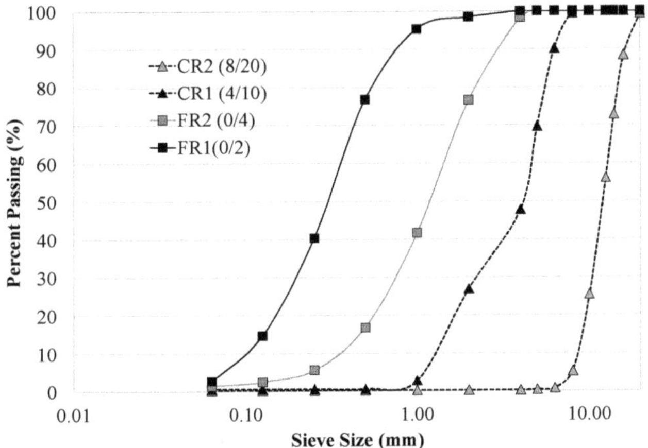

Figure 3. The grading distribution of all the recycled concrete aggregate fractions (FR1, FR2, CR1 and CR2).

The coarse recycled aggregates (CR1 and CR2) contained less than 1.5% of particles under 63 µm classified in the maximum category of f1.5. In addition, the FR1 and FR2 had a filler quantity of less than 3%. Therefore, they were assigned the top category f3.

The density and absorption capacity of all the aggregates fractions, described in Table 2, were determined following UNE-EN 1097-6 specification. All recycled aggregates met the requirements of the Structural Code [36] at 7% (established by the concrete Structural Code). It should be clarified that FR1 0/2 sand was always mixed with FR2 0/4 sand (using 20% 0/2 and 80% 0/4). Therefore, the mixture fraction met the limit set by the structural code of 7% absorption capacity. Furthermore, as mentioned above, all the used recycled aggregates were found to be well within the absorption limitation established by international regulations for recycled aggregates to be used in non-structural concrete. For example, 10% in Hong Kong and 20% in The International Union of Laboratories and Experts in Construction Materials, Systems and Structures (RILEM, Paris, France) German and Norwegian specifications [37].

Table 2. Dry density and absorption capacity of RCA.

	FR1-0/2	Desv	FR2-0/4	Desv	CR1-4/10	Desv	CR2-8/20	Desv
Dry Density (kg/dm^3)	2.12	0.06	2.29	0.05	2.24	0.03	2.30	0.01
Absorption (%)	7.8	0.04	5.6	0.46	6.1	0.05	5.6	0.25

The per cent of acid-soluble sulphate and water-soluble chlorides salts were determined according to the EN 1744-1 specification. The limit established by the Structural Code for aggregates used in the manufacture of concrete is 0.8% and 0.05% of acid-soluble sulphates and soluble chlorides, respectively. Table 3 summarises the obtained values. All recycled aggregates met the standard requirement.

Table 3. Chemical analysis of recycled aggregates.

	FR1-0/2	FR2-0/4	CR1-4/10	CR2-8/20
Acid soluble Sulphate (%)	0.13	0.11	0.23	0.16
Water-Soluble Chloride salts (%)	0.005	0.0014	0.007	0

The shape factor and Los Angeles abrasion coefficient of coarse recycled aggregate fractions were determined following UNE-EN 933-3 and UNE-EN 1097-2 specifications, respectively. The shape factor of the recycled concrete aggregate fraction was 6%, below 35% (maximum value allowed by the Structural Code). According to the Los Angeles coefficient, RCA obtained a maximum of 36%, less than 40%, a limiting value for structural concrete aggregates. Following the EN 12,620 specification, the shape factor and Los Angeles abrasion coefficient are classified as Fl_{15} and LA_{40}, respectively.

As mentioned above, the recycled aggregates were water cleansed. Consequently, after the washing process, all RCA aggregate fractions had a high moisture content in oversaturated conditions. Therefore, in order to control de water amount in concrete mixture, the moisture of aggregates was determined before their use in concrete production. Many researchers recommend the employment of recycled aggregates in highly humid conditions to produce concrete. However, it should not be saturated, as this could negatively affect the interface transition zone [38,39]. Nevertheless, in this case, due to the water washing process, the recycled aggregates were employed in wet conditions, sometimes even oversaturated. According to the obtained properties of RCA, the water-washed industrial treatment process guaranteed high quality RCA, adequate to be used in concrete. Other recycled aggregates treatments [40], only applied in laboratory scale, achieved also improvements in their physical properties. However, recycled aggregates which are required to be in a dry state or with medium humidity grade have been found up to be difficult to combine with the industrial water-washed treatment.

2.2. *Methods*

Two phases of concrete production were conducted. Phase 1 included experimental laboratory work. For pavement-based layer construction, adequate mix proportions of concrete with 100% fine and coarse recycled aggregates (RAC concrete) were designed. The concrete was required to have acceptable workability (6–8 cm of slump test) 30 min after casting for adequate concrete placement. In addition, the produced concrete should have a minimum of 20 MPa compressive strength (strength in cylindrical specimens defined by Structural Code) after 28 days of curing. In phase 2, a RAC concrete pavement-base layer with a minimum of 20 MPa compressive strength (in cylindrical specimens) was built in Barcelona's city centre using an adequate mix proportion; the fresh and hardened properties were determined.

2.2.1. Laboratory Work, Phase 1: Mix Proportions and Test Procedure

All the concretes were produced employing 100% of fine and coarse RCA. 300 kg, 280 kg and 270 kg of cement were used in different mix proportions (see Table 4) to determine the minimum amount of cement needed to achieve adequate properties. In addition, the effective water/cement ratio was defined to establish the concretes' mix proportions. While the effective water/cement ratio 0.55 was defined for the concretes produced with 300 kg and 285 kg of cement per m^3 of concrete, the value of 0.52–0.53 was defined for concrete produced with 270 kg of cement. The day before producing each concrete, all fraction of recycled aggregates were introduced into the oven at 100 °C, and the aggregates' humidity was determined to calculate the water amount to be added for concrete production and control its effective water/cement ratio. The average humidity values (and its standard deviation) of the FR1, FR2, CR1 and CR2 recycled aggregates were 12.2% (0.7%), 6.7% (0.4%), 6.1% (1%) and 4.6% (0.8%), respectively. The fine fractions (FR1 and FR2) were oversaturated when concretes were produced. Consequently, the water present on the surface was considered part of effective water for concrete production. However, the coarse recycled aggregates (CR1 and CR2) had a high humidity but were not saturated. In this case, their effective absorption capacity was calculated (determined by submerging them in water for 20 min), and the determined water amount was added to the concrete to be absorbed by the CR1 and CR2 in order to maintain a constant effective water/cement ratio [5]. The multifunctional (P) and superplasticiser (S) chemical admixtures were used in different percentages to achieve the desired workability of 6–8 cm slump value at 30 min or later after casting.

Table 4. Mix proportions of concrete mixtures. The values are given as weight (in dry condition) over the volume of concrete production (kg/m^3).

	RAC-300	RAC-285-1	RAC-285-2	RAC-270-1	RAC-270-2	RCA-270-3
CEM II A-L 42.5 R	300	285	285	270	270	270
Efective w/c ratio	0.55	0.55	0.55	0.52	0.52	0.53
Total w/c ratio	0.91	0.91	0.90	0.90	0.89	0.94
Water	165	156.75	156.75	140.4	140.4	143.1
CR2 8/20 mm	682.6	694.3	694.3	713.2	713.2	710.8
CR1 4/10 mm	273.6	278.3	278.3	285.9	285.9	284.9
FR2 0/4 mm	589.9	600	600	616.4	616.4	614.3
FR1 0/2 mm	188.9	192.1	192.1	197.4	197.4	196.7
S * (%)	1	1	1	1.2	1	1.3
P * (%)	1.5	2	2.5	2.5	2.7	3
		WORKABILITY (slump test in cm)				
t = 0 min	21.5	17	20	20	11	22
t = 30 min	7	-	3.5	-	6	18
t = 60 min	-	-	-	-	-	6.5

* The amount of S and P used in concrete production is defined as per cent of the cement weight.

All the produced concretes' workability and slump value were determined following the UNE-EN 123350-2:2020 specifications. The slump values were determined immediately after concrete casting (t = 0 min) and 30 min or 60 min after concrete casting (t = 30 min or t = 60 min). Between 0 min (immediately after concrete casting) and 30 min or 60 min, the concrete mixture was kept in the mixer, which was stopped and covered with a plastic sheet until the test time elapsed. The concrete was then mixed for one minute before the slump was again measured.

The concrete specimens were produced and cured following UNE-EN 12,390-2:2001 regulations and manually compacted using a steel rod. The concrete specimens were then covered with a plastic sheet and air-cured for the first 24 h.

After 24 h of casting, the concrete specimens were demoulded and stored in the humidity room at 22 °C and 95% humidity until tested. In the concretes' hardened state, the compressive strength at 7, 28 and 56 days were determined following UNE-EN 12390-3:2020 specifications using cylindrical specimens of diameter Ø100 × 200 mm in length. In addition, the physical properties at 28 days were determined following UNE-EN 12390-7:2020 specifications using cubic specimens of 100 × 100 × 100 mm. Drying shrinkage was measured using concrete prismatic specimens of 70 × 70 × 285 mm, exposed to 20 ± 2 °C and relative humidity 55 ± 5% for 56 days following UNE-EN 12390-16:2020 specifications. Each result represented the average of three measurements.

2.2.2. Pavement-Base Construction

In May 2021, a 500 m long pavement-base layer with a 30-cm thickness (highlighted in green in Figure 4) in Passeig de Colom (PC) in Barcelona, Spain was constructed. Plain concrete with a minimum of 20 MPa compressive strength was manufactured using FR1, FR2, CR1 and CR2 fractions. In addition, more than 815 m^3 of washed recycled aggregates were employed in pavement-base layer construction.

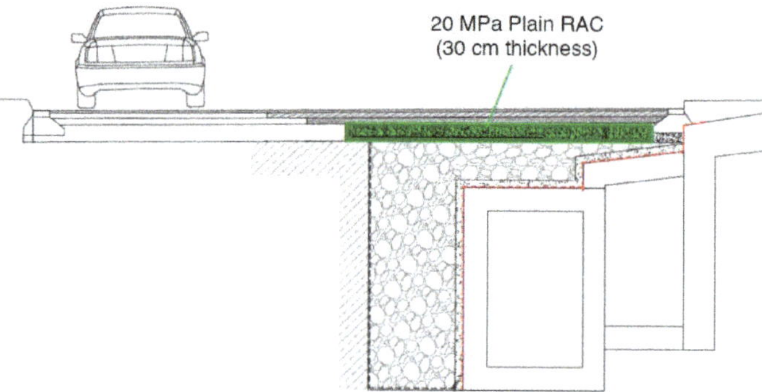

Figure 4. Section of pavement-base built using 100% recycled aggregate.

Table 5 describes the three-mix proportions employed for pavement-base layer construction. The humidity of the aggregates was determined before producing the concretes, except for the first day RAC-300-PC1 concrete. The humidity was determined after concrete production and was higher than initially estimated. The average humidity values (and its standard deviation) of the FR1, FR2, CR1 and CR2 recycled aggregates were 21.5% (1.2%), 9.9% (0.4%), 6.4% (0.2%) and 4.9% (0.5%), respectively. The fine recycled aggregates (FR1 and FR2) had higher-humidity concretes produced in the laboratory. The concrete mixture used for pavement-base layer construction was designed with 300 kg of cement and an effective water/cement ratio of 0.52–0.53 to assure the minimum strength of 20 MPa. However, because the humidity of the aggregates used in RCA-300-PC1 concrete production was higher than the estimated values, the RAC-300-PC1 concrete was produced with an effective water/cement ratio of 0.59. The rest of the pavement-base layer production was constructed using the RCA-300-PC2 and RCA-300-PC3 concretes, with effective water/cement ratios of 0.52 and 0.53, respectively.

Table 5. Mix proportions of the concrete produced. The values are given as weight (in humid conditions) over the volume of concrete production (kg/m^3).

	RAC-300-PC1	RAC-300-PC2	RAC-300-PC3
CEM II A-L 42.5	300	300	300
TOTAL water	101.5	88.0	97.4
Efective w/c ratio	0.59	0.52	0.53
FR1 0/2 mm	246.4	253.3	244.8
FR2 0/4 mm	688.8	708.1	702.3
CR1 4/10 mm	274.1	281.8	280.5
CR2 8/20 mm	665.3	683.9	691.5
S ** (%)	1.05 + 0.2	1.05 + 0.2	1.05 + 0.2
P (%)	1	1	1

** S and P: % of admixture for the cement weight. The second quantity of superplasticizer (+0.2) was added directly to the truck on-site to increase the workability of the concrete for placement in suitable conditions since the concrete arrived at the site one hour after its manufacture.

All the concretes' workability and slump values were determined on-site (see Figure 5a–c).

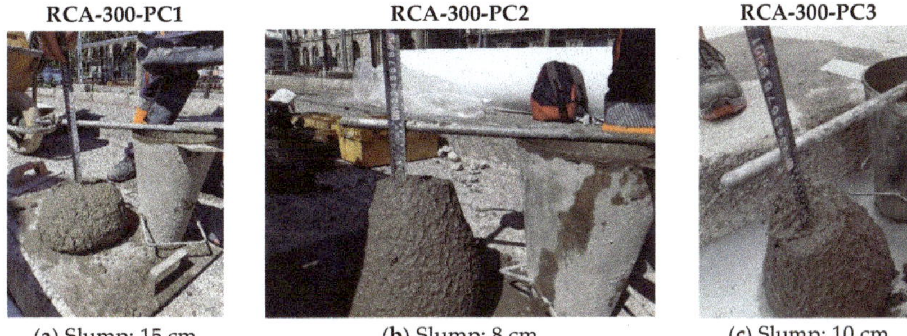

(**a**) Slump: 15 cm (**b**) Slump: 8 cm (**c**) Slump: 10 cm

Figure 5. On-site slump test determination.

The concrete mixtures were placed on-site directly from the truck (see Figure 6a). After compaction using a needle vibrator (see Figure 6b), the admixture of acrylic resin-based evaporation reducer, curing improvement admixture, was added to the concrete surface (see Figure 6c) to guarantee an adequate curing process. Figure 6 summarises the concrete laying process.

(**a**) (**b**) (**c**)

Figure 6. On-site concrete laying (**a**), compaction (**b**) and curing process (**c**) of the concrete.

Concrete specimens were fabricated to characterise the produced concretes. The concrete samples were manually compacted using a steel rod. The specimens were then covered with a plastic sheet and air-cured for the first 24 h at the worksite. After 24 h of casting, the concretes specimens were moved to the university laboratory, demoulded and stored in the humidity room until the required testing. The mechanical properties of compressive and flexural strength were determined. The compressive strength at 7, 14, 28 and 56 days was determined for cylindrical specimens of Ø100 mm × 200 mm and cubic specimens of 150 × 150 × 150 mm for 28 days. In addition, the flexural strength was determined using prismatic samples of 100 × 100 × 400 mm at 28 days of curing. The physical properties at 28 days were determined using cubic specimens of 100 × 100 × 100 mm. Finally, the durability properties of drying shrinkage and depth of penetration of water under pressure were determined. The drying shrinkage was determined using 70 × 70 × 285 mm specimens. Cylindrical specimens of Ø100 × 200 mm were used to calculate the penetration depth, following UNE-EN 12,390−8:2020 specification. Each result was recorded as the average of three measurements.

3. Results

3.1. Laboratory Work: Phase 1

3.1.1. Workability of Produced Concrete

Table 4 shows the workability achieved of all the concretes produced. The RCA-300 concrete achieved an initial slump value of 21.5 cm and adequate laying properties with 7 cm of slump value after 30 min of production. It was produced with an effective and total water/cement ratios of 0.55 and 0.91, respectively. In addition, 1% and 1.5% of admixtures S and P, respectively, were used for concrete production.

The concrete produced with 285 kg of cement (RCA-285) with S 1% and P 2.5% achieved an initial slump value of 20 cm. However, after 30 min of casting, the slump value dropped to 3.5 cm, too dry to lay it on-site under normal conditions. The RCA-285 concrete was also produced with an effective water/cement ratio of 0.55, but with a little less cement. Consequently, the water amount employed was also lower than that used for RCA-300 production. Thus, the RCA-285 concrete would initially require a slightly higher S admixture and a somewhat higher slump value.

The concretes produced with a lower cement amount (RAC-270) were produced with a lower effective w/c ratio (0.52–0.53). Consequently, they needed more admixtures to achieve a good slump value. The RCA-270-1 concrete using S 1% and P 2.7% achieved a slump value of 6 cm at 30 min. The RCA-270-3 concrete with S 1.3% and P 3% achieved a slump test of 6.5 cm at 60 min. Therefore, an amount slightly higher than 1% S admixture (reached 1.3%) together with up to 3% P was necessary for an adequate slump value at 30 or 60 min after casting.

3.1.2. Hardened Properties of Concrete

The hardened properties were determined only in concretes with an initial slump value ≥20 cm. Table 6 shows the mechanical and physical properties obtained by the produced concretes. All the produced concretes achieved the required compressive strength of 20 MPa at 28 days, reaching similar values to those obtained by other researchers [41]. While the RCA-300 concrete achieved a strength of 29.7 MPa at 28 days, the RCA-285-2 concrete achieved a strength value of 30.4 MPa. However, as shown in Table 4, the RCA-285-2 concrete failed to reach an adequate slump value.

Table 6. The mechanical and physical properties of concretes produced in the laboratory.

	Compressive Strength (MPa)			**Flexural Strength (MPa)**	**Dry Density (kg/dm^3)**	**Absorption (%)**	**Accesible Porosity (%)**
Concrete Type	7 Days	28 Days	56 Days	28 Days	28 Days	28 Days	28 Days
RCA-300	25.5 (0.8)	29.7 (0.6)	30.7 (0.5)	3.86 (0.02)	2.10	6.06	12.74
RCA-285-2	25.9 (0.2)	30.4 (1.0)	32.3 (1.0)	-	2.11	5.65	11.90
RCA-270-1	22.4 (0.4)	26.0 (0.8)	26.9 (0.9)	-	2.06	5.97	12.30
RCA-270-3	19.9 (1.8)	26.4 (0.6)	27.2 (0.6)	-	2.09	6.83	14.29

In addition, the RAC-270 concrete, produced with 270 kg of cement, achieved a compressive strength of 26 MPa at 28 days, with an acceptable slump value (see Table 4), thus being adequate for the defined application. However, it should be noted that the RCA-270-3 concrete achieved the lowest strength value at seven days and the highest standard deviation, probably due to the use of P 3% admixture. Nevertheless, the compressive strength increase from 7 days to 28 days was 34% (due to effective of 3% P [5]) compared to the other concrete, which rose to 16.5% during the same period. All the produced concretes achieved a similar compressive strength increase of 3–6% from 28 to 56 days of curing. In addition, the standard deviation of compressive strength values was low in all the concretes. The flexural strength was only determined in RCA-300, guaranteeing that this concrete would be adequate for a pavement-base application [42].

A study of the physical properties achieved, showed that all the produced concretes had a lower dry density and higher absorption capacity than conventional concrete and concrete using only coarse RCA [43,44]. It was also determined that they achieved similar densities to those concretes produced with 100% coarse recycled mixed aggregates [45]. In addition, the RCA-270-3 achieved the highest porosity, as it combined the lowest cement amount of 270 kg, a higher water/cement ratio than RCA-270-1 and the highest P admixture. However, the obtained values were shown to be similar to values achieved by various researchers' concretes, which were produced with 100% coarse and fine aggregates [41]. Consequently, they are acceptable for non-structural plain concrete pavement-base applications [42].

Figure 7 shows the obtained drying shrinkage values (Figure 7a) and the mass loss (Figure 7b) of the produced four concretes. The obtained values in the four concretes were alike as they employed similar amounts of recycled aggregates and total water in all the produced concretes. The value of all the concretes reached −1000 $\mu\epsilon$ in 56 days. In addition, according to ACI [46], the typical drying shrinkage values were −200 to −800 in conventional concrete when a high water/cement ratio was used. The drying shrinkage of concrete with recycled aggregates is always higher than that produced with natural aggregates due to the reduced restraint of recycled aggregate and the high water content [33], reaching to high shrinkage values when 100% of fine and coarse aggregates are employed [47]. However, most of the research works carried out up to date have only analysed concretes using recycled coarse aggregates [48]. In addition, Figure 7b shows that all the concrete mixtures achieved a mass loss % of 4.25% after 56 days of drying test exposure.

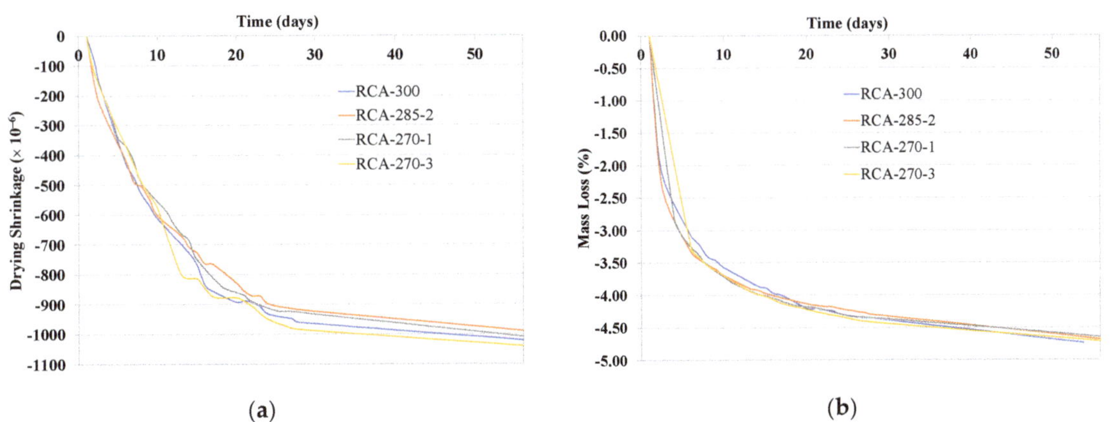

Figure 7. (**a**) drying shrinkage ($\times 10^{-6}$) and (**b**) mass loss (%) of the concrete samples.

Figure 8 shows the relationship between weight loss % and shrinkage value of the four concretes. The concrete RCA-270-1 and RCA-270-3 produced with the highest volume of recycled aggregates suffered a higher mass loss with very low shrinkage value. In addition the RCA-270-3, produced with the highest total w/c ratio, suffered more than 3% mass loss with a shrinkage value of −120 $\mu\epsilon$. However, all the concretes achieved similar last shrinkage and total mass loss (%) values.

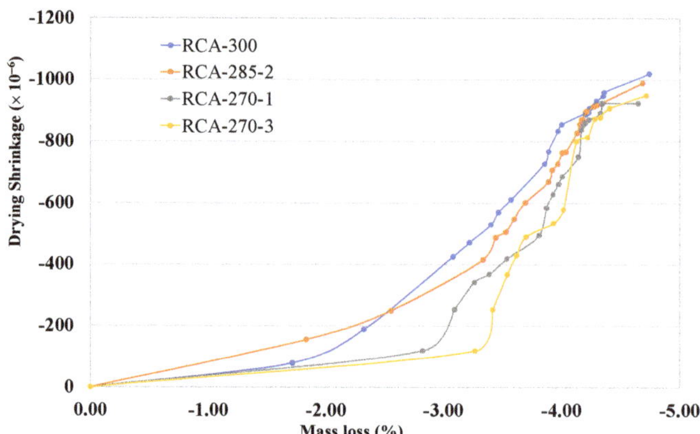

Figure 8. Relationship between mass loss (%) and drying shrinkage value.

3.2. *Pavement-Base Construction*

3.2.1. Workability of Concrete Produced

Figure 5 shows the slump values for the three concrete mixtures produced. All concrete included 1% S and 1% P in the production process. The recycling plant and concrete producer were located 25 km from Barcelona. However, heavy morning traffic at the entrance to Barcelona caused huge delays. Under those conditions, the trip could take up to 50 min. When the truck arrived, all the concrete mixtures had a slump lower than 5 cm. Consequently, to increase the concrete workability, it was necessary to add 0.20% S to the concrete mixture. After mixing the concrete for 10 min, slump values of 15 cm, 8 cm and 10 cm were determined for the RCA-300-PC1, PCA-300-PC2, and RCA-300-PC3 concretes, respectively (Figure 5). The RCA-300-PC1 was produced with the highest water/cement ratio, achieving the highest slump. The PCA-300-PC2 and RCA-300-PC3 concretes were made with a similar effective w/c ratio and reached a similar adequate slump value.

3.2.2. Hardened Properties of Concrete

Table 7 shows the results obtained for the pavement concrete. The RCA-300-PC1 concrete achieved a lower strength due to a higher effective water/cement ratio (see Table 5) in its production than the RCA-300-PC2 and RCA-300-PC3 concrete. However, all concretes achieved the minimum required 20 MPa and a low standard deviation value. The water-washed recycled aggregates had a positive effect, reaching a low standard deviation for the mechanical results.

Table 7. The mechanical properties of produced pavement-base concrete. The standard deviation of the obtained results is described between the brackets.

	Compressive Strength (MPa)			Flexural Strength (MPa)
	7 Days	28 Days *	56 Days	28 Days
RCA-300-PC1	19.0 (0.2)	21.8 (0.2)	23.9 (0.6)	3.6 (0.3)
RCA-300-PC2	21.2 (0.1)	25.7 (0.7)	27.9 (0.7)	3.6 (0.5)
RCA-300-PC3	23.6 (0.4)	30.1 (0.6)	31.5 (1.4)	3.5 (0.2)

* Specimens of different shapes were tested. However, all the results were evaluated as cylindrical specimens.

The flexural strength of the concretes was similar in all concretes produced, with a value of 3.6 MPa. Other researchers [41] obtained a similar value, which could also be valid for higher pavement requirements limited to 3.5 MPa [42]. The flexural strength not only depends on the effective w/c ratio used but also on the effectiveness of the bonding zone between the cement paste and the aggregate, on the quality of the aggregate to adhere to the cement paste [31,32]. In this case, all the concretes were manufactured with the same quantity and quality of aggregates.

Table 8 describes the physical properties of concrete and the depth of water penetration under the pressure of the produced concretes. The concretes reached low densities, caused by the density of the recycled aggregates used in 100% of the aggregate volume in the mixture [27]. In addition, the water absorption capacity was also high and higher in concrete produced with a higher water/cement ratio (RCA-300-PC1) [27,41]. It must be noted that since there was no presence of contaminants in the aggregates, and the aggregates and the cement used had good compatibility, these properties did not present any disadvantage for the durability of this plain concrete.

Table 8. The physical properties and depth of water penetration under pressure in produced concrete.

	Dry Desity (kg/dm^3)	Absorption (%)	Depth of Penertation (cm)	
			Average	Max
RCA-300-PC1	2.02	8.37	5.9	6.3
RCA-300-PC2	2.04	7.72	1.6	2.1
RCA-300-PC3	2.07	7.15	1.6	2.1
Structural Code requirement	-	-	3	5

According to the Structural Code, the maximum and average water penetration depth under pressure should be 5 cm and 3 cm, respectively. However, the RCA-300-PC1 reached a higher penetration depth than the code required (see Table 8 and Figure 9a–c) caused by a concrete mixture with a water/cement ratio that was too high. The other two concretes, RCA-300-PC2 and RCA-300-PC3, employed in all the pavements volume except the first casting, achieved adequate properties in concretes for structural code requirements. However, this concrete was designed as a pavement-base layer (non-structural element). Consequently, it did not need to achieve the requirements of structural concrete.

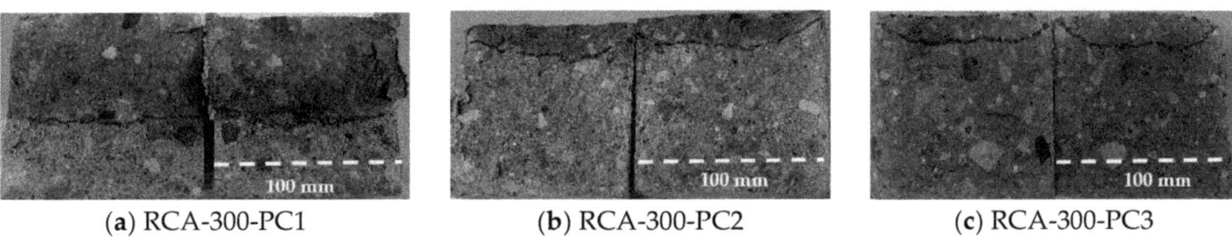

(**a**) RCA-300-PC1 (**b**) RCA-300-PC2 (**c**) RCA-300-PC3

Figure 9. Depth of penetration of water under pressure of three mixtures (**a**–**c**).

Figure 10 shows the obtained drying shrinkage values (Figure 10a) and the mass loss (Figure 10b) of the specimens produced in three different mixtures. As described above, the drying shrinkage was high due to the high water content in the 100% of coarse and fine recycled aggregate employed, concrete production [33]. In addition, the mass loss was 4% in concretes. The obtained values were very similar to those of the laboratory.

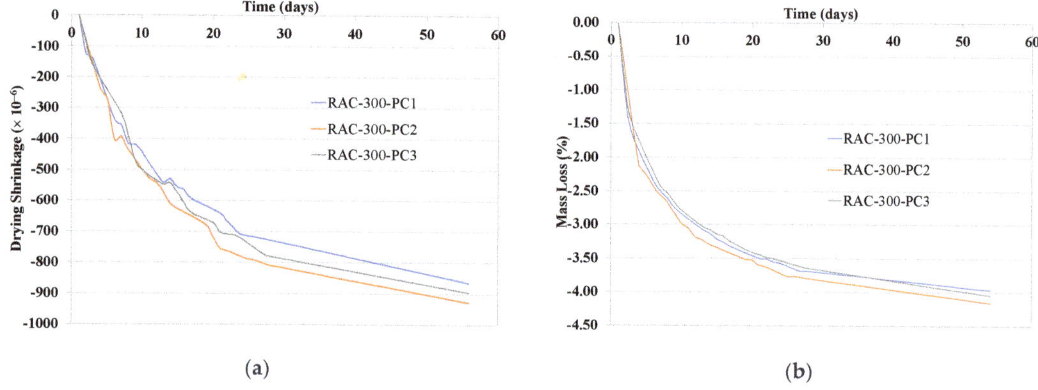

Figure 10. (**a**) drying shrinkage ($\times 10^{-6}$) and (**b**) mass loss (%) of the concrete samples.

The relationship between the mass loss % and shrinkage value of concrete mixtures of the three concrete mixtures was similar. The three concretes achieved similar total shrinkage and total mass loss (%) values (see Figure 11).

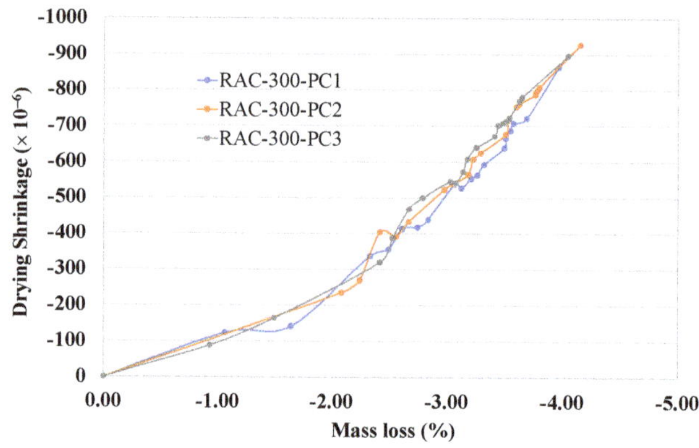

Figure 11. Relationship between mass loss (%) and drying shrinkage value of concrete mixtures of pavement.

4. Discussion

4.1. Workability

The obtained results determined that the workability was difficult to maintain over a sustained period of time when the concrete was produced using C-RCA and F-RCA aggregates. Several researchers [49] reached a similar conclusion regarding the packing effect of recycled sand, surface roughness of the particles and loss of chemical admixtures efficiency due to mortar adherence.

Laboratory analysis determined that an initial slump >20 cm was needed to achieve a good (6–8 cm) slump value of 30 or more minutes after concrete production. In addition, the combination of both admixtures, the high water-reducing superplasticiser (S) and multifunctional admixture (P), were necessary to achieve adequate workability. The RCA-300 concrete achieved sufficient but limited workability using 1% S and 1.5% P within the admixtures producer's recommendations. However, reducing the amount of cement and the effective water/cement ratio employed in RCA-285 and RCA270 concretes resulted in the need for a higher admixture to achieve adequate workability. RCA-270-3 concrete with S 1.3% and P 3% achieved acceptable slump values of 6.5 cm at 60 min after casting.

The multifunctional admixture (P) was required for use in a higher volume than that defined by its manufacturers to achieve adequate workability after 30 min of production. Tahar et al. [49] found that bad retention rheology was observed when the concrete was made with 100% F-RCA. In addition, they also used double the volume of admixtures in recycled concrete than conventional concrete to achieve workability similar to conventional concrete. However, a high amount of P admixture could retard the setting time in the winter season and should be validated.

In the pavement-based layer construction (industrial-scale), the concrete producers used a limited 1% S and 1% P for concrete production. The concrete mixtures had a dry consistency when the trucks arrived at the worksite. Consequently, a 0.2% S admixture was added into the ready-mix truck before concrete laying, achieving adequate workability in all the concretes. However, a higher S plus P was required to offset the concrete mixer driver's likelihood of adding water to the concrete to achieve a workable mixture. Gress et al. [12] also described that the recycled pavement concrete mixtures (using 100% C-RCA and up to 25% F-RCA) generally exhibited reduced workability due to the inherent angularity, rough surface texture, and high absorption characteristics of the RCA, being necessary to use water reducers, or fly ash to improve workability. Sadati and Khayat [9] described that did not have problem with placement, finishing and tinning at the job site, which included casting 450 m^3 of concrete in a length of 300 mm when the concrete produced with 40% of C-RCA had a slump value of 35 mm.

4.2. Hardened State Properties

All the concretes produced in the laboratory achieved the required 20 MPa for the application. Initially, 300 kg of cement/m^3 of concrete was used, which reached adequate compressive and flexural strength with a low standard deviation. Moreover, the concrete produced with 270 kg of cement per m^3 also achieved a higher compressive strength than the minimum required for the application (20 MP). This concrete is the most acceptable from an environmental point of view. RCA-270 concrete employed 100% of recycled aggregates (fine and coarse), achieving circularity by re-introducing the construction and demolition waste as recycled aggregates into the production process. In addition, the over-exploitation of raw materials is reduced, maximising material life cycles [3]. Moreover, the RCA-270 concrete used the lowest amount of cement. Consequently, it also has the lowest contribution to global warming potential (GWP) by employing a lower amount of cement than any other concrete [50].

For industrial use, the pavement-base layer was produced using 300 kg of cement per m^3 of concrete. Although the concrete mix proportion was validated in the laboratory before being used in pavement construction, the first mixture (RCA-300-PC1) was produced using an effective water/cement ratio of 0.59, which proved to be too high in order to achieve the minimum strength required. When the recycled aggregates were used under high humidity, an effective water/cement ratio of 0.53–0.52 was required.

The density and absorption values obtained in laboratory and pavement construction concretes were comparable. They achieved lower and higher values, respectively, than those usually achieved by conventional concrete [27] and concrete produced using only C-RCA or F-RCA [27,33]. However, the depth of water penetration under pressure was below the maximum limit (except RCA-300-PC1 concrete made with a water/cement ratio of 0.59). Thomas et al. [51] concluded that permeability depended on the effective water/cement ratio.

In addition, the drying shrinkage values were high due to the low stiffness of RCA and high water accumulation by the RCA aggregates due to their high water absorption capacity [5,33]. The concretes produced in the laboratory and pavement construction achieved values similar to 950 microstrains drying shrinkage and 4.25% mass loss. As a consequence of the high strain value obtained, an admixture of acrylic resin-based evaporation reducer for concrete surfaces was applied (see Figure 6c) to reduce cracking risk. Sadati and Khayat [9] also described that increasing the w/c ratio increased the shrinkage value of pavement concretes. They too sprayed a liquid curing compound on the finished surfaces after finishing.

The high drying shrinkage limited the use of RAC in structural applications due to two effects: (1) excessive shrinkage deformation could endanger the safety of the structure because shrinkage of concrete plays a significant role in the design of the service limit state of structural members [33,48,52]; and (2) higher shrinkage may cause cracks. This could affect the overall performance of concrete and enable the ingress of harmful substances, resulting in the corrosion of structural concrete reinforcement [5]. However, the plain concrete pavement-base this risk does not occur as it is not reinforced.

5. Conclusions

The following conclusions are based on the results of our study of concrete made with 100% coarse and fine RCA for a pavement-base layer:

- The water-washed RCA achieved type A category, aggregates for concrete, adequate for pavement-base layer construction. In addition, the coarse and fine RCA fractions achieved an absorption capacity lower than 7%, and their physical and mechanical properties and chemical components fulfilled the requirements for use in structural concrete.
- The properties of the pavement-base layer concrete produced on the worksite were similar to those obtained in the laboratory. Consequently, the laboratory results were validated for high-scale production.
- All recycled concrete achieved adequate initial workability. However, to achieve an extended slump value (minimum for 30 min), a value of 3% multifunctional admixture plus 1–1.2% superplasticiser must be used in concrete production.
- On the worksite, a lower chemical admixture was employed, causing a dry consistency, and requiring 0.20% superplasticiser.
- All concretes achieved the required strength of 20 MPa. The concrete produced with 270 kg of cement per m^3 and a water/cement ratio of 0.53 resulted in the best properties with the lowest environmental impact.
- The concretes presented a high absorption capacity. However, the water penetration depth under pressure value was below the maximum limit established by the structural code when the recycled concretes were produced with a water/cement ratio of 0.53 independent of the amount of cement used.
- The drying shrinkage was high due to the employment of 100% of recycled aggregate concrete. Although there was no cracking in the concrete measured in the laboratory specimens, it is possible that the concrete could suffer high strain in low humidity days. However, pavement-base case is a non-structural plain concrete and consequently, problems with deformation and cracking to ingress harmful substances, resulting in corrosion cannot happen.

The water-washed recycled aggregates do not include filler or any chemical components. In addition, they have good physical and mechanical properties for use in structural concrete production. The employment of 100% coarse and fine RCA in concrete production causes a considerable increase in drying shrinkage. This could increase deformation, significantly reducing the service life of structural members and the risk of cracking, enabling the ingress of harmful substances. However, non-structural elements do not have those limitations, and they allow the massive volume of recycled aggregates validated in this case study to be employed. The authors believe it is a straightforward, simple way to re-circulate construction and demolition waste.

Author Contributions: Conceptualization, M.E. and M.Á.P.; methodology, M.E., M.K. and C.G.; validation, M.E., M.K., C.G. and M.Á.P.; investigation, M.E., M.K. and C.G.; resources, M.E. and M.Á.P.; writing—original draft preparation, M.E.; writing—review and editing, M.E.; project administration, M.E.; funding acquisition, M.E. and M.Á.P. All authors have read and agreed to the published version of the manuscript.

Funding: This research received no external funding.

Institutional Review Board Statement: Not applicable.

Informed Consent Statement: Not applicable.

Data Availability Statement: Not applicable.

Acknowledgments: The authors thank Ajuntament de Barcelona (Barcelona city council), Bimsa and SBS Simón i Blanco for their courage and support in working with recycled aggregate concrete. The first author also wants the staff of the Laboratory of Technology of Structures and Materials "Lluis Agulló" of the UPC for their support.

Conflicts of Interest: The authors declare no conflict of interest.

References

1. Tam, V.W.Y.; Soomro, M.; Evangelista, A.C.J. A review of recycled aggregate in concrete applications (2000–2017). *Constr. Build. Mater.* **2018**, *172*, 272–292. [CrossRef]
2. European Statistics-Eurostat. *Energy, Transport and Environment Statistics 2020 Edition*; Publications Office of the European Union: Luxembourg, 2020; ISBN 9789276207375.
3. European Commission. Communication from the Commission to the European parliament, the council, the European Economic and Social Committee and the Committee of the regions. Next steps for a sustainable European future European action for sustainability, Strasbourg, France, 2016. Available online: https://eur-lex.europa.eu/legal-content/EN/TXT/?uri=COM%3A2016%3A739%3AFIN (accessed on 6 January 2022).
4. Plaza, P.; del Bosque, I.S.; Frías, M.; de Rojas, M.S.; Medina, C. Use of recycled coarse and fine aggregates in structural eco-concretes. Physical and mechanical properties and CO2 emissions. *Constr. Build. Mater.* **2021**, *285*, 122926. [CrossRef]
5. Neville, A.M.; Neville, A.M. *Properties of Concrete*; Pearson Education Limited: London, UK, 2011; ISBN 9780273755807.
6. Renan Sorato, S.P.; Venkateswarlu, B. Recycled aggregate concrete; an overview. *J. Struct. Eng.* **2016**, *18*, 67–75.
7. Nwakaire, C.M.; Yap, S.P.; Onn, C.C.; Yuen, C.W.; Ibrahim, H.A. Utilisation of recycled concrete aggregates for sustainable highway pavement applications; a review. *Constr. Build. Mater.* **2019**, *235*, 117444. [CrossRef]
8. Selvam, M.; Debbarma, S.; Singh, S.; Shi, X. Utilization of alternative aggregates for roller compacted concrete pavements—A state-of-the-art review. *Constr. Build. Mater.* **2021**, *317*, 125838. [CrossRef]
9. Sadati, S.; Khayat, K. Field performance of concrete pavement incorporating recycled concrete aggregate. *Constr. Build. Mater.* **2016**, *126*, 691–700. [CrossRef]
10. Cuttell, G.D.; Snyder, M.B.; Vandenbossche, J.M.; Wade, M.J. Performance of Rigid Pavements Containing Recycled Concrete Aggregates. *Transp. Res. Rec. J. Transp. Res. Board* **1997**, *1574*, 89–98. [CrossRef]
11. Nassar, R.; Soroushian, P. Use of recycled aggregate concrete in pavement construction. *J. Solid Waste Technol. Manag.* **2016**, *42*, 137–144. [CrossRef]
12. Gress, D.L.; Snyder, M.B.; Sturtevant, J.R. Performance of Rigid Pavements Containing Recycled Concrete Aggregate. *Transp. Res. Rec. J. Transp. Res. Board* **2009**, *2113*, 99–107. [CrossRef]
13. Silva, R.V.; de Brito, J.; Dhir, R.K. Use of recycled aggregates arising from construction and demolition waste in new construction applications. *J. Clean. Prod.* **2019**, *236*, 117629. [CrossRef]
14. Poon, C.S.; Chan, D. The use of recycled aggregate in concrete in Hong Kong. *Resour. Conserv. Recycl.* **2007**, *50*, 293–305. [CrossRef]
15. Zhang, H.; Zhao, Y. Performance of Recycled Aggregate Concrete in a Real Project. *Adv. Struct. Eng.* **2014**, *17*, 895–906. [CrossRef]
16. Xiao, J.; Wang, C.; Ding, T.; Nezhad, A.A. A recycled aggregate concrete high-rise building: Structural performance and embodied carbon footprint. *J. Clean. Prod.* **2018**, *199*, 868–881. [CrossRef]
17. Chen, W.; Jin, R.; Xu, Y.; Wanatowski, D.; Li, B.; Yan, L.; Pan, Z.; Yang, Y. Adopting recycled aggregates as sustainable construction materials: A review of the scientific literature. *Constr. Build. Mater.* **2019**, *218*, 483–496. [CrossRef]
18. Pedro, D.; de Brito, J.; Evangelista, L. Influence of the use of recycled concrete aggregates from different sources on structural concrete. *Constr. Build. Mater.* **2014**, *71*, 141–151. [CrossRef]
19. Etxeberria, M.; Mari, A.; Vázquez, E. Recycled aggregate concrete as structural material. *Mater. Struct.* **2006**, *40*, 529–541. [CrossRef]
20. Kirthika, S.K.; Singh, S.; Chourasia, A. Alternative fine aggregates in production of sustainable concrete—A review. *J. Clean. Prod.* **2020**, *268*, 122089. [CrossRef]
21. López-Gayarre, F.; Serna, P.; Domingo-Cabo, A.; Serrano-López, M.; López-Colina, C. Influence of recycled aggregate quality and proportioning criteria on recycled concrete properties. *Waste Manag.* **2009**, *29*, 3022–3028. [CrossRef]
22. Miren, E.; Enric, V.; Marí, A.R. Microstructure analysis of hardened recycled aggregate concrete. *Mag. Concr. Res.* **2006**, *58*, 683–690.
23. Hansen, T.C. *Recycling of Demolished Concrete and Masonry*; E&FN Spon: London, UK, 1992.
24. Khatib, J.M. Properties of concrete incorporating fine recycled aggregate. *Cem. Concr. Res.* **2005**, *35*, 763–769. [CrossRef]

25. Bogas, J.A.; de Brito, J.; Ramos, D. Freeze-thaw resistance of concrete produced with fine recycled concrete aggregates. *J. Clean. Prod.* **2016**, *115*, 294–306. [CrossRef]
26. Lavado, J.; Bogas, J.; de Brito, J.; Hawreen, A. Fresh properties of recycled aggregate concrete. *Constr. Build. Mater.* **2020**, *233*, 117322. [CrossRef]
27. Nedeljković, M.; Visser, J.; Šavija, B.; Valcke, S.; Schlangen, E. Use of fine recycled concrete aggregates in concrete: A critical review. *J. Build. Eng.* **2021**, *38*, 102196. [CrossRef]
28. Sasanipour, H.; Aslani, F. Durability properties evaluation of self-compacting concrete prepared with waste fine and coarse recycled concrete aggregates. *Constr. Build. Mater.* **2019**, *236*, 117540. [CrossRef]
29. Tobori, N.; Hosoda, T.; Kasami, H. Properties of Superplasticized Concrete Containing Recycled Fine Aggregate. In *International Conference on High performance Concrete, and Performance and Quality of Concrete Structures*; Farmington Hills: Gramado, RS, Brazil, 1999.
30. Pereira, P.; Evangelista, L.; de Brito, J. The effect of superplasticizers on the mechanical performance of concrete made with fine recycled concrete aggregates. *Cem. Concr. Compos.* **2012**, *34*, 1044–1052. [CrossRef]
31. Evangelista, L.; de Brito, J. Durability performance of concrete made with fine recycled concrete aggregates. *Cem. Concr. Compos.* **2010**, *32*, 9–14. [CrossRef]
32. Wang, Y.; Liu, F.; Xu, L.; Zhao, H. Effect of elevated temperatures and cooling methods on strength of concrete made with coarse and fine recycled concrete aggregates. *Constr. Build. Mater.* **2019**, *210*, 540–547. [CrossRef]
33. Zhang, H.; Wang, Y.; Lehman, D.E.; Geng, Y.; Kuder, K. Time-dependent drying shrinkage model for concrete with coarse and fine recycled aggregate. *Cem. Concr. Compos.* **2019**, *105*, 103426. [CrossRef]
34. Pacheco-Torgal, F.; Ding, Y.; Colangelo, F.; Tuladhar, R.; Koutamanis, A. *Advances in Construction and Demolition Waste Recycling. Management, Processing and Enviromental Assessment*, 1st ed.; Elsevier: Amsterdam, The Netherlands, 2020.
35. H-ZERO. 2021. Available online: https://www.hercalzero.es/h-zero-planta-rcd/ (accessed on 6 January 2022).
36. Relaciones con las Cortes y Memoria Democrática; Ministry of Presidential. Codigo Estructural, España (Structural Code-Concrete). 2021. Available online: https://www.boe.es/eli/es/rd/2021/06/29/470 (accessed on 6 January 2022).
37. de Brito, J.; Agrela, F.; Silva, R.V. Legal Regulations of Recycled Aggregate Concrete in Buildings and Roads. In *New Trends in Eco-efficient and Recycled Concrete*; Elsevier: Amsterdam, The Netherlands, 2019; pp. 509–526. [CrossRef]
38. Poon, C.; Shui, Z.; Lam, L. Effect of microstructure of ITZ on compressive strength of concrete prepared with recycled aggregates. *Constr. Build. Mater.* **2004**, *18*, 461–468. [CrossRef]
39. Evangelista, L.; Gueded, M. Microstructural Studies on Recycled Aggregate Concrete. In *New Trends in Eco-Efficient and Recycled Concrete*; Elsevier: Amsterdam, The Netherlands, 2019; pp. 425–451.
40. Kazmi, S.M.S.; Munir, M.J.; Wu, Y.-F.; Patnaikuni, I.; Zhou, Y.; Xing, F. Effect of recycled aggregate treatment techniques on the durability of concrete: A comparative evaluation. *Constr. Build. Mater.* **2020**, *264*, 120284. [CrossRef]
41. Revilla-Cuesta, V.; Faleschini, F.; Zanini, M.A.; Skaf, M.; Ortega-López, V. Porosity-based models for estimating the mechanical properties of self-compacting concrete with coarse and fine recycled concrete aggregate. *J. Build. Eng.* **2021**, *44*, 103425. [CrossRef]
42. Ministerio de Fomento, E. *PG-3 Pliego de Prescripciones Técnicas Generales Para Obras de Carreteras y Puentes*; Ministry of Public Works: Madrid, Spain, 2015. Available online: https://apps.fomento.gob.es/CVP/handlers/pdfhandler.ashx?idpub=ICW020 (accessed on 6 January 2022).
43. Etxeberria, M.; Vázquez, E.; Mari, A.; Barra, M. Influence of amount of recycled coarse aggregates and production process on properties of recycled aggregate concrete. *Cem. Concr. Res.* **2007**, *37*, 735–742. [CrossRef]
44. Thomas, J.; Thaickavil, N.N.; Wilson, P. Strength and durability of concrete containing recycled concrete aggregates. *J. Build. Eng.* **2018**, *19*, 349–365. [CrossRef]
45. Etxeberria, M.; Fernandez, J.M.; Limeira, J. Secondary aggregates and seawater employment for sustainable concrete dyke blocks production: Case study. *Constr. Build. Mater.* **2016**, *113*, 586–595. [CrossRef]
46. Mcdonald, D.B.; Brooks, J.J.; Burg, R.G.; Daye, M.A.; Gardner, N.J.; Novak, L.C. *Report on Factors Affecting Shrinkage and Creep of Hardened Concrete Reported by ACI Committee 209*; American Concrete Institute (ACI): Farmington Hills, MI, USA, 2014; pp. 1–12.
47. Pedro, D.; de Brito, J.; Evangelista, L. Structural concrete with simultaneous incorporation of fine and coarse recycled concrete aggregates: Mechanical, durability and long-term properties. *Constr. Build. Mater.* **2017**, *154*, 294–309. [CrossRef]
48. Mao, Y.; Liu, J.; Shi, C. Autogenous shrinkage and drying shrinkage of recycled aggregate concrete: A review. *J. Clean. Prod.* **2021**, *295*, 126435. [CrossRef]
49. Tahar, Z.-E.; Ngo, T.-T.; Kadri, E.H.; Bouvet, A.; Debieb, F.; Aggoun, S. Effect of cement and admixture on the utilization of recycled aggregates in concrete. *Constr. Build. Mater.* **2017**, *149*, 91–102. [CrossRef]
50. de Souza, A.M.; de Lima, G.E.S.; Nalon, G.H.; Lopes, M.M.S.; Júnior, A.L.D.O.; Lopes, G.J.R.; Olivier, M.J.d.A.; Pedroti, L.G.; Ribeiro, J.C.L.; de Carvalho, J.M.F. Application of the desirability function for the development of new composite eco-efficiency indicators for concrete. *J. Build. Eng.* **2021**, *40*, 102374. [CrossRef]
51. Thomas, C.; Setién, J.; Polanco, J.A.; Alaejos, P.; De Juan, M.S. Durability of recycled aggregate concrete. *Constr. Build. Mater.* **2013**, *40*, 1054–1065. [CrossRef]
52. Silva, R.; de Brito, J.; Dhir, R. Prediction of the shrinkage behavior of recycled aggregate concrete: A review. *Constr. Build. Mater.* **2015**, *77*, 327–339. [CrossRef]

sustainability MDPI

Article

Properties of Concrete with Recycled Aggregates Giving a Second Life to Municipal Solid Waste Incineration Bottom Ash Concrete

Aneeta Mary Joseph [1,2], Stijn Matthys [1] and Nele De Belie [1,*]

[1] Magnel-Vandepitte Laboratory for Structural Engineering and Building Materials, Ghent University, 9052 Ghent, Belgium; aneetamary.joseph@ugent.be (A.M.J.); stijn.matthys@ugent.be (S.M.)
[2] Strategic Initiative Materials (SIM vzw), Project ASHCEM within the Program "MaRes", 9052 Ghent, Belgium
* Correspondence: nele.debelie@ugent.be

Abstract: Economic and environmental factors call for increased resource productivity. Partial or full replacement of Portland cement by wastes and by-products, and natural aggregates by construction and demolition wastes, are two prominent routes of achieving circular economy in construction and related industries. Municipal solid waste incineration (MSWI) bottom ashes have been found to be suitable to be used as a supplementary cementitious material (SCM) after various treatments. This paper reports a brief literature review on optimum use of recycled aggregates in concrete and an experimental study using replacement of natural aggregate by demolished concrete having MSWI bottom ash as partial replacement of Portland cement, and compares its properties to that of completely natural aggregate concrete. Additional water was added as a compensation for the water absorption by the recycled aggregate during the first 30 min of water contact during concrete mixing. Also the fine fraction of crushed concrete (<250 μm) was removed to reduce the ill-effects of using recycled aggregate. The replacement of aggregates was limited to 23% by weight of natural aggregate. The results prove environmentally safe and comparable performance of concrete including recycled aggregate with bottom ash to that of natural aggregate concrete.

Keywords: municipal solid waste incineration bottom ash; recycled concrete aggregate; recycled aggregate concrete; supplementary cementitious material; circular economy

Citation: Joseph, A.M.; Matthys, S.; De Belie, N. Properties of Concrete with Recycled Aggregates Giving a Second Life to Municipal Solid Waste Incineration Bottom Ash Concrete. *Sustainability* **2022**, *14*, 4679. https://doi.org/10.3390/su14084679

Academic Editors: Anibal C. Maury-Ramirez, Jaime A. Mesa and Marc A. Rosen

Received: 17 January 2022
Accepted: 8 April 2022
Published: 13 April 2022

1. Introduction

Municipal solid waste incineration bottom ash is a residue generated as a result of incineration of waste that is not feasible or economical to be recycled. In Europe, around 17.6 million tonnes of MSWI bottom ashes are generated per year [1]. In Denmark, Netherlands, France and Germany, the utilisation is around 98%, 67%, 72% and 65% respectively. In Belgium, 15% of bottom ash generated is utilised in Belgium itself, mainly as aggregate in road construction, 35% is utilised outside Flanders, in the Netherlands and Germany, and the remaining 50% is landfilled after stabilisation [2–4]. However, it also has potential to be used in applications with higher value such as cement replacement. Obstacles for its use are presence of elemental aluminium, heavy metals, salts etc. and various treatments have been proven to be beneficial to make it suitable as a supplementary cementitious material [5–7].

The various aspects of effective utilisation of MSWI bottom ash for building materials were previously reviewed and published in [6]. Furthermore, an elaborate study was conducted to investigate utilisation of treated MSWI bottom ash as supplementary cementitious material. Pre-treatment methods to reduce ill-effects of MSWI bottom ash while used in concrete were devised and an optimised concrete mix containing processed MSWI ashes as 20% of the binder has been designed and its performance in terms of mechanical, durability and environmental properties was studied and is reported [8,9].

In order to establish circular economy, it is also important to utilise the concrete after its service life. The annual demand for aggregates is currently around 50 and 2.7 billion tonnes worldwide and in Europe respectively [10,11]. Around 450 million tonnes of C&D wastes are generated annually in EU and only 28% is recycled [12]. Use of recycled aggregates, reduces mining of natural aggregates and its related environmental problems, and is an answer partly to the shortage of aggregates. So, to prove that also concrete containing MSWI bottom ash as a supplementary cementitious material can find a second life as recycled aggregates at the end of its service life, concrete samples containing MSWI bottom ash as SCM were crushed and used as recycled aggregate. The mechanical and durability properties of the recycled aggregate concrete (RAC) were compared to that of natural aggregate concrete (NAC) and are reported in this paper.

2. Use of Recycled Concrete Aggregates as Replacement for Natural Aggregate

Literature encompasses various studies on use of recycled aggregates (RA) such as recycled concrete aggregate (RCA), recycled masonry aggregate (RMA) and mixed recycled aggregate (MRA) as replacement for natural aggregate (NA). However, RCA is focussed on here. The properties of concrete with recycled aggregate are affected by various factors. Factors relating to aggregates include water absorption of the aggregate, saturation level of aggregate at the time of mixing, grading and size of aggregate, constituents of the aggregate and source and crushing process of the aggregate [13,14]. Factors related to mix design and mixing of concrete include water-cement ratio, presence of mineral additions, use of chemical admixtures, curing age, recycled aggregate content and mixing procedure [15,16].

Water absorption of RCA is an indicator of its porosity and will influence the strength of the RAC. Aggregates with high water absorption partially absorb the mixing water and reduce workability and effective water-cement ratio, thereby affecting the fresh (mainly workability) and hardened concrete properties (mainly shrinkage) [17]. RCAs have high water absorption compared to NA [18], and the resulting decrease of free water in the mix up to a certain level could increase the strength, however, an excessive decrease in free water can deteriorate the strength of concrete due to lower hydration of cement particles and poor workability of the mix compared to conventional concrete.

The volume of voids among coarse aggregate particles is filled up by fine aggregates and cement paste, which affects the density, the amount of cement mortar and performance of concrete. Instead of replacing simply a part of natural aggregate by RA, the aggregates need to be graded according to size fraction. This gives rise to a better packing of the matrix resulting in better mechanical and durability properties [19]. A study showed that increasing the maximum size of aggregates increases the strength of resulting concrete. However, this limits the use of aggregates to large members and mass concreting [20].

The constituents of RA have high impact on its properties. RA sometimes contains wood, asphalt, plastic, glass etc. The bond between these materials and cement paste is very low and that affects the mechanical and durability properties of the resulting concrete [21]. Residual mortar content in RA, and its condition also has an important effect on its properties [22,23]. Higher content of residual mortar in aggregates increases porosity, water absorption, and may affect associated properties. Strength and other hardened concrete properties decrease. Residual mortar content also has an effect on the shape parameters. It gives more angular shape and rough texture to the recycled concrete aggregate, and this significantly affects particle packing. Increase in angularity leads to increase in void content, and thus leads to higher paste demand [24].

The type of aggregate in the original concrete is also found to have an effect. Concrete with RA originating from concrete with pebbles used as coarse aggregate was found to have higher compressive strength than that with crushed rock as aggregates [25].

Properties of parent concrete determine the properties of the recycled aggregate to a great extent. RA generated from high strength concrete has high crushing strength, low porosity, low water absorption etc. [26,27]. In a study damage sensitivity of aggregates to recycling processes was assessed [28]. Three levels of crushing processes were conducted. Level 1 included primary crushing with jaw crusher and impact crusher and then separation into fine and coarse fraction. Level 2 and Level 3 consisted of an additional mechanical grinding step. While double crushing of the source concrete (Level 1) considerably reduced the density of the cracks in the coarse aggregate by eliminating the particles with microdefects and irregular voids, processing up to Level 3 introduced negligible new cracks. Only a very minor amount of cracking could be detected in adhered mortar or at the interfacial transition zone. Consequently, the recycled concrete aggregates obtained at each stage of the recycling process did not show any loss of integrity. Beyond this, extending of the recycling process up to Level 3 efficiently increased the physical performance of the concrete aggregate by reducing the adhering mortar [28].

Water to cement ratio of concrete is a major parameter influencing strength and workability. Actual water-cement ratio should be fixed taking into consideration the water absorption of aggregates. Use of pre-saturated aggregates affects the structure of the ITZ, which can be improved by use of mineral additions. Partially replacing fine aggregates by mineral additions was also found to improve the strength of RAC. Addition of superplasticizer can compensate for loss of workability and also facilitates reduction of water-cement ratio to improve the strength. Increase in content of RA usually decreases the strength of concrete depending on its quality. General agreement is that replacement of less than 30% does not alter the strength considerably [29].

3. Strategies to Improve Properties of Recycled Aggregate Concrete

Many problems associated with recycled aggregates are caused by cracked and adhered mortar on the surface of old aggregates. It is the presence of this mortar that leads to the increase in water absorption of the RCA and hence an increased water demand for the concrete made with RCA to maintain the workability. It leads to reduced strength and durability of the recycled aggregates and of the resulting concrete made with recycled aggregate. There are various options to reduce the ill-effects. These methods can be broadly classified into removal of the cement paste around the aggregates, strengthening the mortar around aggregate, and others related to mix design and mixing of concrete [30].

Removal of mortar can be done by mechanical grinding and sieving, acid treatment [13,31–33], heat treatment [30,33] or microwaving. Treatment of RCA in a Los Angeles abrasion machine for 300 revolutions with 12 charges and then sieving on 4.75 mm sieve decreased water absorption of aggregates by 32.3% [33]. Heating concrete to around 300 °C makes the mortar brittle and easier to remove by mechanical treatment [30]. Heating to high temperature and then suddenly quenching in water decreases the amount of mortar adhered to the aggregate and allows it to be removed easily by sieving or mechanical grinding [33,34]. Microwave heating makes concrete brittle which results in lower fracture energy. This makes it easier for the mortar fraction to be removed by mechanical grinding [35].

Strengthening of mortar can be done by carbonation, calcite deposition, mixing with pozzolan, cement slurry or polymer emulsion. Carbonation of $Ca(OH)_2$ which produces calcium carbonate leads to reduction in porosity and thus improvement in quality of recycled aggregate. However, carbonation for long time at high concentration of CO_2 leads to carbonation of C-S-H which increases the porosity of cement paste. Therefore, the carbonation treatment should be conducted at optimum CO_2 concentration [36–38]. Immersion of RAs in lime water before carbonation can introduce more carbonatable calcium, and it was proven effective in improving tensile strength of the resulting concrete [39]. Immersion of recycled aggregates in lime water even without carbonation was shown to improve the properties [40]. Another recent innovation investigated using recycled fine powder slurry along with carbonation to treat RCAs and it was found beneficial in reducing the porosity [41]. Various strains of bacteria, especially of the bacillus group, precipitate calcium carbonate, and can be used to reduce porosity. *Sporosarcina pasteurii* (*Bacillus pasteurii*) [42] and *Bacillus sphaericus* [43] have been successfully studied for this purpose. Accelerated calcite precipitation by submerging RCA in water and dissolving CO_2 in the water at high pressure was also studied [44]. Cement slurry [45], alkali activated binders [46] and various polymers have been tested to seal the pores and hence to reduce the water absorption of recycled aggregate. Polyvinyl alcohol (PVA) [47,48], silane or siloxane based polymers [49], sodium silicate [45], paraffin etc. were used effectively at optimum concentrations [30]. Spraying pozzolan slurry on recycled aggregate strengthens the cracks in the aggregate by reaction of the pozzolans with $Ca(OH)_2$ in the waste concrete forming C-S-H [30].

A special mixing technique called two stage mixing technique (TSMA) was studied as well [50]. In TSMA first the recycled aggregates are mixed with cement paste which forms a coating on the surface of the aggregate and fills up cracks and voids, before the actual concrete mixing takes place [50]. This mixing procedure was also found to decrease the creep of resulting concrete [51]. Replacing part of the fine aggregate by mineral addition can be used as a strategy to increase the compressive strength of recycled aggregate concrete [14]. Addition of basalt fibre in the range of 0.1–0.5% was found to increase flexural and splitting tensile strength of RAC with marginal increase in compressive strength [32]. Equivalent mortar volume method is yet another method in which quantity and quality of cement paste and aggregate in the RA is accounted for, and both the coarse aggregate and fresh paste content of the mix are adjusted accordingly to achieve the same total mortar volume as a companion mix with the same specified properties but made entirely with coarse natural aggregates of similar properties to the coarse natural aggregate contained in RCA [52].

To avoid problems due to high water absorption of RA, the pre-saturation method or the water compensation method can be used. It was reported that not more than 90% of the total water absorption of the aggregate is absorbed during the first two hours after mixing, so not all of the measured water absorption of the aggregates needs to be added as compensation [24]. In the pre-saturation method, RAs are saturated prior to mixing, and in the water compensation method, additional water required for absorption by RAs is supplied during mixing of concrete. A study confirmed that concrete produced by the pre-saturation method had slightly inferior fresh and hardened properties compared to that produced by the water compensation method. The authors attributed this observation to the absence of nailing effect in pre-saturated aggregates, which is the penetration of cement paste in superficial pores in aggregates [53]. Various other studies reported higher strength for concrete made with dry RAs [54,55]. A microstructural study of RAC using microtomography reported that using RCA in dried state during mixing resulted in RCA absorbing water and releasing air bubbles inside the new concrete matrix as long as the concrete remained in the fresh state. Consequently, a macropore formed surrounding the RCA boundaries [56]. No systematic difference in compressive strength and carbonation resistance between concrete produced with dry and pre-saturated aggregates was observed [17].

These factors need to be incorporated into the quality control of concrete at various stages and they are summarised in Figure 1. Maximum utilisation of RCA can be made possible by preserving and improving its quality right from the demolition up to the mixing of concrete.

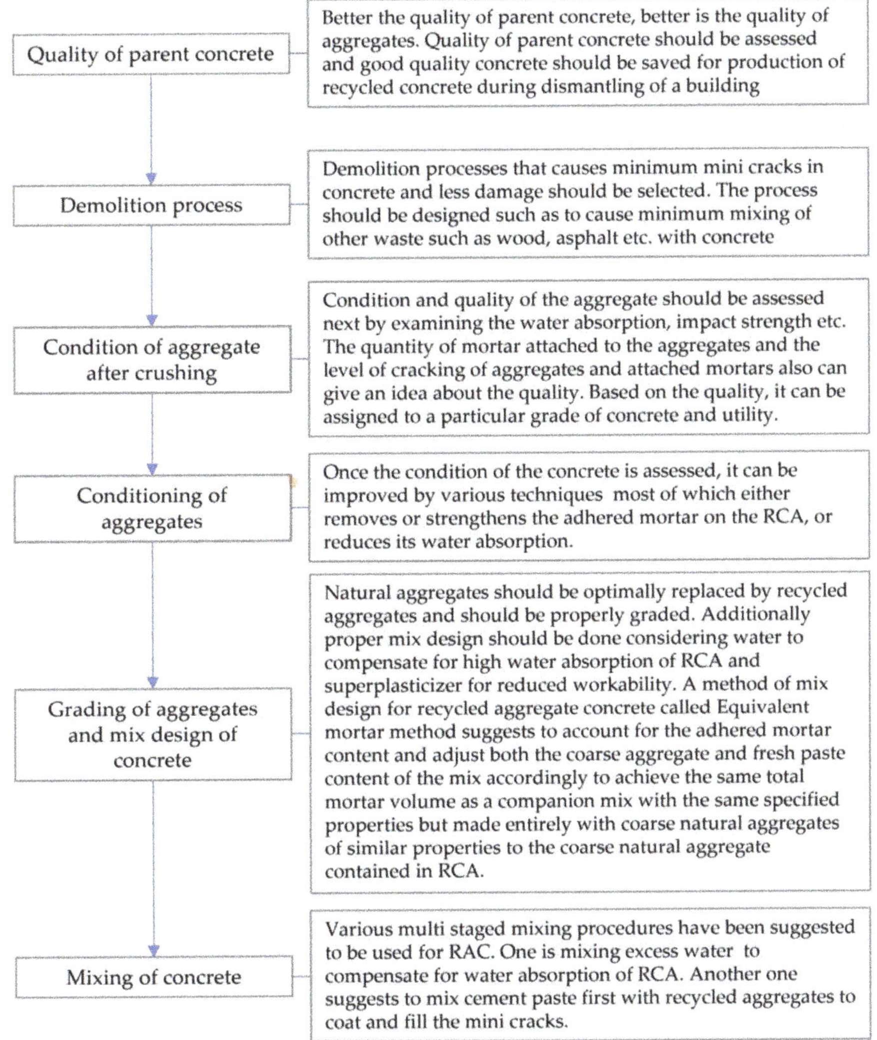

Figure 1. Steps in the quality control of recycled aggregate concrete.

4. Aim of the Study

Although the improvement and application of RCA for use in new concrete has been extensively studied, as shown in the literature overview given above, the introduction of novel SCMs in the parent concrete could affect the properties of the RCA and new RAC. The aim of this study was to show that the application of MSWI bottom ashes as SCM in concrete, will not affect its suitability to have a second life as RCA after demolition of the concrete structure. Therefore, RCA were made from parent concrete containing MSWI bottom ashes as a fraction of the binder. New RAC made with those RCA was tested for its mechanical properties (strength, modulus of elasticity, creep, shrinkage), durability (air permeability, carbonation resistance), and possible negative effects on the environment by leaching of hazardous compounds.

5. Materials and Methods

The experimental study described in this paper is part of a broader study on effective utilisation of MSWI bottom ashes in cement and concrete. Previous parts of the study on pre-treatment of MSWI bottom ashes can be found in our paper by Joseph et al. [5] and results on mechanical, durability and environmental properties of the bottom ash concrete can be found in our article by Alderete et al. [8]. Presence of hazardous elements in the bottom ash is a primary concern for its use in building materials. The bottom ashes used in the study were first processed at the incineration site to extract valuable metals and to render them non-hazardous. The processing steps consisted of washing, sieving, crushing, magnetic and eddy current separation etc. Additionally the ashes were milled and remaining metals were separated. The fraction of bottom ash between size 2 and 6 mm was selected (called 2/6NB), and its chemical composition along with that of the cement CEM I 52.5R used in the parent concrete, determined by XRF, can be found in Table 1.

Table 1. Chemical composition of cement and bottom ash used in the parent concrete and of the cement used in the RAC. Estimated relative error is ±10%.

Oxides	CEM I 52.5N (%)	CEM I 52.5R (%)	2/6NB (%)
CaO	67.80	68.80	18.40
SiO_2	18.10	18.30	43.90
Al_2O_3	4.34	3.52	10.20
Fe_2O_3	2.60	2.97	9.87
SO_3	4.06	4.05	2.26
MgO	1.41	1.02	2.49
P_2O_5	-	0.14	1.43
Cl	0.07	0.03	0.47
Na_2O	-	-	6.92
K_2O	0.78	0.89	1.23
ZnO	0.04	0.01	0.85
TiO_2	0.24	0.12	0.98
CuO	0.02	0.01	0.27
BaO	0.02	0.01	0.14
PbO	0.01	0.00	0.11
MnO		0.04	0.15

Various mixes were cast as trial to obtain a mix design with optimum replacement of cement by milled 2/6NB targeting similar strength as that of a mix with only CEM I 52.5N as binder (the chemical composition of this cement is also shown in Table 1). In these trial mixes, the water to binder ratio ranged from 0.35 to 0.45 and the cement replacement rate ranged from 10 to 20%. Additionally, CEM I 52.5R was used in the mixes with bottom ashes, instead of CEM I 52.5N in the Portland cement mix, to compensate for delay in strength gain when cement replacement is done. Four of these trial mixes were selected to produce RCA and their composition is shown in Table 2. The compressive strength at 2, 7 and 28 days is shown in Figure 2.

The recycled aggregate was manufactured in the lab by crushing equal amounts of concrete remnants left from compressive strength testing of the four trial mixes A, B, C and D, in a laboratory jaw crusher, Matest A092, and sieving out the fraction with size <250 μm since the fines affect the quality of aggregate the most. The crusher had 16 jaws with minimum and maximum opening width of 5 and 15 mm respectively. A mix of crushed particles from the four concretes of the trial mixes was used as recycled aggregate. In this study, a comparison is made between a reference natural aggregate concrete NAC (identical to Mix 1 in Alderete et al. [8]) and a recycled aggregate concrete, designated as RAC. Taking the composition of the NAC and the obtained particle size distribution of the recycled aggregate as a starting point, the recycled concrete mix design has been made by replacing part of the natural aggregates 0/4 and 4/16 as such that the overall particle size distribution remained nearly the same. The particle size distributions of the aggregates used in the study are shown in Figure 3. Extra water was provided to the mix to account for the water absorption of the recycled aggregates during the first 30 min. Total water absorption of recycled aggregates was determined after soaking for 24 h. However, to assess the mentioned extra water for pre-wetting of the recycled aggregates, rather the water absorption during the first 30 min was considered, to avoid the negative effects of the pre-saturation method as discussed in Section 3. This corresponds to the approximate time required for mixing, placing and compacting of concrete. The mix compositions of NAC and RAC are given in Table 3. Overall, 23% (by weight) of the natural aggregates have been replaced, and 27% in reference to the 0/4 and 4/16 fractions of NAC. To obtain similar workability, the amount of superplasticiser was slightly increased, as suggested in the literature (see Section 2).

Table 2. Mix composition of parent concrete.

	Mix A	Mix B	Mix C	Mix D
w/b	0.35	0.35	0.4	0.4
% cement replacement	10	15	10	15
Gravel 4/16, kg/m^3	1056	1055	1029	1028
Sand 0/4 kg/m^3	573	572	556	558
Sand 0/1 kg/m^3	287	286	279	279
CEM I 52.5R kg/m^3	333	314.5	333	314.5
2/6NB kg/m^3	37	55.5	37	55.5
Water kg/m^3	129.5	129.5	148	148
Plasticiser—Sika Viscocrete 1035 kg/m^3	8.4	8.4	3.7	3.7

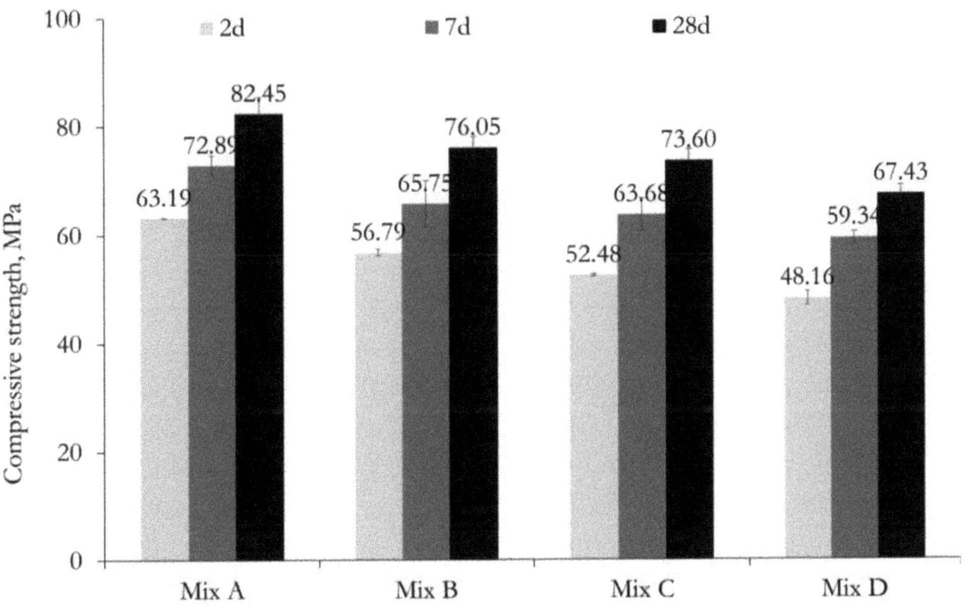

Figure 2. Strength of parent concrete with processed MSWI bottom ash used as cement replacement.

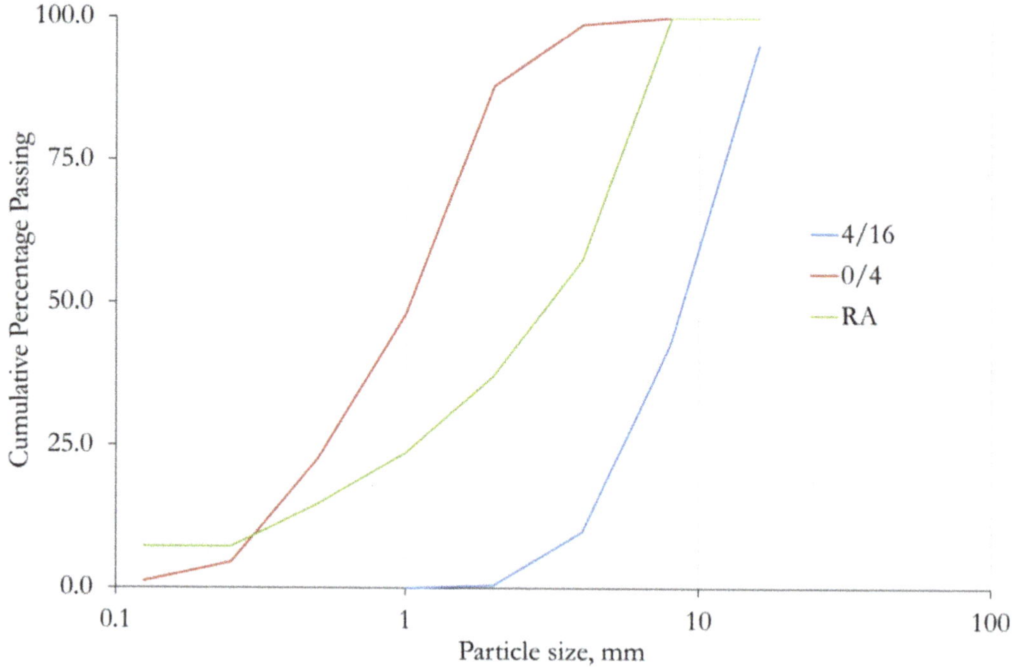

Figure 3. Sieve analysis of aggregates used in NAC and RAC.

Table 3. Composition of mixes used in the study (kg/m^3).

	NAC	RAC
Sand 0/1	271	271
Sand 0/4	541	378.8
Rolled gravel 4/16	997	698
RA		401.2
CEM I 52.5 N	360	360
Water	173	174 + 18 for pre-wetting
Superplasticiser	7	8

The mixing procedure of NAC was as follows: first cement, fine and coarse aggregates were dry-mixed for a minute; then the water was added and the mixing continued for 2 more minutes, finally the plasticiser was added and mixed for 1 extra minute. For RAC, the following procedure was followed. First the aggregates with pre-wetting water were mixed for 1 min. Then binder (CEM I 52.5N) and remaining water was added, and the mixing continued for 2 more minutes; finally the plasticiser was added and mixed for 1 extra minute. All the concrete samples were demoulded 1 day after casting and cured in a climate room at a temperature of (20 ± 2) °C and a relative humidity (RH) higher than 95% until 28 or 90 days of age, after which they were cut and subjected to preconditioning and testing.

Compressive strength tests were carried out on cubes with side length of 150 mm according to the standard NBN B 15-220 (1990) [57]. Three samples per mix were tested at 2, 7, 28, and 91 days of age. The test for modulus of elasticity of concrete was performed on cylinders of 150 mm diameter and 300 mm height after 28 days of curing according to Belgian standard NBN B 15-203 (1990) [58]. During the test, the concrete sample was subjected to a pressure of one third of compressive strength for three times, and then up to failure. While loading, the load and the deformations of the measurement bases of the specimen were recorded electronically. The value of the modulus of elasticity was calculated as the slope of the ascending branch of the last loading-unloading cycle.

Creep testing was conducted according to standard NBN 15-228 [59] on four specimens of size 150 × 150 × 500 mm, two for total creep and two for basic creep. The specimens were cured until 28 days of age in a wet chamber. Then two specimens for basic creep were covered with aluminium tape, and mechanical deformeter points (type demec) were fixed on all four. The other two specimens were tested uncovered. Specimens were loaded in compression to 30% of their compressive strength; the load was maintained constant with pressure vessels and the strain was measured regularly during 14 days using the mechanical deformeter strain gauges of type demec with a base length of 200 mm. Shrinkage of concrete was measured according to standard NBN 15-216 [60] on four specimens of size 150 × 150 × 600 mm after a curing period of 24 h in a wet chamber. After curing, two specimens for basic shrinkage were covered with aluminium tape and mechanical deformeter points (type demec) were fixed on all four sides, both covered and uncovered, similar to the creep specimens. Then strain was measured regularly using the deformeter.

Samples for air permeability tests were cut from slabs of size 400 mm × 400 mm × 100 mm. Four cylinders of 150 mm diameter were cored out of the slab and the middle part was cut out. For carbonation, cylinders of diameter 80 mm and 48 mm height were cut out from cubes of side 100 mm. After the curing period, the samples for air permeability, capillary imbibition rate and carbonation were cut and immersed in water for 72 h, followed by drying at 50 °C till the mass change measured at 24 h intervals was less than 0.1%. The lateral surfaces of samples for carbonation were covered with epoxy resin. After drying the specimens were kept covered in polythene bags for 4 weeks for the moisture to redistribute in the specimens and to have a homogenous relative humidity in the pores.

Air permeability of the specimens was measured by Torrent permeability tester which is a non-destructive method to measure air permeability of concrete according to Swiss standard SIA 262/1:2013. After 4 weeks of homogenisation, the surface moisture content of the samples was measured using a screed moisture meter PCE PMI-4 that measures moisture content by calcium carbide method [61]. The air permeability apparatus was calibrated and the permeability was measured right away. Both pre-conditioning and measurement were carried out in a temperature and humidity-controlled room at 20 °C and 60% RH.

Specimens for carbonation testing were exposed to 1% CO_2 in a conditioned chamber at 20 °C and a RH of 60%. Further, the carbonation depth was measured using phenolphthalein indicator. After different exposure times, specimens were taken out of the carbonation chamber and were split into two halves longitudinally and sprayed with 1% phenolphthalein solution. The carbonated region remains uncoloured and the uncarbonated region turns purple. Carbonation depth is measured at 8 points and its average is reported.

To further prove that the recycled concrete is safe for use, leaching was tested with the two-step shake test according to CMA/2/II/A9.4 [62] and the leachate was characterised using inductively coupled plasma mass spectrometry (ICP-MS). Both NAC and RAC concrete samples after 90 days were crushed first using a compression testing machine. Then the pieces were collected and crushed further with a hammer, and the fraction <4 mm was collected by sieving. In the first step, the sample is shaken at a liquid to solid ratio (L/S) of 2 L/kg for 6 h, and in the second step at 8 L/kg for 18 h. The cumulative L/S is 10 L/kg. The pH value and the composition of leachate was determined after leaching.

6. Results and Discussion

The 28 d compressive strength of concrete used for making recycled aggregate ranged from 67 to 82 MPa as shown in Figure 2. The minimum strength class of this parent concrete was C35/45. The average water absorption of the recycled aggregates at 24 h was obtained to be 6.1% (by weight). This order of magnitude is similar to what is typically reported for various kinds of recycled aggregates. The water absorption of natural aggregates typically lies below 1%, and that of recycled aggregate ranges between 2–20% depending on residual mortar content, cracking etc. [14]. The compressive strength of concrete in which natural aggregates were replaced for about 23% with RA is shown in Figure 4. The strength of RAC improved slightly with respect to NAC. The increase of strength at 28 days is around 16% and 9% at 90 days. Medium compressive strength concrete made with 25% of recycled coarse aggregates is reported to achieve the same mechanical properties as that of conventional concrete employing the same quantity of cement and the equal effective w/c ratio [16]. Here, the strength increase can be attributed to a combination of the following factors. The compressive strength of parent concrete used for production of the RCA ranged from 67 to 82 MPa, similar to or somewhat higher than that of NAC. Also, the fine part of RA which is less than 250 μm in size was removed by sieving. The grading of concrete particle sizes was optimised according to Fuller's curve. Some water absorption by the recycled aggregates would still occur after 30 min, and hence the w/c effective of the RAC would still be lower than of NAC. This could also be an important reason for the higher strength. A major drawback of water absorption of the recycled aggregate is the reduced workability of concrete. Workability is also affected by the roughness and angular shape of recycled aggregates. This could lead to higher amount of compaction pores that could reduce the strength [63]. To mitigate this effect, additional water was added and also the workability was adjusted by using superplasticizer. The slump value of NAC and RAC was 188 and 175 mm respectively which belongs to the same slump class. The reduced effective water to binder ratio combined with similar slump will have resulted in better compressive strength of RAC.

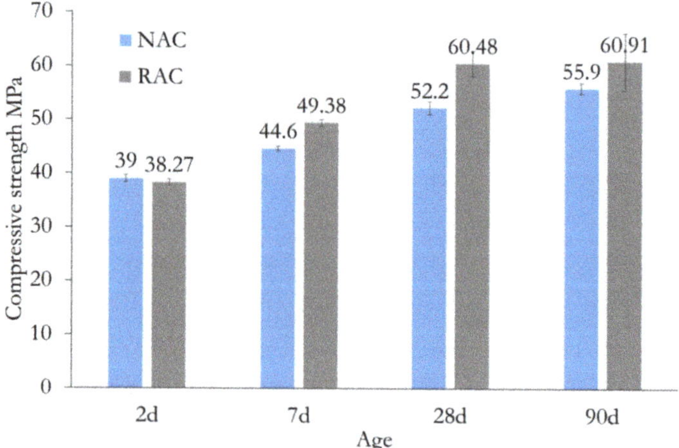

Figure 4. Compressive strength of NAC and RAC at 2, 7, 28 and 90 days.

The modulus of elasticity value was significantly reduced by addition of recycled aggregate as shown in Figure 5. The reduction is around 17% which is similar as reported in literature. In literature, the modulus of elasticity is reported to be greatly reduced by the use of recycled aggregate; it can reach 45% of the modulus of elasticity of corresponding conventional concrete. This percentage reduction varies based on the percentage substitution. The 45% reduction was reported at 100% substitution, and a more moderate reduction of 15% was mentioned at 30% substitution [64].

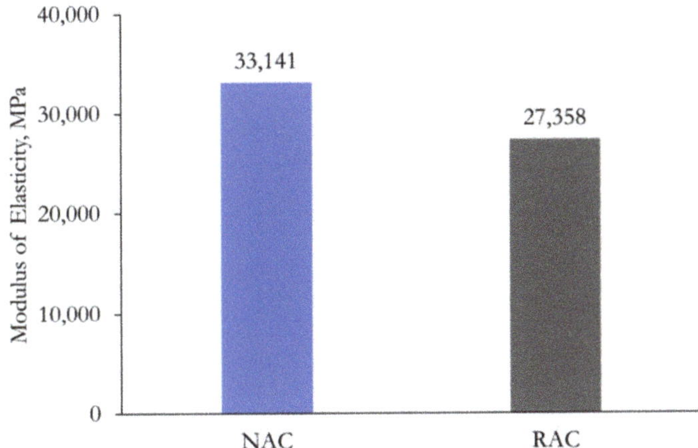

Figure 5. Modulus of elasticity of NAC and RAC.

Total creep and basic creep of NAC and RAC are shown in Figure 6. It can be seen that both total and basic creep are higher for RAC. This could be attributed to the lower modulus of elasticity of the RAC. The difference in creep between RAC (with RCA) and NAC (reference) is more pronounced for the basic creep than for the total creep, meaning that the drying creep is more favourable. Total creep is highly affected by the permeability of concrete. Higher permeability results in faster loss of moisture, and this results in faster rise of strain. The value of air permeability of concrete NAC and RAC determined after 28 days of curing and pre-conditioning is shown in Figure 7. Both these values fall under the classification of moderate permeability (0.1–1 $\times$ 10^{-16}). However, RAC has a lower value of air permeability compared to NAC, that results in a smaller difference between total creep and basic creep. The shrinkage values of both NAC and RAC are similar as can be seen in Figure 8. Shrinkage is highly affected by the binder, and similar shrinkage values could be attributed to the same binder and water to binder ratio used in both the mixes.

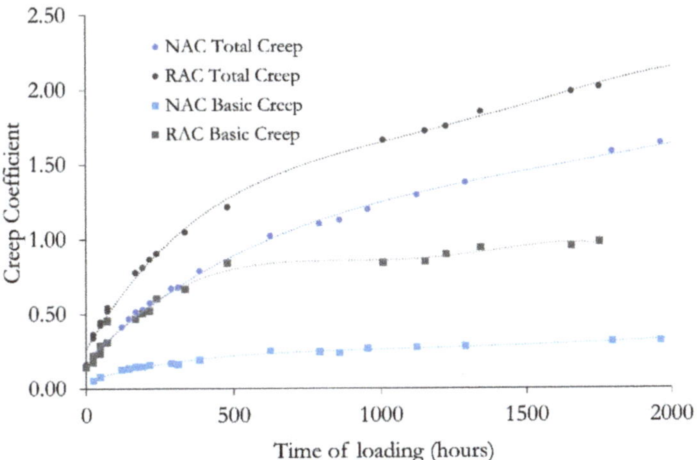

Figure 6. Creep coefficient of NAC and RAC.

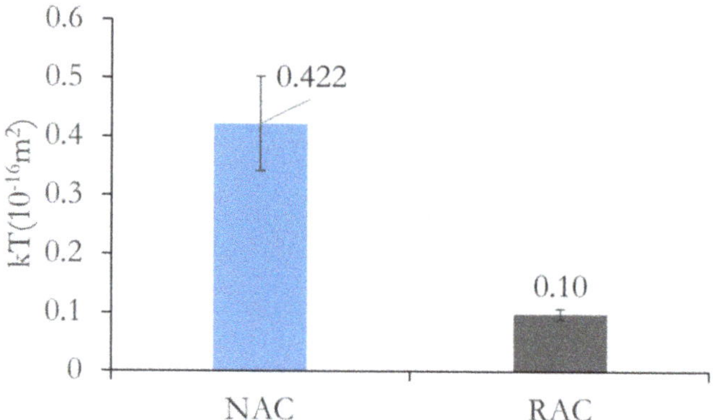

Figure 7. Torrent air permeability values of NAC and RAC after 28 days of curing.

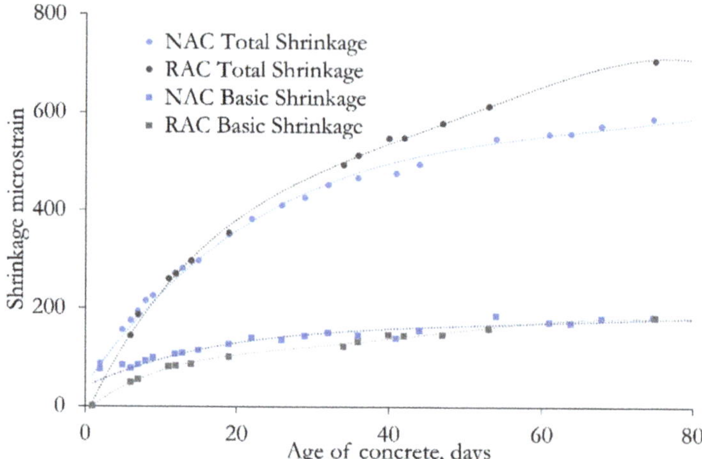

Figure 8. Shrinkage strain of NAC and RAC.

The evolution of carbonation depth with time is shown in Figure 9a, and the carbonation coefficient in Figure 9b. It can be seen that recycled concrete has lower carbonation depth than conventional concrete, and the carbonation coefficient decreases by around 45%. Carbonation depth is mainly affected by air permeability and buffer capacity. Residual mortar content in the RCA which is not carbonated adds to the buffer capacity of RA concrete. Lower carbonation depth of RAC would be a combined effect of both higher buffer capacity and lower permeability. In addition to all the above aspects, the better quality of RAC can partly be attributed to use of recycled aggregate from known parent concrete with a strength class superior to that of the recycled aggregate concrete at moderate replacement levels.

The results of the two-step shake test conducted on crushed concrete, and the limits prescribed by VLAREMA (Decree of the Flemish Government establishing the Flemish regulation on the sustainable management of material cycles and waste) are shown in Table 4. VLAREMA limits indicate values for use of bottom ash in unshaped elements (e.g., road subbase). However, in this application, bottom ash is used in conjunction with concrete which dilutes the content of heavy metals, and also, the cement hydration process encapsulates much of the leachable elements which results in negligible amount of leached heavy metals. Only Cr was detected in concrete mixes, but the concentrations are well within the limits for use in even unbound applications and the leaching from the RAC is comparable to that from the NAC.

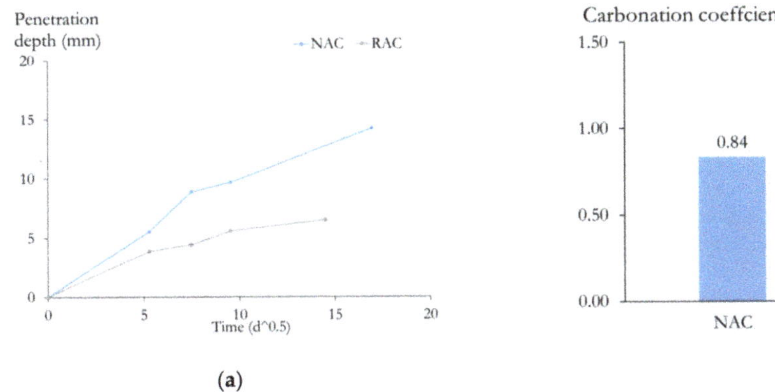

(a) (b)

Figure 9. (**a**) Evolution of carbonation depth with time and (**b**) carbonation coefficient of NAC and RAC.

Table 4. Heavy metal content and pH in eluates from leaching test and VLAREMA limits.

	NAC (mg/kg)	RAC (mg/kg)	VLAREMA Limits (mg/kg)
pH	12.5	12.51	-
As	0.00	0.00	0.8
Cd	0.00	0.00	0.03
Cr	0.17	0.20	0.5
Cu	0.00	0.00	0.5
Pb	0.00	0.00	1.3
Ni	0.00	0.00	0.75
Zn	0.00	0.00	2.8
Hg	0.00	0.00	
Cl	52.0	47.62	430
SO_4	20.3	26.09	540

7. Conclusions and Perspectives

Optimum use of recycled aggregate in concrete can be made possible by taking various steps that can facilitate its preservation and use. The presence of adhered mortar on the surface of crushed concrete aggregate generally increases water absorption and degrades the quality of the recycled aggregate and consequently the fresh and hardened properties of concrete made from it. Recycled aggregate is generally considered inferior to natural aggregate in terms of technical properties. However, it can be used in moderate replacement ratios when proper measures are taken, reducing the environmental footprint of the final product without affecting the properties. Here, the experimental study showed that the replacement of 23% of NA by RCA from concrete with MSWI bottom ashes as SCM, produced concrete with comparable properties. The application of MSWI bottom ashes as SCM in the parent concrete, did not affect its suitability to have a second life as RCA. A decrease in properties was mostly noticed in terms of modulus of elasticity of the concrete. Compressive strength and pore structure improved due to presence of RA. Here, the recycled aggregate being from a parent concrete of superior strength class improved many of the properties of concrete. Further to that, removal of the fine fraction <250 μm in size and addition of water for compensation of water absorption will have added to improvement of workability and other properties. Proper guidelines regarding demolition techniques that preserve the quality of aggregates and techniques to improve the quality of each kind of recycled aggregate can thus increase its utilisation. Life cycle analysis should be conducted specific to the usage of the RA to better estimate the environmental benefits, since availability and impact of both natural aggregate and recycled aggregate varies locally. Production and treatment procedures could add to the environmental load for recycled aggregate [65,66]. The fines generated from sieving the aggregates could be used as a raw meal additive for manufacture of Portland cement production, which is already demonstrated in a previous study (up to ~15%) [67], thus leaving no waste behind ensuring circular economy.

Author Contributions: Formal analysis, A.M.J.; Funding acquisition, S.M. and N.D.B.; Investigation, A.M.J.; Methodology, A.M.J., S.M. and N.D.B.; Project administration, S.M.; Resources, S.M. and N.D.B.; Supervision, S.M. and N.D.B.; Writing—original draft, A.M.J.; Writing—review & editing, N.D.B. All authors have read and agreed to the published version of the manuscript.

Funding: This research is a part of ASHCEM project (grant number: IWT.150076) funded by Strategic Initiative Materials (SIM), Flanders and Flanders Agency for Innovation and Entrepreneurship(VLAIO).

Institutional Review Board Statement: Not applicable.

Informed Consent Statement: Not applicable.

Data Availability Statement: Data will be made available upon reasonable request.

Acknowledgments: The authors thank Natalia Alderete and other partners of ASHCEM project (grant number: IWT.150076), which in itself is a part of the bigger program MaRes aimed at creating and demonstrating an operational, flexible toolbox to recover metals and valorise the residual matrix into building materials funded by SIM (Strategic Initiative Materials in Flanders) and VLAIO (Flanders Innovation & Entrepreneurship). The financial support from the foundations for this study is gratefully appreciated.

Conflicts of Interest: The authors declare no conflict of interest.

References

1. Blasenbauer, D.; Huber, F.; Lederer, J.; Quina, M.J.; Blanc-Biscarat, D.; Bogush, A.; Bontempi, E.; Blondeau, J.; Chimenos, J.M.; Dahlbo, H.; et al. Legal situation and current practice of waste incineration bottom ash utilisation in Europe. *Waste Manag.* **2020**, *102*, 868–883. [CrossRef] [PubMed]
2. Leefmilieu, N.E.; OVAM; VITO. *Monitoringsysteem Duurzaam Oppervlaktedelfstoffenbeleid*; Department Leefmilieu, Natuur & Energie/OVAM/VITO: Brussel, Belgium, 2013.
3. Kaza, S.; Yao, L.; Bhada-Tata, P.; Van Woerden, F. *What a Waste 2.0*; World Bank Publications: Washington, DC, USA, 2018; ISBN 9781464813290.
4. Beylot, A.; Villeneuve, J. Environmental impacts of residual Municipal Solid Waste incineration: A comparison of 110 French incinerators using a life cycle approach. *Waste Manag.* **2013**, *33*, 2781–2788. [CrossRef] [PubMed]
5. Joseph, A.M.; Snellings, R.; Nielsen, P.; Matthys, S.; De Belie, N. Pre-treatment and utilisation of municipal solid waste incineration bottom ashes towards a circular economy. *Constr. Build. Mater.* **2020**, *260*, 120485. [CrossRef]
6. Joseph, A.M.; Snellings, R.; Van den Heede, P.; Matthys, S.; De Belie, N. The Use of Municipal Solid Waste Incineration Ash in Various Building Materials: A Belgian Point of View. *Materials* **2018**, *11*, 141. [CrossRef]
7. Alam, Q. *Valorization of Municipal Solid Waste*; Eindhoven University of Technology: Eindhoven, The Netherlands, 2019.
8. Alderete, N.M.; Joseph, A.M.; Van den Heede, P.; Matthys, S.; De Belie, N. Effective and sustainable use of municipal solid waste incineration bottom ash in concrete regarding strength and durability. *Resour. Conserv. Recycl.* **2020**, *167*, 105356. [CrossRef]
9. Joseph, A.M. *Processed Bottom Ash Based Sustainable Binders for Concrete*; Ghent University: Ghent, Belgium, 2021.
10. Group, F. Global Demand for Construction Aggregates to Exceed 48 Billion Metric Tons in 2015. Available online: https://www.concreteconstruction.net/business/global-demand-for-construction-aggregates-to-exceed-48-billion-metric-tons-in-2015_o (accessed on 25 October 2021).
11. European Aggregate Association. Annual Review 2017–2018. In *A Sustainable Industry for a Sustainable Europe*; Union Europeene Producteurs de Granulats (UEPG): Brussels, Belgium, 2017.

12. CEMBUREAU Construction & Demolition Waste. Available online: https://cembureau.eu/policy-focus/sustainable-construction/construction-demolition-waste/ (accessed on 24 December 2021).
13. Tam, V.W.Y.; Tam, C.M.; Le, K.N. Removal of cement mortar remains from recycled aggregate using pre-soaking approaches. *Resour. Conserv. Recycl.* **2007**, *50*, 82–101. [CrossRef]
14. de Brito, J.; Saikia, N. *Recycled Aggregate in Concrete: Use of Industrial, Construction and Demolition Waste*; Springer Science & Business Media: Berlin/Heidelberg, Germany, 2009; Volume 1, ISBN 9781447145394.
15. Lin, Y.; Tyan, Y.; Chang, T.; Chang, C. An assessment of optimal mixture for concrete made with recycled concrete aggregates. *Cem. Concr. Res.* **2004**, *34*, 1373–1380. [CrossRef]
16. Etxeberria, M.; Vázquez, E.; Marí, A.; Barra, M. Influence of amount of recycled coarse aggregates and production process on properties of recycled aggregate concrete. *Cem. Concr. Res.* **2007**, *37*, 735–742. [CrossRef]
17. Leemann, A.; Loser, R. Carbonation resistance of recycled aggregate concrete. *Constr. Build. Mater.* **2019**, *204*, 335–341. [CrossRef]
18. Eckert, M.; Oliveira, M. Mitigation of the negative effects of recycled aggregate water absorption in concrete technology. *Constr. Build. Mater.* **2017**, *133*, 416–424. [CrossRef]
19. Bui, N.K.; Satomi, T.; Takahashi, H. Improvement of mechanical properties of recycled aggregate concrete basing on a new combination method between recycled aggregate and natural aggregate. *Constr. Build. Mater.* **2017**, *148*, 376–385. [CrossRef]
20. Li, T.; Xiao, J.; Zhu, C.; Zhong, Z. Experimental study on mechanical behaviors of concrete with large-size recycled coarse aggregate. *Constr. Build. Mater.* **2016**, *120*, 321–328. [CrossRef]
21. Del Bosque, I.F.S.; Zhu, W.; Howind, T.; Matías, A.; de Rojas, M.I.S.; Medina, C. Properties of interfacial transition zones (ITZs) in concrete containing recycled mixed aggregate. *Cem. Concr. Compos.* **2017**, *81*, 25–34. [CrossRef]
22. Geng, Y.; Wang, Y.; Chen, J. Creep behaviour of concrete using recycled coarse aggregates obtained from source concrete with different strengths. *Constr. Build. Mater.* **2016**, *128*, 199–213. [CrossRef]
23. Tu, T.Y.; Chen, Y.Y.; Hwang, C.L. Properties of HPC with recycled aggregates. *Cem. Concr. Res.* **2006**, *36*, 943–950. [CrossRef]
24. Mamirov, M.; Hu, J.; Cavalline, T. Geometrical, physical, mechanical, and compositional characterization of recycled concrete aggregates. *J. Clean. Prod.* **2022**, *339*, 130754. [CrossRef]
25. Zhou, C.; Chen, Z. Mechanical properties of recycled concrete made with different types of coarse aggregate. *Constr. Build. Mater.* **2017**, *134*, 497–506. [CrossRef]
26. Koper, A.; Koper, W.; Koper, M. Influence of raw concrete material quality on selected properties of recycled concrete aggregates. *Procedia Eng.* **2017**, *172*, 536–543. [CrossRef]
27. Andal, J.; Shehata, M.; Zacarias, P. Properties of concrete containing recycled concrete aggregate of preserved quality. *Constr. Build. Mater.* **2016**, *125*, 842–855. [CrossRef]
28. Nagataki, S.; Gokce, A.; Saeki, T.; Hisada, M. Assessment of recycling process induced damage sensitivity of recycled concrete aggregates. *Cem. Concr. Res.* **2004**, *34*, 965–971. [CrossRef]
29. Manufacturers, A.C. Recycled Aggregates. Available online: https://www.cement.org/learn/concrete-technology/concrete-design-production/recycled-aggregates (accessed on 25 October 2021).
30. Shi, C.; Li, Y.; Zhang, J.; Li, W.; Chong, L.; Xie, Z. Performance enhancement of recycled concrete aggregate-A review. *J. Clean. Prod.* **2016**, *112*, 466–472. [CrossRef]
31. Wang, L.; Wang, J.; Qian, X.; Chen, P.; Xu, Y.; Guo, J. An environmentally friendly method to improve the quality of recycled concrete aggregates. *Constr. Build. Mater.* **2017**, *144*, 432–441. [CrossRef]
32. Katkhuda, H.; Shatarat, N. Improving the mechanical properties of recycled concrete aggregate using chopped basalt fibers and acid treatment. *Constr. Build. Mater.* **2017**, *140*, 328–335. [CrossRef]
33. Pandurangan, K.; Dayanithy, A.; Prakash, S.O. Influence of treatment methods on the bond strength of recycled aggregate concrete. *Constr. Build. Mater.* **2016**, *120*, 212–221. [CrossRef]
34. Al-Bayati, H.K.A.; Das, P.K.; Tighe, S.L.; Baaj, H. Evaluation of various treatment methods for enhancing the physical and morphological properties of coarse recycled concrete aggregate. *Constr. Build. Mater.* **2016**, *112*, 284–298. [CrossRef]
35. Bru, K.; Touzé, S.; Bourgeois, F.; Lippiatt, N.; Ménard, Y. Assessment of a microwave-assisted recycling process for the recovery of high-quality aggregates from concrete waste. *Int. J. Miner. Process.* **2014**, *126*, 90–98. [CrossRef]
36. Zhang, J.; Shi, C.; Li, Y.; Pan, X.; Poon, C.S.; Xie, Z. Influence of carbonated recycled concrete aggregate on properties of cement mortar. *Constr. Build. Mater.* **2015**, *98*, 1–7. [CrossRef]
37. Shi, C.; Wu, Z.; Cao, Z.; Ling, T.C.; Zheng, J. Performance of mortar prepared with recycled concrete aggregate enhanced by CO2 and pozzolan slurry. *Cem. Concr. Compos.* **2018**, *86*, 130–138. [CrossRef]
38. Tam, V.W.Y.; Butera, A.; Le, K.N. Carbon-conditioned recycled aggregate in concrete production. *J. Clean. Prod.* **2016**, *133*, 672–680. [CrossRef]
39. Kazmi, S.M.S.; Munir, M.J.; Wu, Y.F.; Patnaikuni, I.; Zhou, Y.; Xing, F. Influence of different treatment methods on the mechanical behavior of recycled aggregate concrete: A comparative study. *Cem. Concr. Compos.* **2019**, *104*, 103398. [CrossRef]
40. Prasad, D.; Pandey, A.; Kumar, B. Sustainable production of recycled concrete aggregates by lime treatment and mechanical abrasion for M40 grade concrete. *Constr. Build. Mater.* **2021**, *268*, 121119. [CrossRef]
41. Bian, Y.; Li, Z.; Zhao, J.; Wang, Y. Synergistic enhancement effect of recycled fine powder (RFP) cement paste and carbonation on recycled aggregates performances and its mechanism. *J. Clean. Prod.* **2022**, *344*, 130848. [CrossRef]

42. Grabiec, A.M.; Klama, J.; Zawal, D.; Krupa, D. Modification of recycled concrete aggregate by calcium carbonate biodeposition. *Constr. Build. Mater.* **2012**, *34*, 145–150. [CrossRef]
43. García-González, J.; Rodríguez-Robles, D.; Wang, J.; De Belie, N.; Morán-del Pozo, J.M.; Guerra-Romero, M.I.; Juan-Valdés, A. Quality improvement of mixed and ceramic recycled aggregates by biodeposition of calcium carbonate. *Constr. Build. Mater.* **2017**, *154*, 1015–1023. [CrossRef]
44. Nam, B.H.; An, J.; Youn, H. Accelerated calcite precipitation (ACP) method for recycled concrete aggregate (RCA). *Constr. Build. Mater.* **2016**, *125*, 749–756. [CrossRef]
45. Alqarni, A.S.; Abbas, H.; Al-Shwikh, K.M.; Al-Salloum, Y.A. Treatment of recycled concrete aggregate to enhance concrete performance. *Constr. Build. Mater.* **2021**, *307*, 124960. [CrossRef]
46. Damrongwiriyanupap, N.; Wachum, A.; Khansamrit, K.; Detphan, S.; Hanjitsuwan, S.; Phoo-ngernkham, T.; Sukontasukkul, P.; Li, L.; Chindaprasirt, P. Improvement of recycled concrete aggregate using alkali-activated binder treatment. *Mater. Struct. Constr.* **2022**, *55*, 11. [CrossRef]
47. Kim, J.H.; Robertson, R.E.; Naaman, A.E. Structure and properties of poly (vinyl alcohol)-modified mortar and concrete. *Cem. Concr. Res.* **1999**, *29*, 407–415. [CrossRef]
48. Kou, S.C.; Poon, C.S. Properties of concrete prepared with PVA-impregnated recycled concrete aggregates. *Cem. Concr. Compos.* **2010**, *32*, 649–654. [CrossRef]
49. Zhu, Y.G.; Kou, S.C.; Poon, C.S.; Dai, J.G.; Li, Q.Y. Influence of silane-based water repellent on the durability properties of recycled aggregate concrete. *Cem. Concr. Compos.* **2013**, *35*, 32–38. [CrossRef]
50. Li, W.; Xiao, J.; Sun, Z.; Kawashima, S.; Shah, S.P. Interfacial transition zones in recycled aggregate concrete with different mixing approaches. *Constr. Build. Mater.* **2012**, *35*, 1045–1055. [CrossRef]
51. Silva, R.V.; de Brito, J.; Dhir, R.K. Comparative analysis of existing prediction models on the creep behaviour of recycled aggregate concrete. *Eng. Struct.* **2015**, *100*, 31–42. [CrossRef]
52. Fathifazl, G.; Abbas, A.; Razaqpur, A.G.; Isgor, O.B.; Fournier, B.; Foo, S. New Mixture Proportioning Method for Concrete Made with Coarse Recycled Concrete Aggregate. *J. Mater. Civ. Eng.* **2009**, *21*, 601–611. [CrossRef]
53. Ferreira, L.; De Brito, J.; Barra, M. Influence of the pre-saturation of recycled coarse concrete aggregates on concrete properties. *Mag. Concr. Res.* **2011**, *63*, 617–627. [CrossRef]
54. Amer, A.A.M.; Ezziane, K.; Bougara, A.; Adjoudj, M. Rheological and mechanical behavior of concrete made with pre-saturated and dried recycled concrete aggregates. *Constr. Build. Mater.* **2016**, *123*, 300–308. [CrossRef]
55. Koenders, E.A.B.; Pepe, M.; Martinelli, E. Compressive strength and hydration processes of concrete with recycled aggregates. *Cem. Concr. Res.* **2014**, *56*, 203–212. [CrossRef]
56. Leite, M.B.; Monteiro, P.J.M. Microstructural analysis of recycled concrete using X-ray microtomography. *Cem. Concr. Res.* **2016**, *81*, 38–48. [CrossRef]
57. Institut Belge de Normalisation. *NBN 15-220;* Concrete Testing-Compressive Strength; Bureau of Standardisation: Brussels, Belgium, 1990.
58. Institut Belge de Normalisation. *NBN B 15-203;* Concrete Testing-Statical Modulus of Elasticity with Compression; Bureau of Standardisation: Brussels, Belgium, 1990.
59. Institut Belge de Normalisation. *NBN B 15-228;* Concrete Testing-Creep; Bureau of Standardisation: Brussels, Belgium, 1990.
60. Institut Belge de Normalisation. *NBN B 15-216;* Concrete Testing-Shrinkage and Swelling; Bureau of Standardisation: Brussels, Belgium, 1990.
61. Suraneni, P.; Azad, V.J.; Isgor, O.B.; Weiss, W.J. Use of fly ash to minimize deicing salt damage in concrete pavements. *Transp. Res. Rec.* **2017**, *2629*, 24–32. [CrossRef]
62. Flemish Regulation on the Sustainable Management of Material Cycles and Waste, F.G. CMA/2/II/A.9.4-Leaching of Inorganic Components with the Two Step Shake Test. Available online: https://reflabos.vito.be/2006/CMA_2_II_A.9.4.pdf (accessed on 25 October 2021).
63. Kaplan, M.F. The effects of the properties of coarse aggregates on the workability of concrete. *Mag. Concr. Res.* **1958**, *10*, 63–74. [CrossRef]
64. Yehia, S.; Helal, K.; Abusharkh, A.; Zaher, A.; Istaitiyeh, H. Strength and Durability Evaluation of Recycled Aggregate Concrete. *Int. J. Concr. Struct. Mater.* **2015**, *9*, 219–239. [CrossRef]
65. Rosado, L.P.; Vitale, P.; Penteado, C.S.G.; Arena, U. Life cycle assessment of natural and mixed recycled aggregate production in Brazil. *J. Clean. Prod.* **2017**, *151*, 634–642. [CrossRef]
66. Kleijer, A.L.; Lasvaux, S.; Citherlet, S.; Viviani, M. Product-specific Life Cycle Assessment of ready mix concrete: Comparison between a recycled and an ordinary concrete. *Resour. Conserv. Recycl.* **2017**, *122*, 210–218. [CrossRef]
67. Schoon, J.; De Buysser, K.; Van Driessche, I.; De Belie, N. Fines extracted from recycled concrete as alternative raw material for Portland cement clinker production. *Cem. Concr. Compos.* **2015**, *58*, 70–80. [CrossRef]

Article

Construction and Demolition Waste (CDW) Recycling—As Both Binder and Aggregates—In Alkali-Activated Materials: A Novel Re-Use Concept

Rafael A. Robayo-Salazar, William Valencia-Saavedra and Ruby Mejía de Gutiérrez *

Composites Materials Group (CENM), Universidad de Valle, Calle 13 #100-00, Cali E44, Colombia; rafael.robayo@correounivalle.edu.co (R.A.R.-S.); william.gustavo.valencia@correounivalle.edu.co (W.V.-S.);
* Correspondence: ruby.mejia@correounivalle.edu.co

Received: 30 June 2020; Accepted: 13 July 2020; Published: 17 July 2020

Abstract: This article demonstrates the possibility of producing alkali-activated materials (AAM) from a mixture of mechanically processed concrete, ceramic, masonry, and mortar wastes, as a sustainable alternative for recycling construction and demolition wastes (CDWs) under real conditions. The addition of 10% Portland cement allowed the materials to cure at room temperature (25 °C). CDW binder achieved a compressive strength of up to 43.9 MPa and it was classified as a general use and low heat of hydration cement according to ASTM C1157. The concrete produced with this cement and the crushed aggregates also from CDW reported a compressive strength of 33.9 MPa at 28 days of curing and it was possible to produce a high-class structural block with 26.1 MPa according to ASTM C90. These results are considered one option in making full use of CDWs as binder and aggregates, using alkaline activation technology thereby meeting the zero-waste objective within the concept of the circular economy.

Keywords: construction and demolition wastes; alkali-activated materials; recycling; binder; recycled aggregates

1. Introduction

Construction and demolition wastes (CDWs) represent a severe environmental pollution problem in most countries, especially underdeveloped ones, due to their inadequate management and low rates of recycling or utilization. Indeed, global CDW generation is alarming, currently representing 25–30% of total solid waste worldwide [1], exceeding the quantity of 3 trillion tons per year [2]. In response to this problem, the "2030 Agenda for Sustainable Development of the United Nations General Assembly", approved in 2015 by the member countries of the UN, established the "Sustainable Cities" model. Within this model, it is planned to reduce the negative environmental impact of cities by 2030, paying special attention to the management and recycling of generated waste [3]. Sustainable cities are thus required to achieve, through technological innovation, substitution of the "linear economy" for a "circular economy", in which waste is incorporated (again and again) in the production processes of new products and/or materials (towards the "zero-waste" objective) [4].

In relation to CDWs, these can be classified into two categories: usable and non-usable waste. Non-usable wastes are those that are contaminated with hazardous wastes, so their use is restricted by environmental regulations. Within the group of usable wastes are mainly concrete, ceramic, masonry, and mortar waste. Generally, these wastes have been treated to produce fine and coarse aggregates. These can be incorporated into mortar and concrete mixtures as a replacement (partial or total) for natural aggregates [5], in addition to being used in geotechnical applications, such as the stabilization of slopes, shallow and deep foundations, granular bases and sub-bases [6]. However, the production of recycled aggregates does not represent a solution for exploiting CDWs globally, mainly due to the low commercial competitiveness they possess when comparing their economic (non-environmental) cost with that of natural aggregates in many cities and countries. This disadvantage is compounded by the lack of policies that control the indiscriminate exploitation of natural quarries, aimed at promoting the massive use of recycled aggregates as a replacement for natural aggregates. For this reason, the search for recycling alternatives and/or applications with greater commercial value, which ensure the real use of CDWs, is a priority worldwide.

One of the newest and most promoted solutions for the use of this type of waste is alkaline activation technology or "geopolymerization" [7]. This technology is related with the reaction of a solid aluminosilicate powder (or precursor) with an alkali activator to produce a hardened binder. There are three types of alkali-activated binders: low-calcium alkali-activated aluminosilicate, high-calcium alkali-activated aluminosilicate, and hybrid, which is produced mixing the aluminosilicate precursor with the activator and Portland cement in a proportion lower than 30%. The product of the reaction is a gel whose structure depends on the type of precursor and activator used. Thus, with high-calcium precursors such as granulated blast furnace slag, the main product is a hydrated calcium aluminosilicate gel (C-A-S-H), similar to the gel generated in the hydration of a Portland cement (gel C-S-H), but with lower proportions of CaO/SiO_2. In the case of low-calcium precursors such as fly ashes or calcined clays, the gel is an amorphous alkaline aluminosilicate (gel N-A-S-H) of three-dimensional structure.

Using this technology, it is possible to take advantage of most of the aluminosilicate materials that make up the CDW, i.e., concrete, ceramic, masonry, and mortar wastes, as precursors of new cementitious materials, from which it is possible to produce mortars, concrete, and manufacture various structural and non-structural elements (applications of greater commercial value). In this regard, in recent years the International Union of Laboratories and experts in Construction Materials, Systems and Structures (RILEM) and ASTM technical committees were formed to work on research and/or standardization of alkali-activated materials or "geopolymers" with a view to formulating standards that promote their introduction in different fields of application within the construction sector [8].

Regarding the use of CDWs as precursors of alkali-activated materials, the few results published so far, as shown in Table 1 are promising and demonstrate the potential of this type of waste to be reused by this technology. However, these reports are very recent and are limited in most cases to obtaining and characterizing pastes, without scaling these results to the level of their application (mortars, concretes, construction elements, and/or applications). Likewise, in most cases, thermal curing treatments were applied between 50 and 90 °C to achieve adequate mechanical strength, which limits their technological transfer. It should be noted that these studies start from the use of the separated residues, that is to say "clean", without considering mixtures or combinations between these materials, as can be seen in Table 1.

The need to develop and validate applications based on combinations of these residues is evident, that allow an integral use and simulate the real conditions in which the CDWs are found (mixed), especially in those countries with deficiencies in the collection, transport, treatment, and disposal of these wastes, together with monitoring and regulation of the waste management process.

Table 1. Articles of alkaline activation of construction and demolition waste (CDW) (database: Scopus/keywords: "alkali activated materials" or "geopolymers" + "construction and demolition wastes"). C.S.: compressive strength (max.), OPC: ordinary Portland cement, MK: metakaolin, FA: fly ash, GBFS: blast furnace slag.

Waste	Addition	Optimal Synthesis Conditions			Application	Ref.	Year	Country
		Activator	Cured	C.S.				
Concrete	20% MK	NaOH + Na_2SiO_3	60 °C (3d)	33 MPa	Paste	[9]	2009	UK
Masonry	—	NaOH + Na_2SiO_3	60 °C (7d)	50 MPa	PasteMortar	[10]	2013	Spain
Concrete	—	NaOH + Na_2SiO_3	90 °C (7d)	13 MPa	Paste	[11]	2015	Greece
Masonry				58 MPa				
Tile				50 MPa				
Masonry	0%	NaOH + Na_2SiO_3	25 °C	54 MPa	Paste	[12]	2016	Colombia
	20% OPC			103 MPa				
Concrete	0%	NaOH + Na_2SiO_3	25 °C	26 MPa	Paste	[13]	2016	Colombia
	30% OPC			34 MPa				
	10% MK			46 MPa				
Concrete	—	NaOH + Na_2SiO_3	80 °C (1d)	8 MPa	Paste	[14]	2016	Greece
Masonry				39 MPa				
Tile				58 MPa				
Masonry	—	NaOH + Na_2SiO_3	50 °C (1d)	—	Coating	[15]	2016	Italy
Ceramic	15% OPC	NaOH + Na_2SiO_3	25 °C	58 MPa	Paste	[16]	2017	Colombia
				25 MPa	Mortar			
Ceramic	5% $Ca(OH)_2$	NaOH + Na_2SiO_3	65 °C (3d)	43 MPa	Mortar	[17]	2017	Spain
Masonry	0%	NaOH	25 °C	7 MPa	Paste	[18]	2017	Colombia
	10% OPC			41 MPa				
	0%	NaOH + Na_2SiO_3		54 MPa				
	20% OPC			103 MPa				
Concrete	0%	NaOH		7 MPa				
	30% OPC			10 MPa				
	0%	NaOH + Na_2SiO_3		26 MPa				
	30% OPC			34 MPa				
Masonry	—	NaOH + Na_2SiO_3	90 °C (5d)	36 MPa	Mortar	[19]	2018	Turkey
Masonry	—	NaOH + Na_2SiO_3	25 °C	42 MPa	Paste	[20]	2018	R. Czech
Masonry	30% GBFS + 10% FA	NaOH + Na_2SiO_3	25 °C	70 MPa	Paste	[21]	2019	Taiwan
Ceramic	30% GBFS + 10% FA	NaOH + Na_2SiO_3	25 °C	60 MPa				

This research demonstrates the possibility of reusing CDW mixtures, as binder and aggregates, in the production of alkali-activated materials. A real CDW sample, taken from a local dump site in Cali (Colombia), consisting of concrete, ceramic, masonry, and mortar wastes, was used as a geopolymer precursor. As an alkaline activator solution, mixtures of hydroxide and sodium silicate were used. The CDW precursor was added with 10% (by weight) of Portland cement (OPC) to promote the curing of the mixtures at room temperature (≈25 °C), thereby obtaining a hybrid cement (CDW binder). Evolution of the heat of hydration and compressive strength of the binder was determined and compared with that of a paste based on 100% OPC. Fine and coarse recycled aggregates were also produced from the CDWs, which were characterized according to the provisions of ASTM standards for use in concrete mixtures. A CDW-based alkaline activated hybrid concrete (AAHC), used as binder and aggregates, was produced and characterized both mechanically and microstructurally, validating from the results obtained the possibility of using alkaline activation technology in obtaining concretes based on high CDW content (binder + aggregates).

Finally, an application of the AAHC was formulated in a precast product—a solid block. The foregoing demonstrated the potential of this concrete in the manufacture of precast materials that meet the specifications and technical standards required for their use and application in the construction sector. It should be noted that this is the first time that results have been reported for alkali-activated materials based on the mixture of concrete, ceramic, masonry, and mortar wastes, in addition to the use of these CDWs in obtaining cement (binder) and aggregates in the same application.

2. Materials and Methods

2.1. Construction and Demolition Wastes (CDW)

The CDW sample was collected locally in Cali, Colombia, and used as raw material (binder and aggregates) for the production of the alkali-activated materials. As seen in Figure 1, the initial CDW sample was composed mainly of concrete, ceramic, masonry, and mortar wastes, and to a lesser extent other usable waste such as: asphalt, wood, metals, polymers, glass, Drywall, paper, and cardboard, all mixed together.

Figure 1. Actual composition of the sample of CDW utilized in this research. Site for final disposal of waste located in the city of Cali, Colombia.

Concrete, ceramic (red and white), masonry, and mortar wastes were selected and separated from the initial CDW sample in order to be used. In this study, the waste separation process was carried out manually. Figure 2 is a schematic representation of the processes and treatments to which the wastes were subjected to obtain the geopolymeric precursor (CDW precursor) and the aggregates (fine and coarse). The coarse recycled aggregate (CRA) was obtained from the coarse crushing (≤25.4 mm) of the concrete waste, while the fine recycled aggregate (FRA) was produced from the fine crushing (≤4.76 mm) of the ceramic (red and white) and mortar wastes. It should be noted that during these crushing processes (coarse and fine), a considerable amount (≈20% by weight) of powder (particulate material less than 150 μm (sieve # 100)) is generated, a by-product that was collected for each type of waste (concrete, ceramic, and mortar). For its part, the masonry waste was finely ground. The CDW precursor was obtained by mixing the masonry waste and the powders generated in the crushing of the concrete, ceramic, and mortar wastes; each type of material dosed at 25% by weight. Finally, the CDW precursor was subjected to a grinding process in a ball mill in order to guarantee its homogenization and achieve a uniform particle size, as showm in Figure 2.

Figure 2. Obtaining the CDW precursor and recycled aggregates (fine and coarse) from the use of concrete, ceramic, masonry, and mortar wastes.

2.2. Characterization of Raw Materials

The chemical composition of the CDW precursor, determined by X-ray fluorescence (XRF) using a Philips MagiX-Pro PW-2440 spectrometer, is presented in Table 2. The aluminosilicate nature (SiO_2 + Al_2O_3 = 58.8%) of the geopolymeric precursor is highlighted, with a SiO_2/Al_2O_3 molar ratio of 7.2. During the obtaining of the alkali-activated materials, GU type (general use) Portland cement (OPC) was added, the chemical composition of which is included in Table 2. The particle size analysis of the CDW precursor, carried out by means of laser granulometry in a Mastersizer-2000 kit from Malvern Instruments, resulted in a mean particle size D (4; 3) of 92.1 µm, as shown in Figure 3. The mineralogical composition of the CDW precursor was determined by means of X-ray diffraction using a PANalytical X'Pert MRD X-ray diffractometer as shown in Figure 4.

Table 2. Chemical composition (FRX) of the raw materials (CDW and OPC).

Material	SiO_2	Al_2O_3	CaO	Fe_2O_3	Na_2O	K_2O	MgO	Others	LOI
CDW precursor	47.6	11.2	21.2	5.9	0.6	1.1	1.1	2.3	9.1
OPC	17.9	3.9	62.3	4.8	0.2	0.3	1.8	4.7	4.1

LOI: loss on ignition.

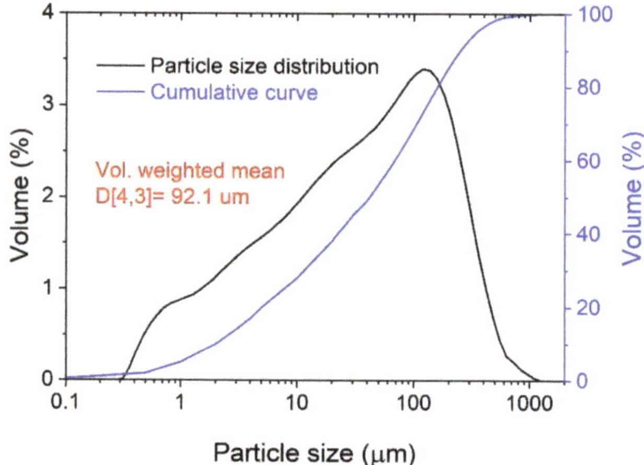

Figure 3. Particle size distribution of the CDW precursor.

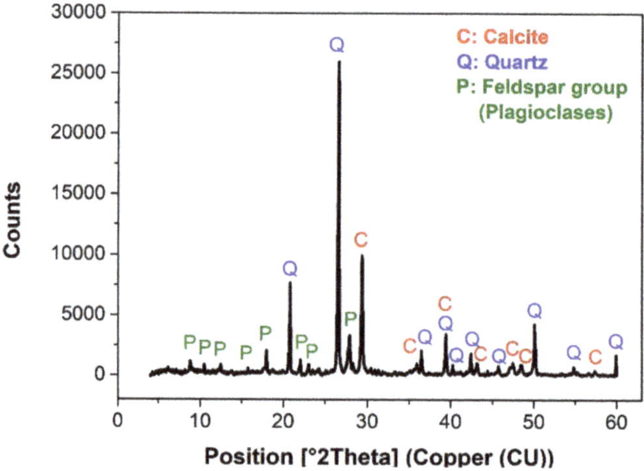

Figure 4. X-ray diffraction of the CDW precursor.

Figure 4 shows the semi-crystalline nature of the CDW precursor, with quartz (SiO_2) (code: 01-083-0539) being the main crystalline phase. Other phases, such as calcite ($CaCO_3$) (code: 01-072-1937) and sodium-calcium feldspars belonging to the plagioclase group (albite-anorthite series) (code: 01-076-0927), were also identified.

As alkaline activator, mixtures of sodium hydroxide (NaOH) and sodium silicate (SS) or "waterglass" (Na_2SiO_3: SiO_2 = 32.09%, Na_2O = 11.92%, H_2O = 55.99%) were used, both of industrial grade.

2.3. Design of Mixture and Production of Alkali-Activated Materials

The procedure followed to obtain the alkali-activated materials from the CDW is summarized in Figure 5. From the combination of the CDW precursor and a solution composed of the alkaline activator (NaOH + SS) and the mixing water it was possible to obtain an alkali-activated cement. The addition of OPC in small proportions (10% by weight with respect to the precursor as a replacement) made it possible to obtain a hybrid cement (CDW binder), which hardened and developed strengths at room temperature (≈25 °C). From the optimization of this CDW binder, it was possible to produce a CDW mortar and an alkali-activated hybrid concrete (AAHC); in both cases the aggregates developed from the CDW were used.

Figure 5. Schematic summary of the methodology developed to obtain the CDW-based alkali-activated materials.

2.3.1. Binder Optimization

For the optimization of the CDW binder, the effect of the alkaline activator content (weight ratios: (NaOH+SS)/(CDW + OPC) and NaOH/SS) on the compressive strength (7 and 28 days of curing) was evaluated. The pastes were obtained in a Hobart mixer with a total mixing time of 5 min. The liquid/solid ratio (L/S) was 0.25. The fresh pastes were molded into 20 mm cubes on each side and vibrated for 30 s on an electric vibrating table to remove trapped air. Subsequently, the molds were covered with a polyethylene film and remained in the laboratory environment for 24 h. After this time, the specimens were removed from the mold and taken to a final curing chamber (≈25 °C) that ensures a relative humidity (RH) of over 80% until reaching the test age. The optimal setting time of the CDW binder was determined according to the procedure described in ASTM C191 (Method B). The evolution of heat (during alkaline activation) and the total heat of reaction (48 h) were evaluated by means of an I-Cal 8000 isothermal calorimeter. For the calorimetric analysis of the CDW binder, a comparison was made with a 100% OPC-based paste (GU type) and a paste based on the mixture 90%CDW+10%OPC+H_2O. Once the alkaline activator content was optimized, the evolution of the compressive strength (1–90 days of curing) was evaluated, using an INSTRON 3369 universal testing machine, which has a capacity of 50 kN force, at a speed of 1 mm/min. For each mix, a minimum of three specimens were tested.

Subsequently, in order to classify the CDW binder according to the specifications of ASTM C1157, the compressive strength was determined at 28 days (≈25 °C) of a mortar produced following the procedure described in the standard ASTM C305. The test specimens were 50.8 mm cubes on each side (ASTM C109).

2.3.2. Characterization of the Recycled Aggregates

Table 3 presents the main characteristics of the fine and coarse recycled aggregates obtained from the CDW sample, which were termed FRA and CRA, respectively. It is highlighted that these aggregates presented high levels of absorption; 12.12% (ASTM C128) for the FRA and 9.17% (ASTM

C127) for the case of the CRA. The high absorption capacity of recycled aggregates is directly related to their nature. It should be noted that the recycled aggregates obtained from concrete or mortar waste contain not only the original aggregates but also hydrated cement paste, which increases its porosity and absorption. In addition, the bulk density and unit weight of FRA and CRA presented a lower value compared to that of natural aggregates due to their higher porosity. The maximum CRA size was 25.4 mm (1 in.) and the FRA fineness modulus was 3.04 (coarse sand) (ASTM C136). The resistance to degradation by abrasion and impact evaluated in the "Los Angeles machine" (ASTM C131) was 33.65% for CRA, being possible to use it in the production of concrete mixtures, taking into account this value and those stipulated by ASTM C131. The organic matter content (ASTM C40) of the FRA was the minimum (organic plate No. 1), allowing its use in mortar and/or concrete mixtures.

Table 3. Characteristics of the recycled aggregates (fine and coarse) obtained from the CDW sample.

Characteristics	Fine Recycled Aggregate (FRA)		Coarse Recycled Aggregate (CRA)	
	Standard	**Result**	**Standard**	**Result**
Bulk density (kg/m^3)	ASTM C128	2029	ASTM C127	2326
Absorption (%)	ASTM C128	12.12	ASTM C127	9.17
Unit weight (kg/m^3)	ASTM C29	1240	ASTM C29	1211
Maximum size (mm)	N/A		ASTM C136	25.4
Fineness modulus	ASTM C136	3.04	N/A	
Organic impurities	ASTM C40	Organic plate No. 1	N/A	
Resistance to degradation (%)	N/A		ASTM C131	33.65

2.3.3. Concretes and Blocks: Production and Characterization

The CDW-based AAHC (binder and aggregates) was produced according to the mix design presented in Table 4. The mix design was based on an adaptation of the "absolute volume" method proposed by ACI 211.1. This method requires knowledge of the volumes occupied by all the components of the mixture. For this, the density of each component, including the alkaline activator solution, was determined. The precursor content (500 kg/m^3) and the liquid/solid ratio (L/S = 0.37) were set to achieve a design compressive strength (28 days) greater than 28 MPa (4000 psi).

Table 4. Design of mixtures of alkali-activated hybrid concrete (AAHC) based on CDW.

Material	Dry Weight (kg)	Density (kg/m^3)	Volume (m^3)
CDW precursor	450	2690	0.167
OPC	50	3100	0.016
Alkaline activator solution *	326	1259	0.259
Coarse recycled aggregate (CRA)	605	2326	0.260
Fine recycled aggregate (FRA)	605	2029	0.298
Total	2036	N/A	1

* Alkaline activator solution: Sodium hydroxide + Sodium silicate (SS) + Mixing water.

AAHC was obtained in a CreteAngle horizontal mixer with a mixing time of 8 min. The slump of the mixtures was verified (75–100 mm) according to ASTM C143. The mixture was poured into cylindrical molds (76.2 mm in diameter) and vibrated for 30 s on an electric vibrating table to remove the trapped air. The cylindrical molds were covered for 24 h with a polyethylene film and after this time the cylinders were removed from the mold. The AAHC specimens were cured at room temperature (≈25 °C) and at a relative humidity (RH) greater than 80% until reaching the test age.

The compressive strength of the AAHC was evaluated according to the ASTM C39 standard in a hydraulic press (ELE International) of 1000 kN capacity. Splitting tensile strength was evaluated according to ASTM C496. The microstructural analysis was performed by a scanning electron microscopy (SEM) technique using a JEOL JSM-6490LV microscope with an acceleration voltage of 20 kV. The specimens were evaluated in the low-vacuum mode. An Oxford Instruments Link-Isis

X-ray spectrometer was coupled to the microscope (EDS). The samples corresponded to pieces of approximately 1 cm^3 extracted from the AAHC (28 days of curing) by means of precision cutting. The samples were encapsulated in epoxy resin and the observation surface was polished.

Finally, from the AAHC, solid concrete blocks were produced which were physically and mechanically characterized according to the provisions of the ASTM C140 standard. In all cases, the data reported in the physical and mechanical tests correspond to the average of three test specimens.

3. Results and Discussion

3.1. Characterization of the CDW-Based Hybrid Cement (Binder)

The effect of the NaOH+SS/CDW+OPC and NaOH/SS ratios on the compressive strength of the CDW binder can be seen in the contour diagrams in Figure 6. The results indicate that the presence of high NaOH+SS/CDW+OPC (>0.35) ratios cause a decrease in compressive strength, both at 7 and 28 days of curing. This indicates that alkaline activator contents (NaOH+SS) above an optimal value affect the mechanical performance of CDW hybrid cement. This effect coincides with that reported by Olivia and Nikraz [22] and Reig et al. [23], who report that the increase in the content of alkaline activator can cause a decrease in compressive strength in this type of system. The NaOH/SS ratio meanwhile did not have a significant effect on compressive strength. In this regard, a horizontal orientation of the contour lines is observed at 28 days of curing, as shown in Figure 6b, which demonstrates the non-significant effect of the NaOH/SS ratio on the compressive strength of the CDW binder. This phenomenon agrees with that reported by Puertas et al. [24], who demonstrated that the NaOH content was not an influential factor in the alkaline activation process of ceramic residues using the NaOH+SS mixture as an activator solution. It should be noted that the optimal type, content and concentration of alkaline activator depends on the raw material used as a precursor (FA, GBFS, MK, etc.) and is not a fixed value for all alkali-activated materials; therefore, the results corresponding to CDW differ from those reported by other authors using other types of precursors. From this analysis it was obtained that the optimal NaOH+SS/CDW+OPC and NaOH/SS ratios for the CDW binder were 0.35 and 0.34, respectively; values that allowed the achievement of compressive strengths of 16 and 31 MPa at 7 and 28 days of curing, respectively.

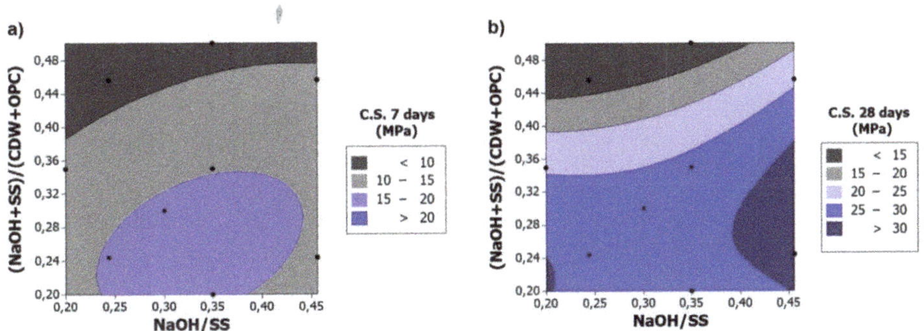

Figure 6. Compressive strength (pastes) at 7 (**a**) and 28 (**b**) days of curing: effect of alkaline activator content and optimization of CDW-based hybrid cement (binder).

Figure 7 represents the evolution of the compressive strength of the optimal CDW binder (NaOH + SS/CDW + OPC = 0.35 and NaOH/SS = 0.34) as a function of curing time (1–90 days) of paste, compared to a reference paste based 100% on OPC. It can be seen how both pastes increase their strength with the evolution of the curing time; noting that, in general, the mechanical performance of the OPC paste is superior to the CDW paste. However, it should be underlined that this difference decreases between 7 and 90 days, a period in which the CDW paste showed a higher strength gain than the OPC paste. The lower compressive strength of CDW paste at an early age (1 and 7 days) is attributable to the low reactivity of the precursor due to its highly crystalline character, as shown in Figure 4. However, from 7 to 28 days, the compressive strength increases significantly (100%), which is attributed to the addition of OPC (10%) that allowed the hardening of the material at room temperature and also accelerated the formation of the alkaline reaction products (C–S–H, C–A–S–H or N,C–A–S–H). The maximum compressive strengths reported by CDW and OPC pastes at 90 days were 43.9 MPa and 49.3 MPa, respectively. These maximum strength values at 90 days (CDW paste = 43.9 MPa vs. OPC paste = 49.3 MPa) represent an increase of 397% and 138% in relation to the strength reported by CDW and OPC pastes at 1 day of curing (CDW paste = 8.8 MPa vs. OPC paste = 20.7 MPa), respectively. This behavior agrees with that reported by Mejia et al. [25], who obtained similar strength increases for an alkali-activated cement based on an FA–GBFS mixture, among other authors [26,27]. In relation to the above, the increase in the strength of hybrid cements is related to the densification of the matrix and the greater formation of "hydrated sodium-calcium aluminum silicates" or (N,C)–A–S–H and C–A–S–H type gels as the curing time increases [28,29].

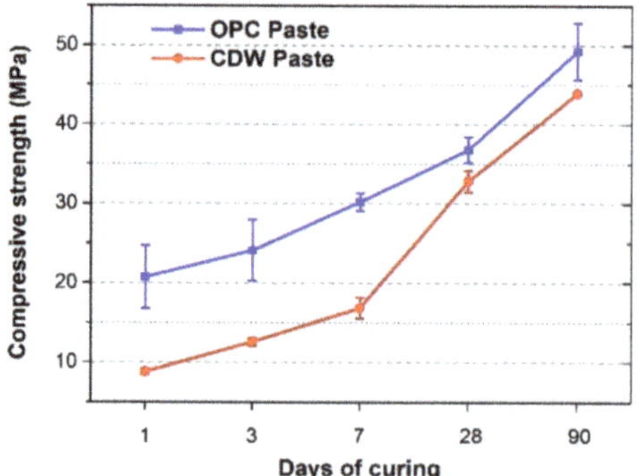

Figure 7. Evolution of compressive strength of the CDW-based hybrid cement (optimal mix): comparison with a 100% OPC-based paste (reference mix).

It should be noted that this is the first time that results have been reported for alkali-activated materials based on a CDW precursor consisting of the mixture of concrete (25%), ceramic (25%), masonry (25%) and mortar (25%) wastes. Regarding the results reported in Table 1, which are related to alkali-activated materials produced from separated or "clean" wastes, it is highlighted that the maximum compressive strength achieved by the CDW binder (43.9 MPa) in this research exceeds the performance of most of these reports, as shown in Table 1. Additionally, the curing conditions of the CDW binder were at room temperature (≈25 °C and RH ≈ 80%), that is, no hydrothermal curing processes (50–90 °C) such as those indicated by other authors were necessary, as shown in Table 1. Likewise, it should be noted that the content of OPC added to the CDW binder was only 10% (by weight), lower than the content (≤40%) of addition or incorporation of calcium-rich materials (OPC, GBFS or $Ca(OH)_2$) or alumino-silicates (FA or MK) previously reported, as shown in Table 1.

Figure 8 shows the results obtained by the calorimetric analysis of CDW paste (90%CDW + 10%OPC + NaOH-SS) compared to OPC paste (100% OPC + H_2O) and a reference paste composed of the mixture 90%CDW + 10%OPC + H_2O. In the case of CDW paste, the maximum peak was identified in the heat evolution curve, as shown in Figure 8, (left), in a reaction time of less than 10 min, reaching a maximum of 12.6 J/g of binder. In this regard, the appearance of a single high intensity peak on the heat evolution curve coincides with that reported by other authors [30–33] for alkali-activated cements based on MK, FA, GBFS, natural pozzolans, and/or their corresponding mixtures.

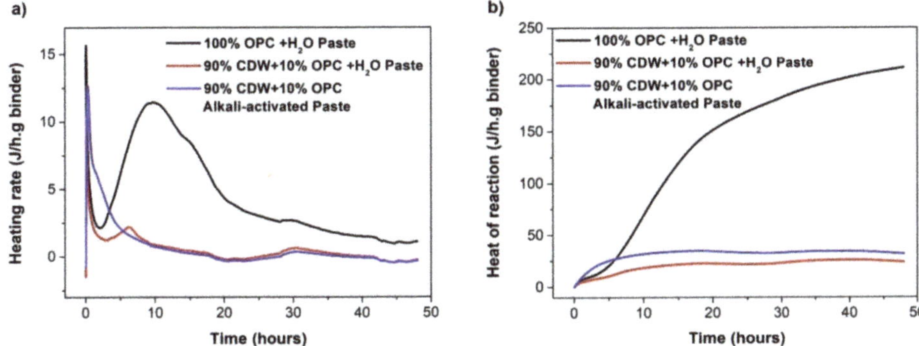

Figure 8. Isothermal calorimetry: reaction kinetics of hybrid cement (binder) based on CDW (optimal mix). Evolution of heat of hydration (**a**) and total heat of hydration (48 h) (**b**).

Likewise, the maximum heat peak reported by the CDW paste (12.6 J/g of cement) was higher than that shown by the 90%CDW + 10%OPC+ H_2O paste (10.4 J/g of cement); behavior that can be associated with the development of alkaline activation reactions in the CDW binder, in accordance with what was reported by Garcia-Lodeiro et al. [34]. Indeed, the total heat of reaction, as shown in Figure 8, (right) of the CDW paste (alkali-activated) and the 90%CDW + 10%OPC + H_2O (hydrated) paste was 31.8 and 23.8 J/h·g, respectively. This shows that the presence of the alkaline activator (NaOH + SS) in the CDW binder promotes a positive effect on the kinetics of the alkaline activation reactions for the 90%CDW+10%OPC mixture, exceeding the heat value by 33.6%, the value of total heat reported by the mixture without alkaline activator (90%CDW + 10%OPC + H_2O paste). The OPC paste, meanwhile, showed the maximum heat peak in the first minutes, as shown in Figure 8, (left), among the pastes studied, with a value of 14.9 J/h·g of cement. This value coincides with the maximum heat range (12.6–36.4 J/h·g) reported by other authors for OPC-based pastes [35–37]. Using the total heat of hydration curve, as shown in Figure 8, (right), it was possible to observe that the CDW paste released a greater amount of heat during the first hours (≤5 h) compared to the OPC paste. However, this behavior was reversed after 5 h. In fact, the OPC paste released a total heat of hydration of up to 211.2 J/h·g of cement; a value that exceeds the total heat reported by the CDW paste by 664.2% (31.8 J/h·g).

Additionally, the initial and final setting times (ASTM C191) of the optimal CDW binder (NaOH + SS/CDW + OPC = 0.35 and NaOH/SS = 0.34) were determined, yielding values of 40 and 52 min, respectively. Likewise, standard mortars with a CDW/sand ratio of 1:2.75 were manufactured, which were subjected to the compressive strength test (ASTM C109) after 28 days of curing, reaching a strength value of 34.6 MPa.

Based on the specifications established by the ASTM C1157 standard, where the compressive strength is one of the most important parameters for the classification of cements based on their performance, the CDW-based hybrid cement obtained in this research could be classified as a general-use cement (GU type), because the mortar made with this binder attained a strength (34.6 MPa) higher than that required by ASTM C1157 (28 MPa). Additionally, based on the isothermal calorimetry results, CDW-based hybrid cement could also be classified as a low-heat-of-hydration cement (type LH) with a total heat of reaction of 31.8 J/g of cement, less than the maximum value required by the standard to achieve this classification (290 J/g of cement).

3.2. Characterization of CDW-Based Hybrid Concrete (AAHC) (As Both Precursor and Aggregates)

Figure 9 presents the internal appearance of the CDW-based AAHC compared to 100% OPC-based conventional concrete. This image highlights, in addition to the adequate (homogeneous) distribution of the CDW aggregates, the brown color of the AAHC. This phenomenon is due to the color of the CDW precursor, mainly due to ceramic waste (red) and masonry waste (red), as shown in Figure 2. In this regard, what in principle could be considered a disadvantage (a color other than the conventional "gray"), becomes an advantage at the application or industrial level. Currently, the demand and production of colored architectural concrete has shown an increase with the modernization of the construction industry [38], leaning on a technological level due to a particular interest in alternative pigments to minerals (waste), of low cost, with greater stability than conventional paints, and even obtaining colored concrete with reflective or thermally insulating properties (cool colored concretes) [39].

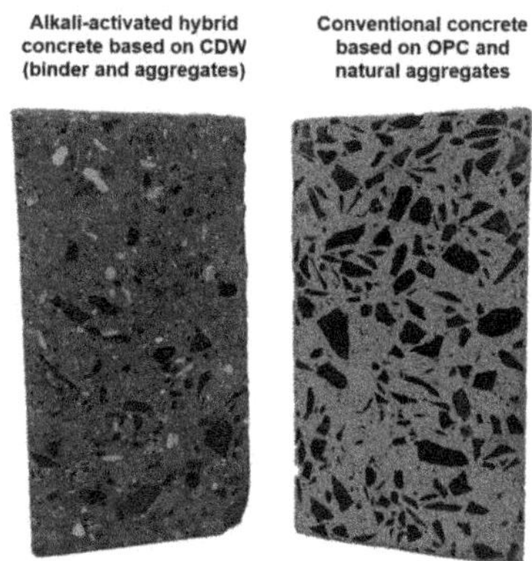

Figure 9. Internal appearance and aggregate distribution of CDW-based hybrid concrete (AAHC) vs. OPC concrete.

The evolution of the compressive strength of the CDW-based AAHC is presented in Figure 10. The values of compressive strength at 1, 3, 7, 28, and 90 days were 8.5, 17.4, 20.5, 33.9, and 42.6 MPa, respectively. A progressive increase in mechanical performance with the course of curing time stands out; this being an increase of 403% considering the values reported between days 1 (8.5 MPa) and 90 (42.6 MPa). Considering the strength limit (17.5 MPa) established by ACI 318 (red line), the CDW–based AAHC reaches its structural classification after 7 days of curing; noting that the CDW-based AAHC curing process was at room temperature (≈25 °C) and under humid conditions (relative humidity (RH) ≈80%) and not under immersion in water as established for OPC-based concretes. Meanwhile, the splitting tensile strength (ASTM C496) of the CDW-based AAHC at 28 days of curing was 3.0 MPa; a value that represents 8.9% of its compressive strength (33.9 MPa) at this same age.

It should be noted that, contrary to the studies reported by other authors, as shown in Table 1, in the field of CDW-based alkali-activated materials (pastes and mortars), the results presented here correspond to the level of concrete. In this sense, and despite the fact that it is not considered correct to compare these results with determined values for pastes and mortars, the level of compressive strength (up to 42.60 MPa at 90 days) reached by the CDW-based AAHC stands out, as shown in Figure 10, highlighting its curing at room temperature (≈25 °C), that is, the non-dependence on additional hydrothermal curing processes. Likewise, the results of compressive strength (28 days) obtained at the level of paste (32.9 MPa), mortar (34.6 MPa), and concrete (33.9 MPa) highlight the binding capacity of the CDW binder, managing to maintain the same resistant level (32–35 MPa) despite the presence of fine aggregates (FRA) and/or coarse recycled (CRA) in the mixture.

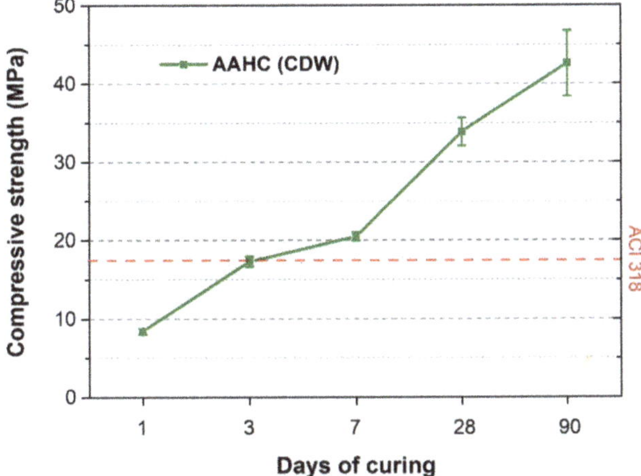

Figure 10. Evolution of compressive strength (≈25 °C) of CDW-based hybrid concrete (AAHC).

Figure 11 represents the SEM-EDS microstructural analysis of the CDW-based AAHC at 28 days of curing. Through this technique it was possible to study the elemental chemical composition found for AAHC at the microstructural level and thus obtain a graphic representation in a SiO_2-Al_2O_3-CaO ternary diagram of the results obtained, as shown in Figure 11.

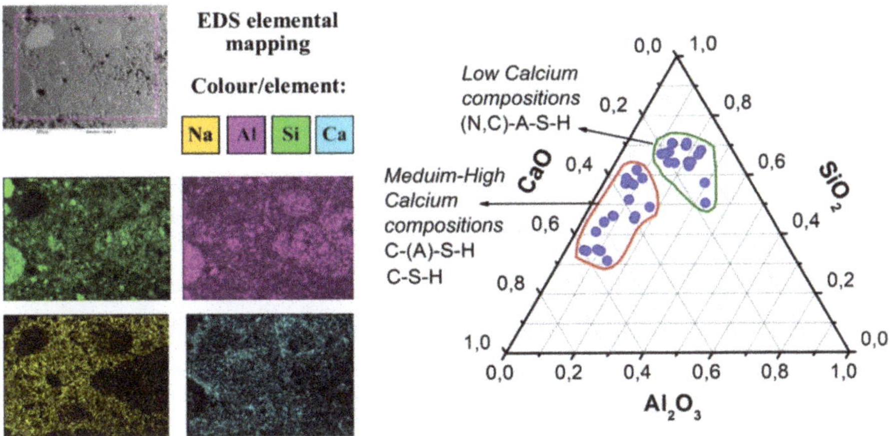

Figure 11. SEM-EDS microstructural analysis (28 days of curing) of AAHC binder.

This ternary diagram represents a total of 35 EDS points analyzed on the AAHC binder. The results obtained are grouped into the regions of "low calcium compositions" and "medium-high calcium compositions". Some authors have managed to correlate these representative areas of the SiO_2–Al_2O_3–CaO ternary diagram with (N,C)–A–S–H (poor in Ca^{2+}), C–S–H, and C–A–S–H type (rich in Ca^{2+}) reaction products for alkali-activated materials and hybrid cements based on other precursors, such as FA, GBFS, and/or natural pozzolans [29,40–45]. An EDS elementary color mapping performed on a specific area of the AAHC, as shown in Figure 11, showed the homogeneous distribution of silica (Si: green), alumina (Al: purple), sodium (Na: yellow), and calcium (Ca: turquoise) in the concrete, distinguishing the CDW aggregates (poor areas in Na) from the CDW binder (areas rich in Si, Al, Na, and Ca).

In addition, the microstructural observation of the AAHC, as shown in Figure 12, demonstrates the obtaining of a homogeneous interface transition zone between the CDW binder and the fine (FRA) and coarse (CRA) aggregates based on CDW, which agrees with the high mechanical performance reported for this material.

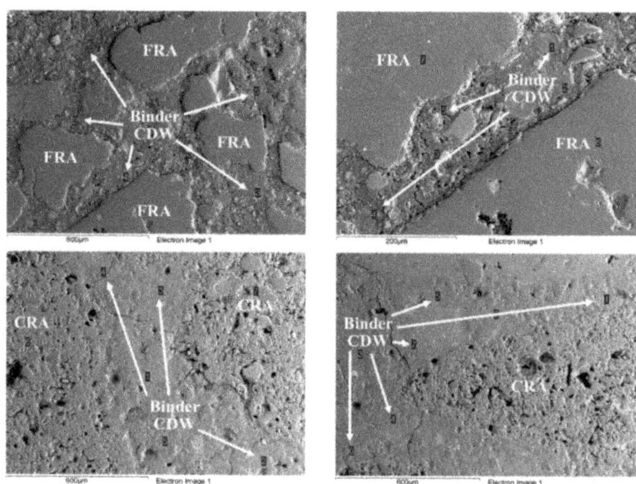

Figure 12. SEM microstructural observation (28 days of curing) of CDW-based hybrid concrete (AAHC).

The mechanical and microstructural results validate the possibility of using alkaline activation technology to obtain structural concretes based on fully utilizing CDW (binder + aggregates).

3.3. *Production and Characterization of a Building Element (Solid Block Type) from the AAHC*

Solid concrete blocks were produced from the AAHC, which were physically and mechanically characterized according to the specifications set out by ASTM C140. Table 5 presents the properties and characteristics of the solid block, which highlights the achievement of a compressive strength at 28 days of 26.1 MPa; a value that exceeds the established lower limit of strength (13 MPa) by 101%. According to ASTM C90 (equivalent to NTC 4026) is classified as "high class structural block". For comparative purposes, the ASTM C1790 standard establishes, in the case of blocks based on the alkaline activation of "fly ash", that the compressive strength must be greater than 24.1 MPa. According to NTC 4026, the maximum percentage of water absorption allowed for medium-weight blocks (1680–2000 kg/m^3) is 15%; a condition that the CDW block meets, with water absorption at 14.4%. It should be noted that the ASTM C140 standard establishes that these limits of water absorption and compressive strength must be met within 12 months after the units are produced. The results obtained (28 days) allowed us to validate the potential use of the CDW-based AAHC in structural precast products suitable for construction applications.

Table 5. Properties and characteristics of the solid block made from CDW-based hybrid concrete (AAHC).

CDW Concrete Block	Properties and Characteristics	Result
	Dimensions (length × width × height)	200 × 100 × 80 mm
	Curing temperature	25 °C
	Compressive strength (28 days)	26.1 MPa
	Rupture modulus (28 days)	3.6 MPa
	Density	1926 kg/m^3
	Water absorption	14.4%

4. Conclusions

The results of this research demonstrate the possibility of producing alkali-activated materials (pastes, mortars, concrete, and precast building elements) based on the mixture of concrete, ceramic, masonry and mortar wastes as a sustainable alternative for the recycling of construction and demolition wastes (CDWs). It should be noted that CDWs were used to obtain both binder and aggregates (fine and coarse), thus achieving a comprehensive application of CDWs. The following conclusions can be drawn from the experimental results.

Alkaline activation of the CDW precursor, using mixtures of hydroxide and sodium silicate, allowed the synthesis of a CDW binder with compressive strengths of up to 43.9 MPa after 90 days of curing at room temperature. In this context, the addition of only 10% OPC allowed the mixtures to cure at room temperature (≈25 °C). The compressive strength achieved by the CDW binder-based mortar was 34.6 MPa. The total heat of reaction of this hybrid cement reached maximum values of 31.8 J/h·g. Based on mechanical performance and total heat of reaction, the CDW binder can be classified as general-use cement (GU type) and low heat of hydration (LH type), meeting the specifications of ASTM C1157.

The CDW concrete achieved compressive strength values of up to 42.6 MPa (90 days of curing (≈25 °C)), demonstrating the possibility of obtaining structural classification concretes according to ACI 318 specifications. The compressive and splitting tensile strengths of CDW concrete were 33.9 and 3.0 MPa, respectively (28 days of curing). Based on the results obtained at the levels of paste (32.9 MPa), mortar (34.6 MPa) and concrete (33.9 MPa), the binding capacity of the CDW binder was demonstrated, managing to maintain the same strength level (32–35 MPa) despite the presence of recycled aggregates (fine and/or coarse). These results further demonstrated that recycled aggregates do not affect the mechanical performance of alkali-activated materials. The SEM-EDS micro-structural analysis carried out on the CDW concrete demonstrated the formation, in the binder phase, of (N,C)-A-S-H (poor in Ca^{2+}), C-S-H, and C-A-S-H (rich in Ca^{2+}) type reaction products, in addition to homogeneous interfacial transition zones, which taken together is consistent with the high mechanical performances reported for alkali-activated materials.

The production and physical-mechanical characterization of a solid concrete block demonstrated the application potential of CDW concrete in the production of precast building elements. This CDW concrete block (1925.8 kg/m^3) reported a compressive strength of 26.1 MPa (28 days), exceeding by 101% the lower strength limit (13 MPa) established by the ASTM C90 standard to be classified as a "high class structural block".

The reported results validate the possibility of using alkaline-activation technology in obtaining alkali-activated materials based on making the fullest use of CDWs (binder + aggregates). This innovative recycling concept is considered a sustainable alternative for the real-life use of CDWs in conditions where the separation of concrete, ceramic, masonry, and mortar wastes is not possible. Furthermore, these results validate the possibility of reusing CDWs to obtain new materials with greater commercial value (cements and building applications) than recycled aggregates (common use;, an advantage that would promote the proper handling and management of this type of waste in the construction sector, towards the goal of "zero waste", complying with the basic principles of the circular economy.

Author Contributions: Conceptualization, R.A.R.-S., W.V.-S., and R.M.d.G.; methodology and investigation, R.A.R.-S. and W.V.-S.; writing—original draft preparation, R.A.R.-S. and W.V.-S.; supervision, writing—review and editing, project administration, R.M.d.G. All authors have read and agreed to the published version of the manuscript.

Funding: This research was funded by Colombian Institute for the Development of Science, Technology, and Innovation (Colciencias), grant number 096-2016 and the Universidad del Valle (Cali, Colombia), grant number CI-21025.

Acknowledgments: R. Robayo-Salazar and W. Valencia-Saavedra thanks the call No. 848-2019 (Minciencias) for the postdoctoral fellowship in Colombia.

Conflicts of Interest: The authors declare no conflict of interest.

References

1. European Commission, Construction and Demolition Waste (CDW). 2016. Available online: http://ec.europa.eu/environment/waste/construction_demolition.htm (accessed on 13 March 2018).
2. Akhtar, A.; Sarmah, A.K. Construction and demolition waste generation and properties of recycled aggregate concrete: A global perspective. *J. Clean. Prod.* **2018**, *186*, 262–281. [CrossRef]
3. Naciones Unidas CEPAL, Agenda 2030 y los Objetivos de Desarrollo Sostenible. Una oportunidad Para America Latina y el Caribe. Available online: http://www.sela.org/media/2262361/agenda-2030-y-los-objetivos-de-desarrollo-sostenible.pdf (accessed on 18 January 2018).
4. Akcil, A.; Agcasulu, I.; Swain, B. Valorization of waste LCD and recovery of critical raw material for circular economy: A review. *Resour. Conserv. Recycl.* **2019**, *149*, 622–637. [CrossRef]
5. Tam, V.W.Y.; Soomro, M.; Evangelista, A.C.J. A review of recycled aggregate in concrete applications (2000–2017). *Constr. Build. Mater.* **2018**, *172*, 272–292. [CrossRef]
6. Vieira, C.S.; Pereira, P.M. Use of recycled construction and demolition materials in geotechnical applications: A review. *Resour. Conserv. Recycl.* **2015**, *103*, 192–204. [CrossRef]
7. Wu, Y.; Lu, B.; Bai, T.; Wang, H.; Du, F.; Zhang, Y.; Cai, L.; Jiang, C.; Wang, W. Geopolymer, green alkali activated cementitious material: Synthesis, applications and challenges. *Constr. Build. Mater.* **2019**, *224*, 930–949. [CrossRef]
8. RILEM Technical Committee 247-DTA, 247-DTA: Durability Testing of Alkali-Activated Materials. Available online: https://www.rilem.net/groupe/247-dta-durability-testing-of-alkali-activated-materials-290 (accessed on 18 January 2018).
9. Lampris, C.; Lupo, R.; Cheeseman, C. Geopolymerisation of silt generated from construction and demolition waste washing plants. *Waste Manag.* **2009**, *29*, 368–373. [CrossRef]
10. Reig, L.; Tashima, M.; Borrachero, M.; Monzó, J.; Cheeseman, C.; Payá, J. Properties and microstructure of alkali-activated red clay brick waste. *Constr. Build. Mater.* **2013**, *43*, 98–106. [CrossRef]
11. Komnitsas, K.; Zaharaki, D.; Vlachou, A.; Bartzas, G.; Galetakis, M. Effect of synthesis parameters on the quality of construction and demolition wastes (CDW) geopolymers. *Adv. Powder Technol.* **2015**, *26*, 368–376. [CrossRef]
12. Robayo, R.A.; Mulford, A.; Munera, J.; De Gutiérrez, R.M. Alternative cements based on alkali-activated red clay brick waste. *Constr. Build. Mater.* **2016**, *128*, 163–169. [CrossRef]
13. Vásquez, A.; Cárdenas, V.; Robayo, R.A.; De Gutiérrez, R.M. Geopolymer based on concrete demolition waste. *Adv. Powder Technol.* **2016**, *27*, 1173–1179. [CrossRef]
14. Zaharaki, D.; Galetakis, M.; Komnitsas, K. Valorization of construction and demolition (C&D) and industrial wastes through alkali activation. *Constr. Build. Mater.* **2016**, *121*, 686–693. [CrossRef]
15. Sassoni, E.; Pahlavan, P.; Franzoni, E.; Bignozzi, M.C. Valorization of brick waste by alkali-activation: A study on the possible use for masonry repointing. *Ceram. Int.* **2016**, *42*, 14685–14694. [CrossRef]
16. Murillo, L.M.; Delvasto, S.; Suárez, M.G. A study of a hybrid binder based on alkali-activated ceramic tile wastes and portland cement. In *Sustainable and Nonconventional Construction Materials Using Inorganic Bonded Fiber Composites*; Elsevier BV: Amsterdam, The Netherlands, 2017; pp. 291–311.

17. Reig, L.; Sanz, M.; Borrachero, M.; Monzó, J.; Soriano, L.; Payá, J. Compressive strength and microstructure of alkali-activated mortars with high ceramic waste content. *Ceram. Int.* **2017**, *43*, 13622–13634. [CrossRef]
18. Robayo-Salazar, R.; Rivera, J.F.; De Gutiérrez, R.M. Alkali-activated building materials made with recycled construction and demolition wastes. *Constr. Build. Mater.* **2017**, *149*, 130–138. [CrossRef]
19. Tuyan, M.; Andiç-Çakir, Ö.; Ramyar, K. Effect of alkali activator concentration and curing condition on strength and microstructure of waste clay brick powder-based geopolymer. *Compos. Part B Eng.* **2018**, *135*, 242–252. [CrossRef]
20. Fort, J.; Vejmelková, E.; Koňáková, D.; Alblová, N.; Čáchová, M.; Keppert, M.; Rovnaníková, P.; Černý, R. Application of waste brick powder in alkali activated aluminosilicates: Functional and environmental aspects. *J. Clean. Prod.* **2018**, *194*, 714–725. [CrossRef]
21. Hwang, C.-L.; Yehualaw, M.D.; Vo, D.-H.; Huynh, T.-P. Development of high-strength alkali-activated pastes containing high volumes of waste brick and ceramic powders. *Constr. Build. Mater.* **2019**, *218*, 519–529. [CrossRef]
22. Olivia, M.; Nikraz, H. Properties of fly ash geopolymer concrete designed by Taguchi method. *Mater. Des.* **2012**, *36*, 191–198. [CrossRef]
23. Reig, L.; Soriano, L.; Borrachero, M.; Monzó, J.; Payá, J. Influence of the activator concentration and calcium hydroxide addition on the properties of alkali-activated porcelain stoneware. *Constr. Build. Mater.* **2014**, *63*, 214–222. [CrossRef]
24. Puertas, F.; Barba, A.; Gazulla, M.F.; Gómez, M.P.; Palacios, M.; Martinez-Ramirez, S. Residuos cerámicos para su posible uso como materia prima en la fabricación de clínker de cemento Portland: Caracterización y activación alcalina. *Mater. Constr.* **2006**, *56*, 73–84. [CrossRef]
25. De Gutiérrez, R.M.; Mejía, J.M.; Puertas, F. Ceniza de cascarilla de arroz como fuente de sílice en sistemas cementicios de ceniza volante y escoria activados alcalinamente. *Mater. Constr.* **2013**, *63*, 361–375. [CrossRef]
26. Puertas, F.; Martínez-Ramírez, S.; Alonso, S.; Vázquez, T. Alkali-activated fly ash/slag cements. *Cem. Concr. Res.* **2000**, *30*, 1625–1632. [CrossRef]
27. Phoo-Ngernkham, T.; Chindaprasirt, P.; Sata, V.; Pangdaeng, S.; Sinsiri, T. Properties of high calcium fly ash geopolymer pastes with Portland cement as an additive. *Int. J. Miner. Met. Mater.* **2013**, *20*, 214–220. [CrossRef]
28. Rivera, J.F.; Mejia, J.M.; De Gutiérrez, R.M.; Suárez, M.G. Hybrid cement based on the alkali activation of by-products of coal. *Rev. Constr.* **2014**, *13*, 31–39. [CrossRef]
29. Garcia-Lodeiro, I.; Fernández-Jiménez, A.; Palomo, A. Variation in hybrid cements over time. Alkaline activation of fly ash–portland cement blends. *Cem. Concr. Res.* **2013**, *52*, 112–122. [CrossRef]
30. Bernal, S.A.; Rodriguez, E.D.; De Gutiérrez, R.M.; Provis, J.L.; Delvasto, S. Activation of Metakaolin/Slag Blends Using Alkaline Solutions Based on Chemically Modified Silica Fume and Rice Husk Ash. *Waste Biomass Valoriz.* **2011**, *3*, 99–108. [CrossRef]
31. Mejia, J.; Rodriguez, E.D.; De Gutiérrez, R.M. Utilización potencial de una ceniza volante de baja calidad como fuente de aluminosilicatos en la producción de geopolímeros. *Ing. Univ.* **2014**, *18*, 309. [CrossRef]
32. De Gutiérrez, R.M.; Robayo, R.A.; Gordillo, M. Natural pozzolan—And granulated blast furnace slag-based binary geopolymers. *Mater. Constr.* **2016**, *66*. [CrossRef]
33. Robayo-Salazar, R.; De Gutiérrez, R.M.; Puertas, F. Study of synergy between a natural volcanic pozzolan and a granulated blast furnace slag in the production of geopolymeric pastes and mortars. *Constr. Build. Mater.* **2017**, *157*, 151–160. [CrossRef]
34. Lodeiro, I.G.; Fernandez-Jiménez, A.; Palomo, A. Hydration kinetics in hybrid binders: Early reaction stages. *Cem. Concr. Compos.* **2013**, *39*, 82–92. [CrossRef]
35. Gruyaert, E.; Robeyst, N.; De Belie, N. Study of the hydration of Portland cement blended with blast-furnace slag by calorimetry and thermogravimetry. *J. Therm. Anal. Calorim.* **2010**, *102*, 941–951. [CrossRef]
36. Xu, H.; Gong, W.; Syltebo, L.; Lutze, W.; Pegg, I.L. DuraLith geopolymer waste form for Hanford secondary waste: Correlating setting behavior to hydration heat evolution. *J. Hazard. Mater.* **2014**, *278*, 34–39. [CrossRef] [PubMed]
37. Xu, Q.; Hu, J.; Ruiz, J.M.; Wang, K.; Ge, Z. Isothermal calorimetry tests and modeling of cement hydration parameters. *Thermochim. Acta* **2010**, *499*, 91–99. [CrossRef]
38. López, A.; Guzmán, G.A.; Di Sarli, A.R. Color stability in mortars and concretes. Part 2: Study on architectural concretes. *Constr. Build. Mater.* **2016**, *123*, 248–253. [CrossRef]

39. Levinson, R.; Akbari, H.; Berdahl, P.; Wood, K.; Skilton, W.; Petersheim, J. A novel technique for the production of cool colored concrete tile and asphalt shingle roofing products. *Sol. Energy Mater. Sol. Cells* **2010**, *94*, 946–954. [CrossRef]
40. Ortega, E.A.; Cheeseman, C.; Knight, J.; Loizidou, M. Properties of alkali-activated clinoptilolite. *Cem. Concr. Res.* **2000**, *30*, 1641–1646. [CrossRef]
41. Lodeiro, I.G.; Macphee, D.; Palomo, A.; Fernandez-Jiménez, A. Effect of alkalis on fresh C–S–H gels. FTIR analysis. *Cem. Concr. Res.* **2009**, *39*, 147–153. [CrossRef]
42. Gebregziabiher, B.S.; Thomas, R.J.; Peethamparan, S. Temperature and activator effect on early—Age reaction kinetics of alkali-activated slag binders. *Constr. Build. Mater.* **2016**, *113*, 783–793. [CrossRef]
43. Herrero, M.J.S.; Fernandez-Jiménez, A.; Palomo, A. Alkaline Hydration of C2S and C3S. *J. Am. Ceram. Soc.* **2015**, *99*, 604–611. [CrossRef]
44. Djobo, J.N.Y.; Tchakouté, H.K.; Ranjbar, N.; Elimbi, A.; Tchadjié, L.N.; Njopwouo, D. Gel Composition and Strength Properties of Alkali-Activated Oyster Shell-Volcanic Ash: Effect of Synthesis Conditions. *J. Am. Ceram. Soc.* **2016**, *99*, 3159–3166. [CrossRef]
45. Robayo-Salazar, R.; De Gutiérrez, R.M.; Puertas, F. Alkali-activated binary concrete based on a natural pozzolan: Physical, mechanical and microstructural characterization. *Mater. Constr.* **2019**, *69*, 191. [CrossRef]

Article

Ecotoxicity of Concrete Containing Fine-Recycled Aggregate: Effect on Photosynthetic Pigments, Soil Enzymatic Activity and Carbonation Process

Diana Mariaková [1,*], Klára Anna Mocová [2], Jan Pešta [1,2], Kristina Fořtová [1], Bhavna Tripathi [3], Tereza Pavlů [1] and Petr Hájek [1]

1 Research Team Architecture and the Environment, University Centre for Energy Efficient Buildings of Czech Technical University in Prague, Trinecka 1024, 273 43 Bustehrad, Czech Republic; jan.pesta@cvut.cz (J.P.); kristina.fortova@cvut.cz (K.F.); tereza.pavlu@cvut.cz (T.P.); petr.hajek@fsv.cvut.cz (P.H.)
2 Department of Environmental Chemistry, Faculty of Environmental Technology, University of Chemistry and Technology, Technicka 5, 166 28 Prague, Czech Republic; klara.mocova@vscht.cz
3 School of Civil and Chemical Engineering, Manipal University Jaipur, Dehmi Kalan, Jaipur 303 007, India; bhavna.tripathi@jaipur.manipal.edu
* Correspondence: diana.mariakova@cvut.cz

Abstract: Recycling of materials such as masonry or concrete is one of the suitable ways to reduce amount of disposed construction and demolition waste (CDW). However, the environmental safety of products containing recycled materials must be guaranteed. To verify overall environmental benefits of recycled concrete, this work considers ecotoxicity of recycled concrete, as well as potential environmental impacts of their life cycle. Moreover, impacts related with carbonation of concrete is considered in terms of durability and influence of potential CO_2 uptake. Concrete containing fine recycled aggregate from two different sources (masonry and concrete) were examined experimentally at the biochemical level and compared with reference samples. Leaching experiments are performed in order to assess physicochemical properties and aquatic ecotoxicity using water flea, freshwater algae and duckweed. The consequences, such as effects of material on soil enzymatic activity (dehydrogenase activity), photosynthetic pigments (chlorophylls and carotenoids), and the carbonation process, are verified in the laboratory and included in the comparison with the theoretical life cycle assessment. As a conclusion, environmental safety of recycled concrete was verified, and its overall potential environmental impact was lower in comparison with reference concrete.

Keywords: recycled concrete; carbonation; life cycle assessment

Citation: Mariaková, D.; Mocová, K.A.; Pešta, J.; Fořtová, K.; Tripathi, B.; Pavlů, T.; Hájek, P. Ecotoxicity of Concrete Containing Fine-Recycled Aggregate: Effect on Photosynthetic Pigments, Soil Enzymatic Activity and Carbonation Process. *Sustainability* **2022**, *14*, 1732. https://doi.org/10.3390/su14031732

Academic Editors: Anibal C. Maury-Ramirez and Jaime A Mesa

Received: 10 December 2021
Accepted: 31 January 2022
Published: 2 February 2022

1. Introduction

Construction and demolition waste (CDW) constituted approximately 35.9% of the total waste production in the EU in 2018. CDW, as one of the highest waste streams, consists of materials like red bricks, mortar, masonry, and concrete, which can be recycled and used as secondary raw materials. This approach reduces not only waste but also the demand for primary resources. However, there is a risk of using recycled materials with content, which is potentially harmful to human health or the environment. Therefore, prior to their use, the ecotoxicity of such materials must be tested using ecotoxicological bioassays and their potential environmental impact should be assessed.

To evaluate the ecotoxicological impact of concrete containing recycled materials, bioassays according to the European law system can be performed. These tests are designed to determine the potential influence of various chemicals or their mixtures, along with the transport from the source to the reservoir. To model this transport, the leachates of considered materials are prepared. However, just the impact caused by bioavailable chemicals can be evaluated using these tests.

Ecotoxicity of construction waste or materials was assessed in previous studies, which were carried out with a simple test design, such as the freshwater algae growth test, seed germination test, crustacean acute assay, marine bacteria bioluminescence test, or the yeast growth test [1–5]. Nevertheless, tests with these organisms are focused on the influence of chemicals on water ecosystems only.

Green plants (algae, aquatic and terrestrial plants) are usually examined not only at morphological level, such as growth rate or yield. Photosynthetic pigments represent the most typical chemicals in plants. Chlorophylls are closely related to primary production, while carotenoids serve as protection against adverse effects of the environment. Both groups of pigments are known to be sensitive to contamination, alkaline pH, and consequently oxidative stress [6–8]. With a significant decrease in chlorophyll, it is likely that plant growth will also decrease. An increase in total carotenoids indicates internal oxidative stress, which can result from lack of nutrients, heavy metal accumulation, and other stresses associated with the formation of reactive oxygen species (ROS) [9].

Impacts on soil ecosystems can be assessed using tests focused on nonspecified microbial communities where a selected metabolic activity is determined. Soil enzymes are produced mainly by bacteria and fungi, and are suitable for the determination of various external effects on the soil microbiota [10]. Various methods for the determination of soil enzymes, such as oxidoreductases, hydrolases, transferases, etc., have been described [10], but the most often found soil enzymes belongs to the group of dehydrogenases (DHA). In contrast to most soil enzymes, DHA are intracellular, and so DHA can be used as an indicator of living (active) cells [11].

Besides ecotoxicological impact, other environmental impacts, such as an impact on climate change, should be assessed. The impact of CO_2 emissions is one of the most discussed issues in the European Union. The EU aims to reduce CO_2 emissions values by 40% by 2030 [12]. Moreover, up to 9% of CO_2 emissions are directly related to the construction industry, and about 3% specifically to concrete [13]. This is also associated with a large amount of energy consumption, which is spent on the construction process (from material production, building the construction, construction life, and also demolition). This amount is estimated to be up to 40% of total energy consumption [14].

On the other hand, one of the beneficial influences of concrete is the absorption of CO_2 during a slow process called carbonation, in which CO_2 reacts with the cement matrix, mainly portlandite. Limit conditions for this reaction are the environment, the amount of carbon dioxide in the air, and the type of concrete (great influence, e.g., porosity) [15–20]. The CO_2 and moisture of the environment neutralize concrete by forming calcium carbonate and reducing alkaline balance, which means that the initial properties are rapidly changing during the carbonation process. During the reaction, the pH values decrease from 12–12.5 to 9, and as a result the protective properties of the material are weakened and a suitable environment appears for the development of corrosion [21]. These effects decrease the quality and possible utilization of concrete, and so the speed of carbonation is used to characterize the concrete quality. Thus, many researchers have stated that the durability of concrete is better with slower carbonation speed [22,23]. However, even concrete with a higher speed of carbonation can be used in some applications. Thus, subsequent absorption of CO_2 by concrete should be assessed as a potential benefit and compared with other environmental impacts in the life cycle of concrete.

Potential environmental impacts caused by recycled concrete can be assessed using the Life Cycle Assessment (LCA) method. The LCA is used to analyze not only the life cycle of the concrete itself, but also material and energy flows between the concrete and the environment, as well as the impact of these flows. In life cycle assessment, it is necessary to take into account the issue of care for the structure at the end of its life, and also benefits such as CO_2 uptake.

This study aims to verify the environmental safety of different types of concrete containing recycled aggregates in two strength classes. Each strength class had its own reference sample (control) with which the recycled mixtures were compared. These environmentally friendly mixtures have been designed with regard to the properties of individual materials and are intended for use in the construction sector, for example, as the foundations of buildings. In addition to the influence of leachates on aquatic plants and invertebrates, this research deals with the determination of photosynthetic pigments and impact on soil enzymes. Recycled concretes were exposed to the carbonation test, to analyze the impact of the environment on the samples. The rate of CO_2 absorption was measured according to the valid Czech standard ČSN EN 12390-12 (73 1302) [24]. Following the gained practical knowledge from laboratory experiments, the theoretical level was evaluated in the form of life cycle analysis.

2. Materials and Methodology

2.1. Materials

This work is based on the solid foundations of previous research, which verified the chemical analysis and ecotoxicity of selected waste materials from different sources [25]. The authors investigated four types of waste materials, and after evaluation and verification, two types were picked and used in this investigation as a substitute for natural aggregate. Natural aggregate concrete (NAC), which contains natural aggregate, was used as a reference sample in both strength classes.

Two types of strength class were tested to compare the properties:

- Strength class I—corresponds to ordinary concrete in strength class C16/20
- Strength class II—corresponds to ordinary concrete in strength class C25/30

In each strength class was the reference sample containing natural aggregate and two types of samples with recycled aggregate. Therefore, there were a total of three samples in each strength class (reference sample and two mixtures with recycled aggregate). Thus, a total of six mixtures were tested (two strength classes of three mixtures each).

The first type of recycled aggregate used originates from masonry structures and contains mainly red bricks, mortar, and plasters (RA4) [25]. It was prepared from reinforcement concrete at the recycling center using the two-step recycling process and used in recycled masonry aggregate concrete (RMAC) in this research. This type of concrete was made in two mixtures with different strength classes (RMAC-I, RMAC-II).

The second type of aggregate used was prepared from reinforcement concrete in the recycling center by the two-step recycling process (RA1) [25]. The crushed and separated recycled aggregate of fraction 16/128 mm from the first step of the recycling process was crushed and sieved into fractions in the second step. Two concrete mixtures containing RA1 were prepared (RCAC-I, RCAC-II).

In general, six concrete mixtures were made and tested in the field of ecotoxicity at the biochemical level with regard to environmental impacts; specifically, a comparison of actual exposure and potential life cycle was examined.

- NAC-I, as a reference concrete sample for strength class C16/20
- RMAC-I, as recycled concrete containing RA4, strength class C16/20.
- RCAC-I, as recycled concrete containing RA1, strength class C16/20.
- NAC-II, as a reference concrete sample for the C25/30 strength class
- RMAC-II, as a recycled concrete containing RA4, strength class C25/30
- RCAC-II, as recycled concrete containing RA1, strength class C25/30.

The tested samples containing recycled aggregates are shown in Figure 1.

Figure 1. Tested samples containing recycled aggregates: (**a**) RMAC I, (**b**) RMAC II, (**c**) RCAC I, and (**d**) RCAC II.

2.2. Methodology

In this research, recycled aggregate was used that has been tested in previous research [25] with the aim of proving the possibility of replacing normally used raw materials in concrete with secondary raw materials. On the basis of the results from previous research, materials were selected and concrete mixtures were designed, which were subsequently exposed to the experiments on the basis of the international standards. All samples were tested according to the valid Czech standards as well.

2.3. *Ecotoxicology*

2.3.1. Chemical and Ecotoxicological Analysis of Leachate

The concrete cubes were leached as described in [26]. The concentrations of Na, Mg, Al, K, Ca, Cr, Mn, Fe, Co, Ni, Cu, Zn, As, Se, Sr, Mo, Cd, Ba, Hg, and Pb were determined in leachates acidified to pH 2.0 using inductively coupled plasma optical emission spectrometry (Integra 6000, GBC, Melbourne, Australia).

Aquatic ecotoxicity tests were performed with non-treated leachates in the concentration range from 510 to 1000 $mL.L^{-1}$, and nutrient-amended leachates diluted 10 times (100 + n $mL.L^{-1}$). The water flea (*Daphnia magna*) acute immobilization test followed the methodology described in [26]. The algal toxicity test using *Desmodesmus subspicatus* and the duckweed (*Lemna minor*) test were conducted according to [25] with minor changes. In algae, growth rate was determined based on optical density measurements at 750 nm using a UV/VIS spectrophotometer UV-1900 (Shimadzu Corporation, Kyoto, Japan).

2.3.2. Determination of Photosynthetic Pigments

In the algal and duckweed test, the total chlorophyll a + b (Chls) and total carotenoid (Cars) content was determined after the exposition and growth rate determination.

First, 10 mL of algal suspension was transferred to 15 mL Falcon tubes and centrifuged (2360× *g*, 10 min, 4 °C). The supernatant was disposed of and 5 mL of 99.5% methanol (Lach-Ner) was added. The samples were homogenized in a vortex homogenizer for 15 s and placed in an ultrasound bath with ice-cooled water for 15 min. The extracts were homogenized again and centrifuged (2360× *g*, 10 min, 4 °C). The absorbance in the supernatants was determined at 470, 653, and 666 nm. Total chlorophylls and carotenoids were calculated according to [27] and expressed as pigment content per unit of algal suspension volume and as the Chls/Cars ratio.

In the duckweed test, the total frond material from a test vessel was transferred to a 15 mL centrifugation tube, covered with 3–8 mL of pure methanol (according to the total frond amount) and placed in the dark and 4 °C for 24–48 h. After extraction, the samples were centrifuged (2360× *g*, 10 min, 4 °C) and the absorbance in the supernatants was determined at 470, 653 and 666 nm. Total chlorophylls and carotenoids were calculated according to [27] and expressed as pigment content per unit of frond area and as the Chls/Cars ratio.

The absorbance was determined using the UV/VIS UV-1900 spectrophotometer (Shimadzu, Kyoto, Japan).

2.3.3. Soil Enzymatic Test

To determine the influence on soil enzymes, leachates were added to Lufa soil 2.4, characterized as clayey loam type (LUFA Speyer, Speyer, Germany). Fifty grams of air-dried soil were properly mixed with 15.3 g of nondiluted and untreated leachate in a sterile glass jar to achieve 70% WHC. Pure distilled water was used as a control sample. The containers were covered with sterile aluminum and placed under stable conditions (20 °C, light cycle 16 h/8 h; 1000 lux). The samples were left without humidity treatment for 56 days. Dry mass content (DM), pH, and soil dehydrogenase activity were determined 7, 28, and 56 days after soil contamination.

For the DM content, approximately 2.5 g was dried at 105 °C for 2 h and weighed. For this measurement, two replicates were prepared. DM was calculated as the fresh mass/dry mass ratio. The soil pH was determined in soil suspensions in 0.01M $CaCl_2$, as described in [28].

Soil dehydrogenase activity (DHA) was determined using triphenyltetrazolium chloride (TTC; Sigma-Aldrich) as a substrate for the reaction. The procedure followed ISO Guideline No. 23753-1 [29] with some adjustments. For each sample, 2.00 ± 0.05 g was transferred to a sterile glass tube and 2 mL of 1% TTC solution in Tris buffer (pH of 7.8) was added. Each sample was prepared in triplicate, plus one blank (2.00 ± 0.05 g of soil, 2 mL of Tris buffer). The samples and blanks were carefully homogenized for 10 s and placed on a dark thermostat (25 °C) for 20 h. After that, each sample was extracted using 10 mL of 99.5% acetone (Lach-ner) and homogenized three times, every 60 min. Finally, the extracts were centrifuged (2360× *g*, 10 min, 4 °C) and the absorbance at 485 nm was determined (UV/VIS spectrophotometer (Shimadzu, Kyoto, Japan). The DHA was expressed as the amount of product formation, i.e., triphenyltetrazolium formazan per soil DM and time. Consequently, the data obtained were compared to the control values and recalculated as % inhibition/stimulation, as described in [26].

2.3.4. Statistical Analysis and Data Evaluation

A one-way ANOVA was performed on all ecotoxicity data sets. Normality was tested using the Shapiro–Wilk test. ANOVA was followed by Tukey's post-hoc test to determine significant differences between samples. The nonparametric Kruskal–Wallis test followed by Dunn's post-hoc test was used when the data did not meet the normal distribution. Ecotoxicity based on EC50 and NOEC values was evaluated according to the scale formulated in a previous study [26]. All statistical analysis was performed using GraphPad Prism, v9.1 (GraphPad Software, San Diego, CA, USA).

2.4. Carbonation Testing Process

Concrete structures need to be durable to ensure that service life is achieved; This plays a significant role in resistance to corrosion. This phenomenon is caused by carbonation; consequently, carbonation behavior is an important attribute to measure.

The simplified carbonation reaction of concrete:

$$Ca(OH)_2 + CO_2 = CaCO_3 + H_2O \tag{1}$$

The Czech standard ČSN EN 12390-12 (73 1302) describes the carbonation resistance of concrete using test conditions that accelerate the rate of carbonation [24]. The method used in this research is inspired by this standard, but the conditions were slightly different.

Czech Standard ČSN EN 12390-12 (73 1302)

This document quantifies the carbonation resistance of concrete. The test conditions used an accelerated the rate of carbonation. The experiment is carried out under controlled exposure of carbon dioxide to an increased level after 28 days of hardening concrete samples. The carbon dioxide concentration should be within ±0.5% by volume of the target value.

For each test, the reference sample of concrete should be used. Samples for one test should be made from one concrete mixture. The concrete cubes are cast and cured for 28 days (in accordance with EN 12390-2 [30]), then placed in a storage chamber with carbon dioxide under normal conditions: 1 013 mbar at 25 °C, temperature 20 ± 2 °C, relative humidity 57 ± 3. In addition, 0.8 g of phenolphthalein powder was dissolved in a solution of 70 mL of ethanol and 30 mL of deionized water. Phenolphthalein was used as an indicator.

After the exposure period, which is 28 days, the carbonation depth is measured at three points on each of the four faces of the cube. To locate these points, the length of the edge is divided into four equal distances. Three samples of each mixture were measured and the mean carbonation depth at time t in mm was calculated as a result.

2.5. Life Cycle Assessment

To analyze the environmental performance of the described mixtures from the perspective of their entire life cycle, the life cycle assessment (LCA) method was applied as an analytical tool [31], which is used primarily to assess the environmental impacts caused by processes throughout the life cycle of a product or service according to the international standards ISO 14 040 and ISO 14 044 [32,33]. According to these standards, the LCA method consists of four steps: definition of goals and scope, inventory analysis, impact assessment, and interpretation.

Taking into account the scope and other conditions for the environmental assessment described in EN 15 804 + A2 for construction products [34], the LCA method was used to evaluate all elementary flows, including the inputs and outputs of materials and energy to the environment in the phases of raw resource production, transport of resources to the facility, production of ready mix concrete, and disassembly of concrete and its disposal in landfill.

2.5.1. System Boundaries and Functional Unit

The environmental impacts of the mixtures were related to the declared unit, which was defined as 1 m^3 of the concrete mixture. The system boundaries of the compared concrete mixtures include raw material supply (cement production, water production, production of primary or recycled aggregate), transport of resources to a facility, mixing of materials, and their transport to site. The phase of use of concrete mixtures was not included according to EN 15 804 + A2. The boundaries of the system also include the end-of-life phase (EoL), which consists of the excavation of concrete in the process of deconstruction, the transportation and demolition of concrete waste in the landfill, and the disposal of waste in the landfill. The investigated system boundaries are described in Figure 2.

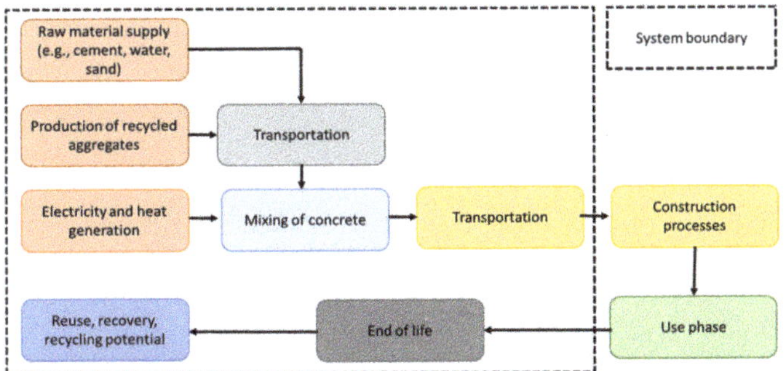

Figure 2. Description of the boundaries of the system.

2.5.2. Life Cycle Inventory

To create the environmental model of the life cycle of the mixtures, GaBi software was used [35]. The mixtures were modelled according to the proportions described in Table 1. To model upstream processes, generic data from the GaBi database were used to describe the environmental impacts of resource production [36]. In addition, the end-of-life processes of concrete were modelled using the mentioned generic data. The energy supply was modelled using the Czech energy mix according to data from the reference year 2016. The transport processes were modelled as transporting on a 50 km distance using a truck trailer (EURO 3, up to 28 t gross weight).

Table 1. Composition of concrete mixes.

Material (kg)	I			II		
	NAC	RMAC	RCAC	NAC	RMAC	RCAC
Cement	260	260	260	300	300	300
Nature Sand	709	-	-	671	-	-
Gravel 4/8	38	-	-	28	-	-
Gravel 8/16	1092	766	949	1139	822	994
Recycled Aggregate 0/4	-	971	843	-	920	800
Water	169	187	186	165	182	181

2.5.3. Influence of Carbonation

As an alternative scenario, the CO_2 uptake potential in concrete was calculated according to EN 16 757 [37]. The expected service life of concrete blocks made of the considered mixtures was assumed to be 50 years. The maximum theoretical uptake of CO_2 was estimated for the cement used as 0.49 kg CO_2/kg of cement. The assumed degree of carbonation was estimated at 0.85 on the basis of the potential future use of concrete as a foundation structure, which will be covered by ground.

2.5.4. Environmental Assessment

To evaluate the impacts of inputs and outputs on the environment, these elementary flows were classified and characterized using the Product Environmental Footprint 3.0 method [38]. This impact assessment method is recommended by the European Commission and uses several environmental indicators [39].

2.5.5. Normalization and Weighting

Taking into account the spectrum of environmental indicators, the results were normalized and weighted to obtain a single score evaluation of the mixtures considered. Normalized values were calculated by dividing the indicators' results by normalized contributions for each indicator according to the normalization data set described in the PEF 3.0 method [39]. Similarly, the weighted values were calculated by multiplying the normalized results using weighting factors. Weighing is used to express the relative importance of each indicator. The data set of the weighing indicators is based on expert opinion and is described in the PEF 3.0 method [39].

3. Results

3.1. *Physicochemical Properties of Concrete Leachates*

Table 2 shows the results of the chemical analysis of the leachates. Mn, Co, Ni, Cu, As, Se, Mo, Cd, Ba, Hg, and Pb were below the detection limit; Cr was found only in NAC I. The main elements found in the leachates were Ca, K, and Na, while the concentration of Mg, Al, Fe, Zn, and Sr was below 0.5 mg.L^{-1}. The chemical composition of the leachates was generally relatively similar. Only Zn content showed different patterns, with the highest content in RMAC I and the lowest content in NAC I. All leachates had similar pH (10.5–10.7), as well as electrical conductivity (162–232 μS.cm^{-2}). The initial pH value decreased to 7.5–8.4 after both dilution and seven-day exposure under the light cycle and 24 ± 1 °C in the duckweed assay (Supplementary Materials, Table S1).

Table 2. Physicochemical properties of leachates.

Element (mg.L^{-1})	I			II		
	NAC	RMAC	RCAC	NAC	RMAC	RCAC
Na	3.04 ± 0.05	3.96 ± 0.09	<2.5	<2.5	4.16 ± 0.17	<2.5
Mg	0.20 ± 0.01	0.29 ± 0.01	0.24 ± 0.01	0.18 ± 0.01	0.20 ± 0.01	0.20 ± 0.01
Al	<0.5	<0.5	<0.5	<0.5	<0.5	<0.5
K	17.27 ± 0.32	15.66 ± 0.63	14.20 ± 0.46	14.03 ± 0.49	19.19 ± 0.32	12.82 ± 0.11
Ca	28.85 ± 0.23	29.44 ± 0.95	24.68 ± 0.49	21.19 ± 0.59	19.75 ± 0.33	22.13 ± 0.49
Cr [1]	<0.5	0	0	0	0	0
Fe	~0.04	~0.08	~0.03	~0.02	~0.04	~0.02
Zn [2]	~0.008	0.182 ± 0.007	0.016 ± 0.001	0.033 ± 0.001	0.055 ± 0.002	0.015 ± 0.001
Sr	<0.03	~0.08	~0.03	~0.03	~0.03	~0.03
pH	10.7 ± 0.1	10.6 ± 0.1	10.7 ± 0.1	10.6 ± 0	10.6 ± 0.1	10.5 ± 0.1
el. conductivity (μS.cm^{-2})	225 ± 12	232 ± 24	191 ± 11	183 ± 6	211 ± 38	162 ± 17

[1] Limit value in waste leachates 7 mg.L^{-1}. [2] Limit value in waste leachates 20 mg.L^{-1}.

3.2. Aquatic Ecotoxicity

Basic ecotoxicity tests performed with water flea, algae, and duckweed showed similar dose-response patterns in all leachates (Tables S2–S4). The duckweed growth rate was the most sensitive endpoint, while the algal growth was the least sensitive.

For most samples, NOEC was found to be 800 mL.L^{-1} in the acute test for algae and water fleas, and 640 mL.L^{-1} in the growth rate of duckweed. Therefore, according to ecotoxicity indexes, all leachates were classified as non-toxic (Table 3).

Table 3. Ecotoxicity assessment of concrete leachates: EC50 with 95% CI (confidence interval) and coefficient of determination (R^2), NOEC values. GR—growth rate; TC—toxicity class [26]; n.c.—not calculable. EC50 and NOEC values are expressed in mL.L^{-1}.

Concrete Mix	Value	Water Flea	Algae GR	Duckweed GR	Toxicity Level
NAC I	EC50	931	>1000	870	
	CI 95%	890–n.c.	-	833–912	
	R^2	0.89	0.80	0.94	
	NOEC	800	800	640	
	TC	NT-1	NT-1	NT-2	Non-toxic
RMAC I	EC50	929	>1000	896	
	CI 95%	894–n.c.	-	838–966	
	R^2	0.96	-	0.86	
	NOEC	800	800	640	
	TC	NT-1	NT-1	NT-2	Non-toxic
RCAC I	EC50	>1000	>1000	911	
	CI 95%	n.c.	-	864–971	
	R^2	0.69	0.77	0.92	
	NOEC	800	800	640	
	TC	NT-1	NT-1	NT-2	Non-toxic
NAC II	EC50	>1000	>1000	844	
	CI 95%	-	-	829–861	
	R^2	0.11	0.82	0.99	
	NOEC	640	800	510	
	TC	NT-2	NT-1	NT-2	Non-toxic
RMAC II	EC50	992	>1000	926	
	CI 95%	976–n.c.	n.c.	909–943	
	R^2	0.94	0.76	0.99	
	NOEC	800	800	640	
	TC	NT-1	NT-1	NT-2	Non-toxic
RCAC II	EC50	>1000	>1000	928	
	CI 95%	-	-	895–966	
	R^2	0.65	0.77	0.95	
	NOEC	800	800	640	
	TC	NT-1	NT-1	NT-2	Non-toxic

3.3. Photosynthetic Pigments

The evaluation of photosynthetic pigments in algae and duckweed was in accordance with observations at the morphological level. As shown in Tables 4 and 5, the pigment ratio (total chlorophyll/total carotenoids) was significantly reduced only in non-diluted leachates in algae, and in 800 and 1000 mL.L^{-1} in duckweed with two exceptions (in NAC-I and RMAC-I diluted to 800 mL.L^{-1}, the change in pigment ratio was not significant). The change in pigment ratio was caused by a decrease in both chlorophylls and carotenoids in algal suspension, where the negative effect of concentrated leachates was more pronounced in chlorophylls than in carotenoids (Figures 3 and 4). The change in the pigment ratio in duckweed was caused by a decrease in total chlorophyll and an increase in total carotenoids at the same time (Figures 5 and 6). The highest carotenoid content per frond was found in duckweed exposed to nondiluted NAC I leachate, which led to the lowest Chls/Cars ratio (Table 5).

Table 4. Total chlorophyll to total carotenoid ratio in algae (mean values ± SD). 100 + n—leachates (100 mL.L^{-1}) amended with nutrients. The letters indicate significant differences between the values (post-hoc test; α = 0.05) within the same column (uppercase) and within the same row (lowercase).

mL.L^{-1}	I						II					
	NAC		RMAC		RCAC		NAC		RMAC		RCAC	
0	A	5.6 ± 0.4	A	5.6 ± 0.4	A	5.6 ± 0.4	A	5.6 ± 0.4	A	5.6 ± 0.4	A	5.6 ± 0.4
640	A	5.5 ± 0.1 a	A	6.0 ± 0.2 a	A	5.4 ± 0.2 a	A	6.2 ± 0.1 a	A	5.4 ± 0.7 a	A	6.3 ± 0.1 a
800	A	5.1 ± 0.2 a	A	5.8 ± 0.8 a	A	5.1 ± 0.2 a	A	5.8 ± 0.3 a	A	5.6 ± 0.1 a	A	5.9 ± 0.1 a
1000	B	1.9 ± 0.1 a	B	1.7 ± 0.2 a	B	1.8 ± 0.1 a	B	2.0 ± 0.1 a	B	1.9 ± 0.2 a	B	1.8 ± 0.2 a
100 + n	A	5.4 ± 0.2 a	A	5.7 ± 0.2 a	A	5.2 ± 0.0 a	A	5.9 ± 0.1 a	A	5.5 ± 0.1 a	A	5.9 ± 0.1 a

Table 5. Total chlorophyll to total carotenoid ratio in duckweed (mean values ± SD). 100 + n—leachates (100 mL.L^{-1}) amended with nutrients. The letters indicate significant differences between the values (post-hoc test; α = 0.05) within the same column (uppercase) and within the same row (lowercase).

mL.L^{-1}	I						II					
	NAC		RMAC		RCAC		NAC		RMAC		RCAC	
0	A	7.9 ± 0.2	A	7.9 ± 0.2	A	7.9 ± 0.2	A	7.9 ± 0.2	A	7.9 ± 0.2	A	7.9 ± 0.2
510	A	7.0 ± 0.3 a	A	7.0 ± 0.5 a	A	7.6 ± 0.6 a	A	7.2 ± 0.2 a	A	7.0 ± 0.3 a	A	8.1 ± 0.4 a
640	A	7.9 ± 1.0 a	A	7.4 ± 0.3 a	A	7.7 ± 0.5 a	A	7.1 ± 0.3 a	A	7.1 ± 0.4 a	A	7.0 ± 0.3 a
800	A	6.9 ± 1.0 a	A	7.0 ± 0.9 a	B	5.1 ± 0.1 b	B	4.6 ± 0.2 b	B	4.4 ± 0.1 b	B	4.4 ± 0.1 b
1000	B	1.5 ± 0.1 b	B	3.8 ± 0.5 a	C	3.9 ± 0.4 a	B	3.2 ± 0.7 a	B	3.8 ± 0.1 a	C	2.8 ± 0.2 ab
100 + n	A	8.2 ± 0.1 a	A	7.7 ± 0.2 a	A	8.6 ± 0.3 a	A	8.1 ± 0.3 a	A	8.3 ± 0.3 a	A	8.1 ± 0.1 a

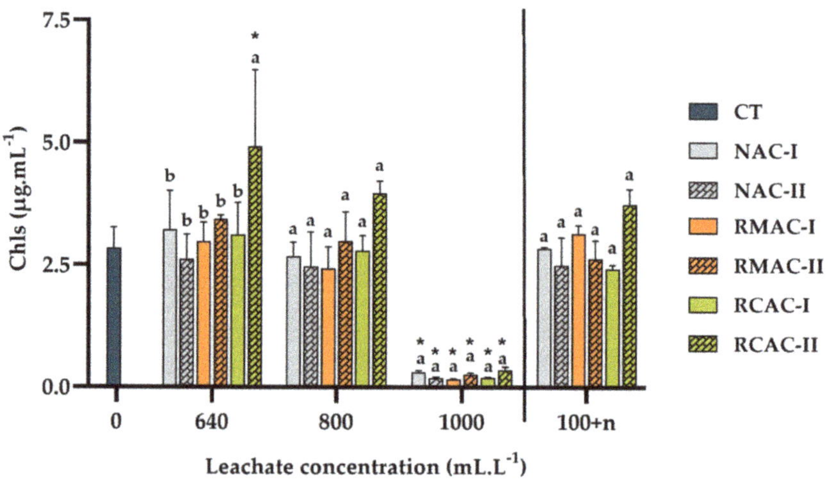

Figure 3. Mean (±SD) total chlorophyll (a + b) content in algal suspension. CT (control)—Bold's Basal medium. 100 + n—leachates (100 mL.L^{-1}) with amended nutrients. Lowercase letters indicate significant differences between samples of a given concentration, and asterisks (*) indicate significant differences between sample and control (post-hoc test; α = 0.05).

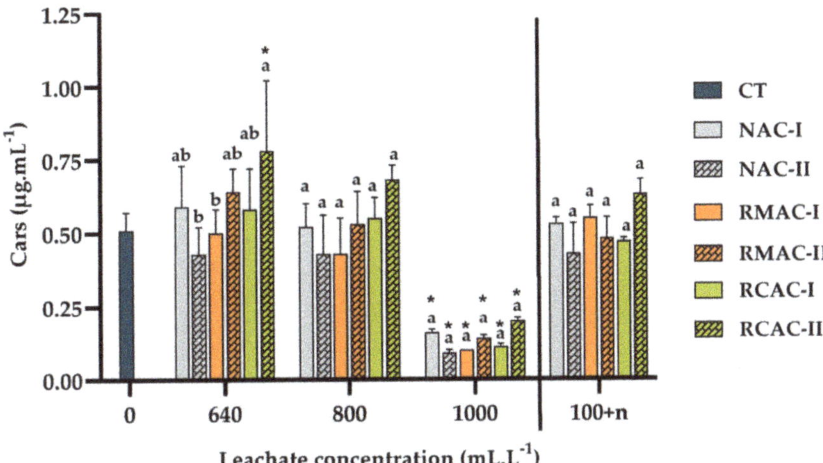

Figure 4. Mean (±SD) total carotenoid content in algal suspension. CT (control)—Bold's Basal medium. 100 + n—leachates (100 mL.L^{-1}) with amended nutrients. Lowercase letters indicate significant differences between samples of a given concentration, and asterisks (*) indicate significant differences between sample and control (post-hoc test; α = 0.05).

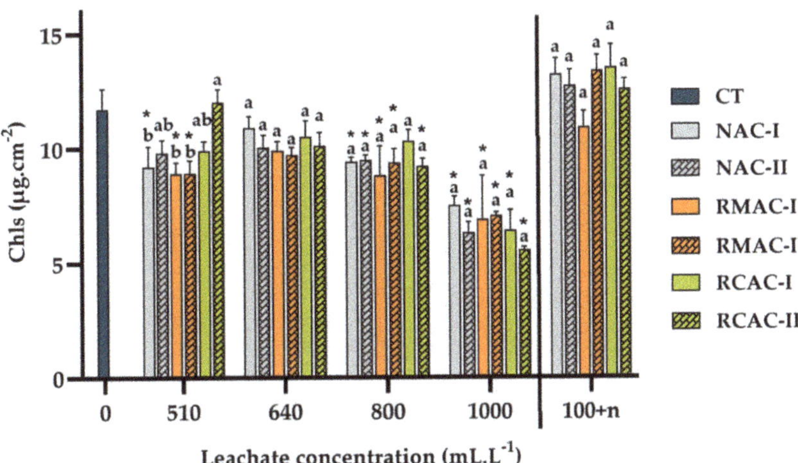

Figure 5. Mean (±SD) total chlorophyll (a + b) content in duckweed. CT (control)—Steinberg medium. 100 + n—leachates (100 mL.L^{-1}) with amended nutrients. Lowercase letters indicate significant differences between samples of a given concentration, and asterisks (*) indicate significant differences between sample and control (post-hoc test; α = 0.05).

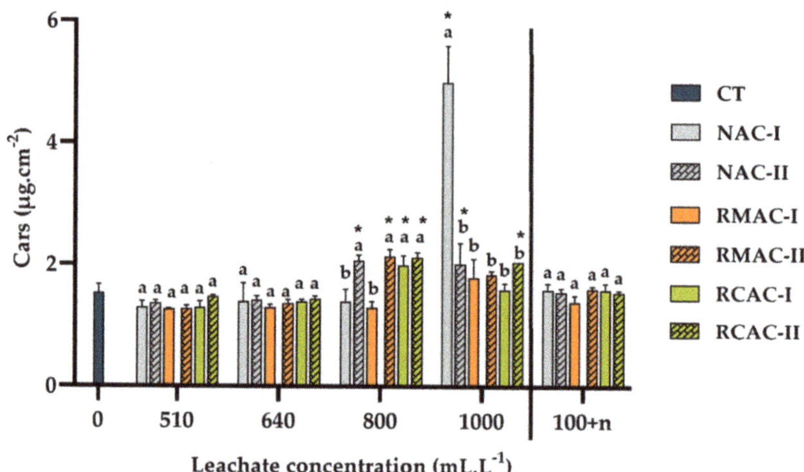

Figure 6. Mean (±SD) total carotenoid content in duckweed. CT (control)—Steinberg medium. 100 + n—leachates (100 mL.L^{-1}) with amended nutrients. Lowercase letters indicate significant differences between samples of a given concentration, and asterisks (*) indicate significant differences between sample and control (post-hoc test; α = 0.05).

3.4. Soil Dehydrogenase Activity

The results of DHA in the soil are summarized in Figure 7. With a few exceptions, the enzymatic activity was slightly stimulated in soils amended with leachates. Stimulation was more pronounced in soils amended with concrete leachates of strength class I. However, the differences among samples, as well as the stimulation, were usually not significant. RMAC II was the only leachate that caused slight inhibition in all measurements, while soils contaminated with NAC I leachate changed their reaction from significant stimulation (−11% and −10% after 7 and 28 days, respectively) to low inhibition (5%) at the end of the exposure. The highest stimulation was observed in soil contaminated with RMAC I leachate after seven days (15%). Generally, it can be said that undiluted leachates did not significantly affect soil microbial activity, or caused a slight increase of up to 15%. The pH of the soil mixtures was relatively similar to that of the control soils (Table S5). The soil pH ranged between 5.7 and 6.0 after seven days and dropped to 5.3–5.6 after 56 days of exposure; therefore, according to the soil pH [29], all samples and the control remained acidic during the whole experiment.

3.5. Carbonation Effect

There are four basic stages of carbonation; most structures reach the maximum of the second stage. The amount of calcium carbonate formed does not completely characterize the carbonation stage [40]. By finding out in what form $CaCO_3$ is present, it is possible to characterize the carbonation process and, at the same time, assess the situation of carbonated concrete. Studies that consider concrete carbonation in general show that concretes of the lower strength class (C16/20) reach deeper carbonation depths compared to the higher strength class (C25/30) [40–42]. This fact is also connected with factors such as porosity and density beside concrete strength [43–45]. Research dealing with carbonation effect has proved that with increasing porosity and density, the carbonation effect is decreased. This phenomenon is also confirmed in this research (Figure 8). The purple-red color adheres to the noncarbon part of the sample, where the concrete is highly alkaline. There was no coloration in places with reduced concrete alkalinity. Mixtures NAC-I, RMAC-I, and RCAC-II have shown deeper penetration compared to the corresponding higher-class concrete (NAC-II, RMAC-II, and RCAC-II). Carbonation depth was determined by image analysis using NIS Elements (v5.20, Laboratory Imaging, Prague, Czech Republic).

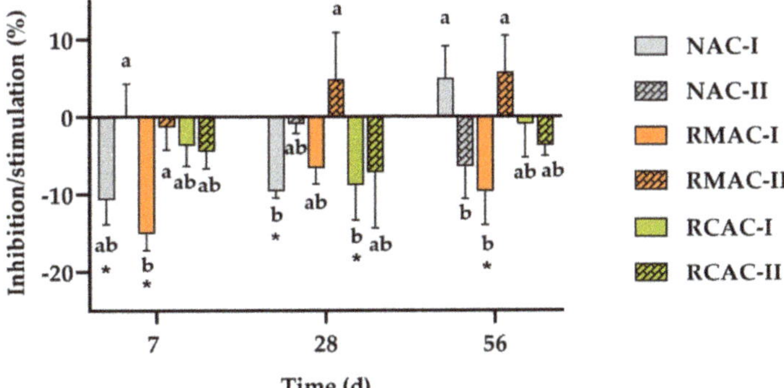

Figure 7. Mean (±SD) inhibition/stimulation of soil dehydrogenase activity measured in soil contaminated with leachates after 7, 28 and 56 days. Different letters indicate significant differences among samples within a given time point. Asterisks (*) indicate significant differences between the sample and control, that is, zero values (post-hoc test; α = 0.05).

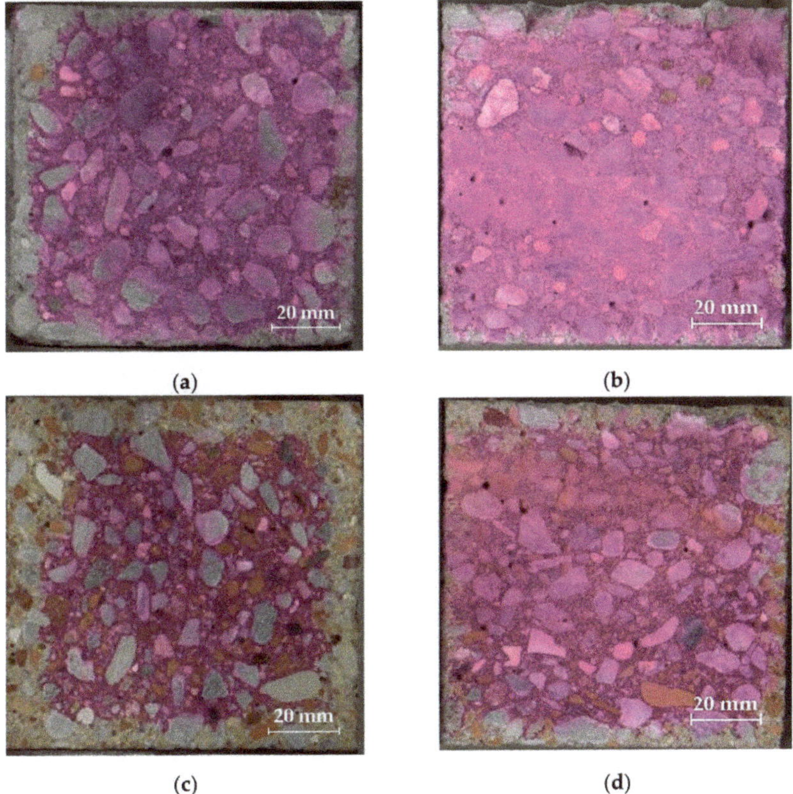

(a) **(b)** **(c)** **(d)**

Figure 8. *Cont.*

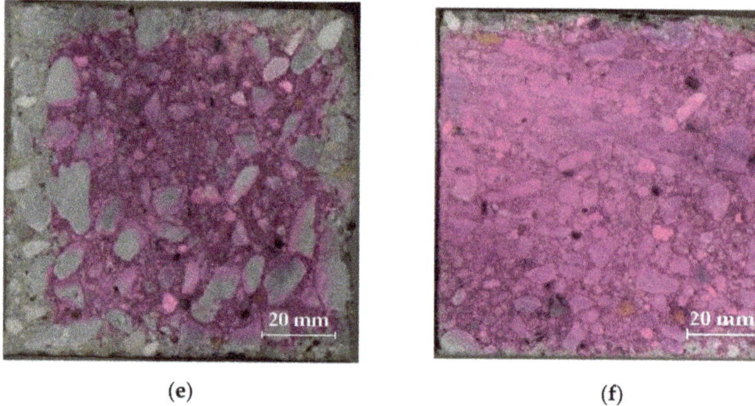

Figure 8. Samples after carbonation test colored with phenolphthalein: (**a**) NAC-I, (**b**) NAC-II, (**c**) RMAC-I, (**d**) RMAC-II, (**e**) RCAC-I, and (**f**) RCAC-II.

The results of the carbonation depth are summarized in Table 6. The NAC-I value 4.41 mm was more than one and a half times higher compared to the same mixture in the higher concrete class NAC-II 2.65 mm. This trend appears similarly in the other mixtures as well, but the ratio increases from 1.6 to 2.9 with RMAC, and to 3.6 with RCAC. In general, the deepest penetration was observed in RMAC in both evaluated grades (10.04 mm and 3.37 mm). However, the RCAC-I was extremely high compared to that of NAC-I. Meanwhile, RCAC-II (with the value 2.45) was almost comparable with NAC-II (2.65).

Table 6. Average carbonation depth results of samples tested containing natural and recycled aggregate.

Mean Carbonation Depth (mm)	I			II		
	NAC	RMAC	RCAC	NAC	RMAC	RCAC
d_1	2.99	12.69	9.66	2.50	1.18	2.56
d_2	6.82	8.41	7.99	5.25	3.74	1.87
d_3	3.66	7.35	6.27	0.34	3.37	4.10
d_4	4.17	11.73	12.14	2.50	6.25	1.30
d_k	4.40 ± 1.45	10.04 ± 2.22	9.01 ± 2.16	2.65 ± 1.74	3.37 ± 1.79	2.45 ± 1.05

3.6. Results of Environmental Assessment

The environmental assessment was performed using the LCA method and the potential environmental impacts were calculated using PEF 3.0. The results of this assessment are given in Table 7.

Taking into account the climate change (total) indicator, which describes the potential impact on one of the key categories, mixtures with natural aggregates cause a higher impact than mixtures with recycled aggregates in the same strength class. Similarly, NAC has a greater impact in most categories. This is affected by the dominant influence of cement. Mixtures in the same strength class are designed with the same amount of cement, so their potential impact is mainly affected by this. However, there is also the influence of the beneficial impact of recycled aggregates, which are used as replacements for natural gravel in the mixture.

In comparison of the two types of recycled aggregates, recycled concrete aggregate has a more beneficial impact than recycled masonry aggregate. This is mainly affected by the higher amount of iron scrap, which can be recycled from concrete structures with steel reinforcement.

Table 7. Results of the selected impact indicators for 1 m^3 of concrete mixtures; the environmental impact assessment was carried out according to the PEF 3.0 method.

	I			II		
	NAC	**RMAC**	**RCAC**	**NAC**	**RMAC**	**RCA**
Acidification (Mole of H+ eq.)	9.96×10^{-1}	8.99×10^{-1}	8.64×10^{-1}	1.06	9.66×10^{-1}	9.34×10^{-1}
Climate Change—total (kg CO_2 eq.)	3.21×10^{2}	2.59×10^{2}	2.21×10^{2}	3.54×10^{2}	2.95×10^{2}	2.59×10^{2}
Climate Change, biogenic (kg CO_2 eq.)	3.62×10^{-1}	3.18×10^{-1}	3.69×10^{-1}	3.88×10^{-1}	3.47×10^{-1}	3.95×10^{-1}
Climate Change, fossil (kg CO_2 eq.)	3.20×10^{2}	2.58×10^{2}	2.20×10^{2}	3.53×10^{2}	2.94×10^{2}	2.58×10^{2}
Climate Change, LULUC (kg CO_2 eq.)	6.12×10^{-1}	6.35×10^{-1}	6.95×10^{-1}	6.29×10^{-1}	6.51×10^{-1}	7.09×10^{-1}
Ecotoxicity, freshwater—total (CTUe)	1.71×10^{3}	1.39×10^{3}	1.57×10^{3}	1.77×10^{3}	1.47×10^{3}	1.65×10^{3}
Eutrophication, freshwater (kg P eq.)	1.07×10^{-3}	7.31×10^{-4}	8.75×10^{-4}	1.12×10^{-3}	7.91×10^{-4}	9.28×10^{-4}
Eutrophication, marine (kg N eq.)	3.42×10^{-1}	3.25×10^{-1}	3.30×10^{-1}	3.59×10^{-1}	3.43×10^{-1}	3.48×10^{-1}
Eutrophication, terrestrial (Mole of N eq.)	3.76	3.59	3.66	3.95	3.78	3.85
Human toxicity, cancer—total (CTUh)	9.03×10^{-8}	4.41×10^{-8}	2.19×10^{-8}	9.24×10^{-8}	4.87×10^{-8}	2.77×10^{-8}
Human toxicity, non-cancer—total (CTUh)	6.75×10^{-6}	5.62×10^{-6}	5.31×10^{-6}	7.12×10^{-6}	6.05×10^{-6}	5.76×10^{-6}
Ionising rad., human health (kBq U235 eq.)	6.49	4.83	6.29	6.99	5.41	6.80
Land use (Pt)	5.57×10^{2}	4.66×10^{2}	5.30×10^{2}	5.88×10^{2}	5.01×10^{2}	5.63×10^{2}
Ozone depletion (kg CFC-11 eq.)	3.73×10^{-7}	2.53×10^{-7}	3.13×10^{-7}	3.85×10^{-7}	2.71×10^{-7}	3.28×10^{-7}
Particulate matter (Disease incidences)	1.12×10^{-5}	6.37×10^{-6}	5.74×10^{-6}	1.20×10^{-5}	7.43×10^{-6}	6.83×10^{-6}
Photochem. ozone form., hum. health (kg NMVOC eq.)	8.52×10^{-1}	7.98×10^{-1}	7.80×10^{-1}	9.02×10^{-1}	8.51×10^{-1}	8.35×10^{-1}
Resource use, fossils (MJ)	2.08×10^{3}	1.31×10^{3}	9.71×10^{2}	2.17×10^{3}	1.44×10^{3}	1.12×10^{3}
Resource use, mineral and metals (kg Sb eq.)	3.05×10^{-5}	6.43×10^{-5}	1.81×10^{-4}	3.30×10^{-5}	5.68×10^{-5}	1.67×10^{-4}
Water use (m^3 world equiv.)	1.27×10^{3}	7.67×10^{2}	8.40×10^{2}	1.31×10^{3}	8.35×10^{2}	9.03×10^{2}

LULUC–Land use and land use change

4. Discussion

4.1. Impact of Chemical Composition on Leachate Ecotoxicity

Except for reference samples, the concentration of leached elements from concrete cubes was significantly lower compared to leaching patterns of homogenized recycled aggregates, as expected (Table 2, [25]). However, the general proportion of leached elements was similar for primary materials and construction applications. Heavy metals which are non-essential for organisms, i.e., hazardous at any concentration (As, Ba, Cd, Hg, Ni, and Pb), were below the detection limit. Ca, Na, and K that belong to the main metals released in concrete leachates [4] are not considered toxic; in fact, quite the opposite, as they are essential mineral macroelements that are included in the culture media for both crustacean and aquatic plants [46–48]. Mg, Fe and Zn represent other mineral nutrients required especially by plants. However, Zn is included in risk metals and therefore has to be analyzed in wastewaters, sludge or waste leachates [49,50]. Moreover, secondary salinization of surface waters and soils caused by increasing concentration of ions including Na^+, Mg^{2+}, Ca^{2+}, K^+ and Fe ions together with climate change is an issue of growing concern [51–53]. In this study, the essential minerals were often below the concentration required in growth media.

The results from ecotoxicity tests indicate that the high growth inhibition/immobilization in original untreated leachates was caused most particularly by lack of nutrients. This can be considered as a favorable result because abundant elements in eluates entering aquatic or terrestrial environment can cause ecological imbalance [54,55].

4.2. Selection of Leaching and Ecotoxicity Testing Design

Various leaching test methods have been reported from batch tests in one stage, percolation tests, and long-term tests with leachant renewal [56]. For the leaching experiment, we have chosen the simple batch design in one stage that was already applied in the previous study [26] to compare the ecotoxic potential of recycled glass waste in the form of homogenized material and its subsequent use in concrete cubes. This 24-h leaching design was also chosen to prevent potential metal sorption on glass vessels, change of the leachate pH in time, biocontamination, as well as potential biodegradation of the leached compounds.

Ecotoxicity tests are usually based on a simple experimental design with acute exposure that provide quick screening of potential environmental risks. However, acute exposure which usually lasts several hours to several days is suitable mainly for detection of larger amounts of hazardous substances affecting living organisms. To detect the potential risk of lower concentrations of toxicants, chronic ecotoxicity tests may be used. Such methods are time-, space- and sample-consuming, and thus can be problematic for routine application. The use of semi-chronic tests provides a suitable solution.

Ecotoxicological impact of concrete leachates is usually tested by a set of two or three aquatic bioassays. In consumers, the most popular test is immobilization of freshwater or marine crustaceans [1–5,57,58]. The embryonic stage of zebrafish eggs (*Danio rerio*) represents another possibility of how to avoid problematic animal models, as the early developmental stage is not protected by regulatory framework [59]. In the inter-laboratory study, tests with zebrafish eggs was applied, but was evaluated as the least sensitive model [57]. Marine luminescent bacteria *Aliivibrio fischeri* (previously *Vibrio fischeri*) is often used in concrete leachate testing [1,57,60] as the test design is simple, short-term (30 min exposure), and easy to perform using modern luminometers [61]. Heisterkamp et al. [57] reported the bacterial luminescent test as the most sensitive for construction product evaluation. Plant models can be examined at both the individual (lethality, necrosis) and population (reproduction) levels, making them semi-chronic tests. At the same time, additional endpoints at the biochemical level [9,60] can be determined. As duckweed and unicellular algae reproduce asexually, they represent genetically homogeneous plant material and have another advantage over seed germination tests [9].

4.3. *Photosynthetic Pigment Ratio as Stress Indicators in Aquatic Plants*

Aquatic plants growing in metal-contaminated waters are able to accumulate heavy metals [62]. Besides the negative effect on plant growth, metal contamination also causes oxidative stress, as reported for duckweed exposed to Cd, Cu, Cr, and Hg [63,64]. Oxidative stress in aquatic plants can be detected by increased activity of antioxidative enzymes, malondialdehyde, or changes in total carotenoids content [64,65]. However, deficiency of essential metals such as Cu also has a negative impact on photosynthetic pigments [66]. Duckweed exposed to heavy metals in industrial wastewater was more seriously affected at the morphological level (growth rate based on the frond number and weight) than in the chlorophyll content [9]. This is in agreement with our results (Table S4, Figure 5).

Another task is to determine how the pigment content is expressed. Calculation per weight unit or frond area may be subject to error in the event that the water content in the fronds differs or the fronds overlap. The effect of heavy metal pollution in wastewaters lead to changes in chlorophyll a and b, and the total carotenoids exceeded the total chlorophyll content in duckweed, which indicated internal oxidative stress [65]. Hence, Chls/Cars ratio can be easily used for comparison among various samples and control. In this study, a significant decrease of Chls/Cars was generally in accordance with significant growth inhibition in duckweed (Table 5, Table S4). Besides, by determination of the pigment ratio, both the actual state of the plant and the prediction of the future plant response can be considered.

Traditional algal assays are often based on indirect estimation of biomass or population growth through cell counting under a microscope, flow cytometry, or optical density measurement [67,68]. These approaches do not take into account the cell size and the cell quality, including colour, i.e., pigment profile.

Direct biomass determination on the cell dry mass basis is usually impossible due to the very low dry matter content. At the same time, the extraction of photosynthetic pigments enables the quantification of algal production at the biochemical level (Chls/mL), and the level of stress pronounced by changes in Cars. Another guideline for measuring aquatic ecotoxicity describes the determination of chlorophyll a in algae using ethanol extraction [69]. However, as summarized in [70], hydrophilic carotenoids are not easily extracted by ethanol. Osorio et al. reported acid-free methanol as a suitable solvent for quantitative extraction for carotenoids in various macro- and micro-algae [71]. For this reason, a similar approach for pigment extraction and measurement as applied in the duckweed assay was chosen in the algal experiment also.

4.4. *Effect of Leachates on Soil Dehydrogenase Activity*

Soil represents an important part of the environment. The balanced functioning of soil is strongly dependent on the soil microbial community. Soils are considered one of the sinks for various kinds of pollutants, including those coming from the construction sector [72]. The release of alkalizing compounds from cement and concrete contributes to the increase of soil pH [73]. Soil pH was reported to be a significant factor influencing the composition of the soil microbiome [74]. Our hypothesis was that the addition of leachates into natural soil would lead to a change in microbial activity in response to metal input. This was observed in most samples, especially seven days after soil contamination (Figure 6). The slight stimulation effect is not surprising, since the total amount of metals leached from concrete was relatively low. Leachate alkalinity also did not affect soil pH significantly, although the pH value decreased slightly over time (Table S5). As the stimulation/inhibition effect of concentrated concrete leachates on DHA was very low (though significant in several cases), addition of diluted leachates was not tested. To our knowledge, there is no study on the addition of concrete leachate to soil. Soil enzymes were not inhibited in soils located near landfills or soils amended with landfill leachates [75,76].

The DHA experiment was performed using only one selected type of an acidic soil material. However, soils located in urban sites vary in physicochemical characteristics [77] and thus may give different results. Furthermore, impact on other components of the soil ecosystem, plants and invertebrates may be also included. The performed type of experiment was the first of its kind due to the untraceable studies in this field. Thus, more research is necessary on terrestrial ecotoxicology of construction products.

4.5. *Impact of the Carbonation Process on Concrete*

The real trigger mechanism is water and oxygen, which means the process of carbonation itself (high CO_2 content) does not cause corrosion. Carbonation is one of the chemical mechanisms that can cause concrete failure, and one of the main factors effecting the process is relative humidity of the environment. In a wet environment (humidity higher than 95%), the carbonation process is inefficient or not going at all [45,78]. However, structures in a very dry environment (relative humidity up to 30%), as well as structures fully immersed in water, show no signs of carbonation or corrosion. This is caused by the absence of oxygen to fill the capillary pores [23]. The definition of the effect of relative humidity on the carbonation process in concrete is an important topic in the scientific field; the research in this area is examined by Matoušek et al. [40]. According to [40], the carbonation process is more intense between 50 and 95% of relative humidity, and between 75 and 95% strongly unsolicited [42]. However, the reduction of concrete alkalinity could be (beside carbon dioxide) caused by nitrogen oxides or sulfur dioxide, which are also pollutants affecting concrete. This scenario could appear with outdoor exposure.

Some studies have also shown refinement of pore structure, but this factor was dependent on the relative humidity. However, the research [78] validates that carbonation of concrete before its utilization could lead to a decrease in water absorption as well. These conclusions are also connected with better durability, e.g., freeze-thaw resistance, which is an important factor for concrete structures in general.

Another factor that affects the carbonation depth could be a higher cement ratio. Studies have shown that carbonation on these samples was negligible [21,79]. This study confirms the prediction that concrete in the lower strength grade has deeper penetration and the extent of carbonation is more significant. However, phenolphthalein as an indicator reveals that the pH level is in fact below 9 (not the real carbonation depth) [79,80].

When dealing with cement, there is also the possibility of using alkali-activated materials. There are studies [81,82] dealing with a high MgO ratio in in alkali-activated slag.

With hydrotalcite as the main secondary product, this can effect and reduce the carbonation process, and this whole case can lead to an increase of the durability of concrete [82].

If focusing purely on carbonation without corrosion, e.g., reinforcement, the process can be considered environmentally beneficial. Carbon dioxide absorption by concrete structures can reduce these emissions. With regard to this theory, it can be said that the recycled concrete that has been investigated in this work will hold more CO_2 than conventional reference concrete in the same strength grade. The usual CO_2 content in the air is 0.03% by volume, depending on the area. In cities, this number could be up to three times higher [42].

In general, based on the results of this research, the investigated recycled concretes can be evaluated as suitable for use in concrete structures that will not have a negative environmental impact higher than similar reference concretes of the same strength class.

4.6. Environmental Assessment of the Alternative Scenario Considering CO_2 Uptake

The alternative scenario describes the potential of concrete mixtures to capture CO_2 as a consequence of carbonation. The approach for this calculation is described in Section 2.5. In this chapter, the assumed factors for the calculation were described to characterize the potential of the mixtures to take up CO_2. The results of the calculation of the total potential uptake are described in Table 8.

Table 8. The potential total CO_2 uptake calculated for concrete cubes (a = 1 m) that have 5 m^2 of the surface below the ground, according to EN 16757.

	NAC I	RMAC I	RCAC I	NAC II	RMAC II	RCAC II
Total CO_2 potential uptake (kg CO_2 per cube)	4.21	4.21	4.21	3.53	3.53	3.53

The calculated uptake contribution can be used as a benefit of the concrete structure, and it can be declared together with the results of the environmental assessment of the entire life cycle. However, assumptions describing expected service life or future utilization or the surface of the cube available for carbonation are highly uncertain. Therefore, the results of this calculation are stated as an alternative scenario which describes the possible use of such concrete. Furthermore, the potential total CO_2 uptake is not considered in comparison with the total impact in the category of climate change, which is mainly influenced by cement production.

Carbonation of concrete also continues after its service life and CO_2 can be absorbed in recycled concrete aggregate. After gridding of recycled concrete to particle size 0–40 mm, the rate of CO_2 can reach even 5.5% of overall CO_2 emissions realized during the life cycle of concrete [83]. The amount of absorbed CO_2 after four months, in which concrete is crushed into the typical size of concrete aggregate, can reach even 20% of the total amount of CO_2 realized during calcination of used cement [84]. A similar result was reported by Yang et al., who calculated the CO_2 uptake during life expectancy of 40 years and recycling span of 60 years as 18–21% of the CO_2 emissions from the production of ordinary Portland cement [85].

4.7. Overall Potential Impact on the Environment

Based on the normalized and weighted results, the overall potential impact can be calculated, and the sums of normalized and weighted results are presented in Figure 9. The highest environmental impact is related to the considered life cycle of NAC II. Mixtures with the same strength class, which were designed with the use of recycled aggregates, cause a smaller potential impact. The same relation is seen among the mixtures designed for the lower strength class.

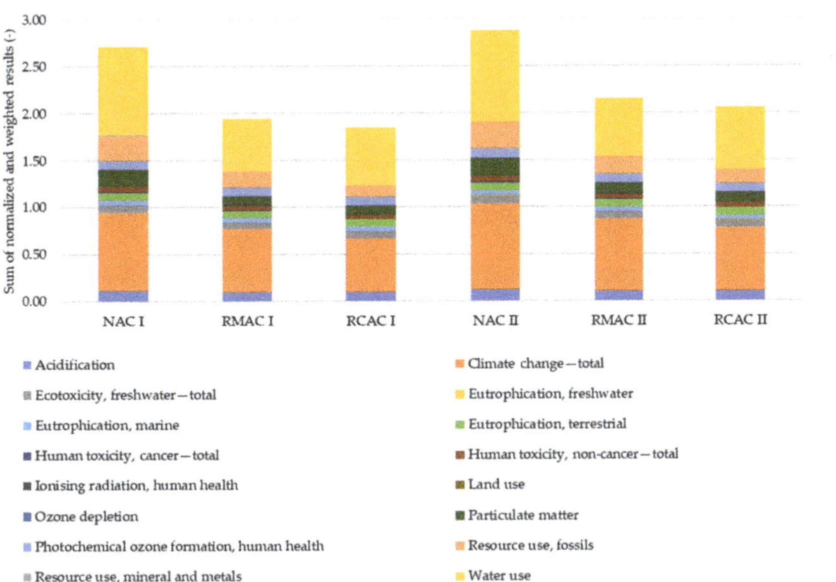

Figure 9. Sum of normalized and weighted results calculated using the PEF 3.0 method.

Regarding the comparison of mixtures containing recycled concrete aggregate and recycled masonry aggregate, the lowest overall impact is reached in the case of RCAC mixtures. The similar conclusion was reported by Marinkovic et al., who, in two scenarios, in which recycled aggregate and natural aggregate concrete were compared, calculated that lower normalized and weighted results of environmental indicators was reached by recycled aggregate concrete [86]. In addition, a study published by Colangelo et al. shows that concrete with 25% recycled aggregates is the best solution from an environmental point of view [87].

The overall impact is significantly affected by the contribution in the water use category. The impact in this category is caused mainly by gravel production, and the production of recycled aggregates has a beneficial impact in this category. This beneficial impact represents the environmental credits, which are connected to the recycling of iron scrap from construction and demolition waste.

Another important contribution to the overall impact is related to the results in the climate change category. The major impact in this category is caused by the production of cement.

5. Conclusions

In this study, the experimental verification of the reaction between concrete and the environment, from the biochemical level up to the mechanical and theoretical levels, was performed. Laboratory leaching experiments that determine the toxic effect of the concrete structure on the environment (water and soil) were combined with evaluation of the environment (air or water) on the concrete structure, through the carbonation process. All of the obtained experimental data were then theoretically compared with results of the life-cycle assessment.

As a conclusion of the observation at both the ecotoxicological and biochemical levels, it is possible to say that all assumptions were confirmed. With a smaller surface, the leachability of both toxic compounds and trace elements also decreases. The effect of concrete leachates on photosynthetic pigment ratio (Chls/Cars) was in accordance with the effect on plant growth. Addition of leachates to natural soil had a very low effect on soil

DHA and did not change soil pH. Hence, from an ecotoxicological point of view, concrete containing fine recycled aggregate does not disturb the balance in the ecosystem and is as nontoxic as reference samples.

At the same time, some types of recycled concrete (mainly RCAC-II) have been proven to reach carbonation depths similar to those of the reference sample, while RMAC-I and RMAC-II showed a deeper penetration of CO_2. In general, it is possible to say that, based on the performed experiments and assumptions from foreign studies, the increasing depth of carbonation with the decreasing strength class was confirmed, regardless of whether it is a reference concrete with natural aggregates or concrete with recycled aggregates.

The potential scenario of CO_2 uptake is evaluated in the LCA, and the captured CO_2 value was evaluated as negligible compared to the value of CO_2 in cement production. However, the assumption of CO_2 capture could be useful given the effort to eliminate environmentally non-friendly materials, such as cement in concrete production, and replace them with waste or recycled materials.

After an overall evaluation of the LCA, recycled concrete (RMAC-I, RCAC-I, RMAC-II, RCAC-II) were evaluated as more environmentally friendly compared to the reference samples (NAC-I, NAC-II). These results will be used as a basis for the subsequent verification of other specific properties of recycled concrete with the aim of implementing them in the industry sector.

Supplementary Materials: The following supporting information can be downloaded at: https://www.mdpi.com/article/10.3390/su14031732/s1, Table S1: pH of the leachates (mean values ± SD) at the end of the duckweed toxicity test (after 7 days of exposition). 100 + n—leachates (100 mL.L^{-1}) amended with nutrients; Table S2: The results of the water flea toxicity tests. Mean (±SD) values of immobilization (%). 100 + n—leachates (100 mL.L^{-1}) amended with nutrients. The letters indicate significant differences between values (post-hoc test; α = 0.05) within the same column (uppercase) and within the same row (lowercase), and the asterisks indicate differences between sample and control (zero values); Table S3: The results of algae toxicity tests. Mean (±SD) values of inhibition/stimulation (%) of growth rate based on optical density at 750 nm. 100+n—leachates (100 mL.L^{-1}) amended with nutrients. Negative values indicate growth stimulation. The letters indicate significant differences between values (post-hoc test; α = 0.05) within the same column (uppercase) and within the same row (lowercase), and the asterisks indicate differences between sample and control (zero values); Table S4: The results of duckweed toxicity tests. Mean (±SD) values of inhibition/stimulation (%) of the growth rate based on the total area of the frond. 100 + n—leachates (100 mL.L^{-1}) amended with nutrients. Negative values indicate growth stimulation. The letters indicate significant differences between values (post-hoc test; α = 0.05) within the same column (uppercase) and within the same row (lowercase), and the asterisks indicate differences between the sample and control (zero values); Table S5: pH (mean values ± SD) measured in soils amended with leachates after 7, 28, and 56 days.

Author Contributions: Conceptualization, D.M.; methodology, K.A.M. and D.M.; validation, T.P., K.A.M., D.M., B.T., K.F. and P.H.; investigation, T.P., K.A.M., D.M., J.P. and K.F.; resources, D.M. and K.A.M.; data curation K.F., K.A.M., J.P. and D.M.; writing—original draft preparation, D.M. and K.A.M.; writing—review and editing, K.F., K.A.M., T.P., P.H., B.T., J.P. and D.M.; visualization, D.M., K.F. and T.P.; project administration, T.P. and P.H. All authors have read and agreed to the published version of the manuscript.

Funding: This work is part of the project "Design of performance-based concrete using sand reclaimed from construction and demolition wastes", LTAIN19205, supported by the Ministry of Education, Youth and Sports of the Czech Republic, through the program INTER-EXCELLENCE.

Institutional Review Board Statement: Not applicable.

Informed Consent Statement: Not applicable.

Data Availability Statement: Data available in a publicly accessible repository.

Conflicts of Interest: The authors declare no conflict of interest.

References

1. Vytlacilova, V. Testing ecological suitability for the utilization of recycled aggregates. *Green Process. Synth.* **2017**, *6*, 225–234. [CrossRef]
2. Rodrigues, P.; Silvestre, J.D.; Flores-Colen, I.; Viegas, C.A.; Ahmed, H.H.; Kurda, R.; De Brito, J. Evaluation of the Ecotoxicological Potential of Fly Ash and Recycled Concrete Aggregates Use in Concrete. *Appl. Sci.* **2020**, *10*, 351. [CrossRef]
3. Rodrigues, P.; Silvestre, J.D.; Flores-Colen, I.; Viegas, C.A.; De Brito, J.; Kurad, R.; Demertzi, M. Methodology for the Assessment of the Ecotoxicological Potential of Construction Materials. *Materials* **2017**, *10*, 649. [CrossRef]
4. Couvidat, J.; Diliberto, C.; Meux, E.; Cotelle, S.; Bojic, C.; Izoret, L.; Lecomte, A. Greening effect of slag cement-based concrete: Environmental and ecotoxicological impact. *Environ. Technol. Innov.* **2021**, *22*, 101467. [CrossRef]
5. Barbosa, R.; Lapa, N.; Dias, D.; Mendes, B. Concretes containing biomass ashes: Mechanical, chemical, and ecotoxic performances. *Constr. Build. Mater.* **2013**, *48*, 457–463. [CrossRef]
6. Naaz, S.; Pandey, S.N. Effects of industrial waste water on heavy metal accumulation, growth and biochemical responses of lettuce (*Lactuca sativa* L.). *J. Environ. Biol.* **2010**, *31*, 273–276.
7. Gao, Z.; Han, J.; Mu, C.; Lin, J.; Li, X.; Lin, L.; Sun, S. Effects of Saline and Alkaline Stresses on Growth and Physiological Changes in Oat (*Avena sativa L.*) Seedlings. *Not. Bot. Horti Agrobot.* **2014**, *42*, 357–362. [CrossRef]
8. Guo, R.; Zhou, J.; Hao, W.; Gu, F.; Liu, Q.; Li, H.; Xia, X.; Mao, L. Germination, Growth, Chlorophyll Fluorescence and Ionic Balance in Linseed Seedlings Subjected to Saline and Alkaline Stresses. *Plant Prod. Sci.* **2014**, *17*, 20–31. [CrossRef]
9. Radić, S.; Stipaničev, D.; Cvjetko, P.; Mikelić, I.L.; Rajčić, M.M.; Širac, S.; Pevalek-Kozlina, B.; Pavlica, M. Ecotoxicological assessment of industrial effluent using duckweed (*Lemna minor* L.) as a test organism. *Ecotoxicology* **2009**, *19*, 216–222. [CrossRef]
10. Schinner, F.; Öhlinger, R.; Kandeler, E.; Margesin, R. (Eds.) *Methods in Soil Biology*; Springer: Berlin/Heidelberg, Germany, 1996; ISBN 9783642646331.
11. Dick, R.P.; Breakwell, D.P.; Turco, R.F. Soil Enzyme Activities and Biodiversity Measurements as Integrative Microbiological Indicators. In *SSSA Special Publications*; Wiley: Hoboken, NJ, USA, 2015; pp. 247–271. ISBN 978-0-89118-944-2.
12. Delreux, T.; Ohler, F. Climate Policy in European Union Politics. In *Oxford Research Encyclopedia of Politics*; Oxford University Press: Oxford, UK, 2019; ISBN 9780190228637.
13. Monteiro, P.J.M.; Miller, S.; Horvath, A. Towards sustainable concrete. *Nat. Mater.* **2017**, *16*, 698–699. [CrossRef]
14. GREEN Solution, s.r.o. *Green Solution s.r.o. Program Předcházení Vzniku Odpadu*; Publisher Ministry of the Environment: Prague, Czech Republic, 2017.
15. Papadakis, V.G.; Vayenas, C.G.; Fardis, M. Experimental investigation and mathematical modeling of the concrete carbonation problem. *Chem. Eng. Sci.* **1991**, *46*, 1333–1338. [CrossRef]
16. Xuan, D.; Zhan, B.; Poon, C.S. Assessment of mechanical properties of concrete incorporating carbonated recycled concrete aggregates. *Cem. Concr. Compos.* **2016**, *65*, 67–74. [CrossRef]
17. Villain, G.; Thiery, M.; Platret, G. Measurement methods of carbonation profiles in concrete: Thermogravimetry, chemical analysis and gammadensimetry. *Cem. Concr. Res.* **2007**, *37*, 1182–1192. [CrossRef]
18. Papadakis, V.G.; Vayenas, C.G.; Fardis, M.N. Fundamental Modeling and Experimental Investigation of Concrete Carbonation. *ACI Mater. J.* **1991**, *88*, 363–373. [CrossRef]
19. Papadakis, V.G.; Vayenas, C.G.; Fardis, M.N. Physical and Chemical Characteristics Affecting the Durability of Concrete. *ACI Mater. J.* **1991**, *88*, 186–196. [CrossRef]
20. Parrott, L.; Killoh, D. Carbonation in a 36 year old, in-situ concrete. *Cem. Concr. Res.* **1989**, *19*, 649–656. [CrossRef]
21. Collepardi, M. The Influence of Slag and Fly Ash on the Carbonation of Concrete. In Proceedings of the Eighth CANMET/ACI International Conferance on Fly Ash, Silica Fume, Slag, and Natural Pozzolans in Concrete, Las Vegas, NV, USA, 23–29 May 2004; Volume 2004, pp. 221–229.
22. Matthews, S. *Design of Durable Concrete Structures*; BRE Trust: Glasgow, UK, 2014. [CrossRef]
23. Chang, C.-F.; Chen, J.-W. The experimental investigation of concrete carbonation depth. *Cem. Concr. Res.* **2006**, *36*, 1760–1767. [CrossRef]
24. ČSN EN 12390-12 (73 1302). *Testing Hardened Concrete—Part 12: Determination of the Resistance of Concrete to Carbonation—Accelerated Carbonation Method*; Office for Technical Standardization, Metrology and State Testing: Prague, Czech Republic, 2020.
25. Mariaková, D.; Mocová, K.A.; Fořtová, K.; Ryparová, P.; Pešta, J.; Pavlů, T. Ecotoxicity and Essential Properties of Fine-Recycled Aggregate. *Materials* **2021**, *14*, 463. [CrossRef]
26. Mariaková, D.; Mocová, K.A.; Fořtová, K.; Pavlů, T.; Hájek, P. Waste Glass Powder Reusability in High-Performance Concrete: Leaching Behavior and Ecotoxicity. *Materials* **2021**, *14*, 4476. [CrossRef]
27. Wellburn, A.R. The Spectral Determination of Chlorophylls a and b, as well as Total Carotenoids, Using Various Solvents with Spectrophotometers of Different Resolution. *J. Plant Physiol.* **1994**, *144*, 307–313. [CrossRef]
28. ISO 10390:2021. *Soil, Treated Biowaste and Sludge—Determination of PH*; ISO: Geneva, Switzerland, 2021; p. 8.
29. ISO 23753-1:2005. *Soil Quality—Determination of Dehydrogenase Activity in Soils—Part 1: Method Using Triphenyltetrazolium Chloride (TTC)*; ISO: Geneva, Switzerland, 2005; p. 5.
30. ČSN EN 12390-2. *Testing Hardened Concrete—Part 2: Making and Curing Specimens for Strength Tests*; Office for Technical Standardization, Metrology and State Testing: Prague, Czech Republic, 2020.

31. Guinée, J. Handbook on life cycle assessment—Operational guide to the ISO standards. *Int. J. Life Cycle Assess.* **2001**, *6*, 255. [CrossRef]
32. International Organization for Standardization. *Environmental Management. Life Cycle Assessment. Principles and Framework; Standard ISO 14040*; International Organization for Standardization: Geneva, Switzerland, 2006; Available online: https://www.iso.org/standard/37456.html (accessed on 27 March 2020).
33. International Organization for Standardization. ISO 14040 Environmental Management-Life Cycle Assessment-Principles and Framework. Available online: https://www.iso.org/standard/38498.html (accessed on 13 December 2019).
34. CEN EN 15804:2012+A2:2019 Sustainability of Construction Works—Environmental Product Declarations—Core Rules for the Product Category of Construction Products. Available online: https://standards.cen.eu/dyn/www/f?p=204:110:0::::FSP_PROJECT:70014&cs=1CFE7BDC38149F238F05C4C13E0E3B4C2 (accessed on 27 November 2020).
35. Life Cycle Assessment LCA Software: Gabi Software. Available online: http://www.gabisoftware.com/ (accessed on 30 October 2020).
36. Kupfer, T.; Baitz, M.; Colodel, C.M.; Kokborg, M.; Schöll, S.; Rudolf, M.; Thellier, L.; Bos, U.; Bosch, F.; Gonzalez, M.; et al. GaBi Database and Modelling Principles. Available online: http://www.gabi-software.com/fileadmin/gabi/Modelling_Principles/Modeling_Principles_-_GaBi_Databases_2020_2.pdf (accessed on 1 December 2020).
37. ČSN EN 16757. *Sustainability of Construction Works—Environmental Product Declarations—Product Category Rules for Concrete and Concrete Elements*; Office for Technical Standardization, Metrology and State Testing: Prague, Czech Republic, 2018.
38. Product Environmental Footprint (PEF) Guide. Available online: https://ec.europa.eu/environment/archives/eussd/pdf/footprint/PEF%20methodology%20final%20draft.pdf (accessed on 7 October 2020).
39. Environmental Footprint 3.0. Available online: https://eplca.jrc.ec.europa.eu//EnvironmentalFootprint.html (accessed on 1 November 2021).
40. Matoušek, M.; Drochytka, R. *Atmosférická Koroze Betonu*; ČKAIT a IKAS: Prague, Czech Republic, 1998.
41. Teplý, B.; Chromá, M.; Rovnanik, P. Durability assessment of concrete structures: Reinforcement depassivation due to carbonation. *Struct. Infrastruct. Eng.* **2010**, *6*, 317–327. [CrossRef]
42. Pašek, J.; Vejvara, L. Karbonatace Betonových a Železobetonových Obvodových Konstrukcí. In Proceedings of the Poruchy a Obnova Obalových Konštrukcií Budov, Vysoké Tatry, Slovakia, 4 April 2004; pp. 86–91.
43. Kurda, R.; de Brito, J.; Silvestre, J.D. Carbonation of concrete made with high amount of fly ash and recycled concrete aggregates for utilization of CO_2. *J. CO2 Util.* **2018**, *29*, 12–19. [CrossRef]
44. Lu, B.; Shi, C.; Cao, Z.; Guo, M.; Zheng, J. Effect of carbonated coarse recycled concrete aggregate on the properties and microstructure of recycled concrete. *J. Clean. Prod.* **2019**, *233*, 421–428. [CrossRef]
45. Liang, C.; Lu, N.; Ma, H.; Ma, Z.; Duan, Z. Carbonation behavior of recycled concrete with CO2-curing recycled aggregate under various environments. *J. CO2 Util.* **2020**, *39*, 101185. [CrossRef]
46. Bold, H.C. The Morphology of *Chlamydomonas chlamydogama*, Sp. Nov. *Bull. Torrey Bot. Club* **1949**, *76*, 101. [CrossRef]
47. Ebert, D.; Zschokke-Rohringer, C.D.; Carius, H.J. Within- and between-population variation for resistance of Daphnia magna to the bacterial endoparasite *Pasteuria ramosa*. *Proc. R. Soc. B Boil. Sci.* **1998**, *265*, 2127–2134. [CrossRef]
48. ISO (International Organisation for Standardization). *2005 ISO 20079 Water Quality—Determination of Toxic Effect of Water Constituents and Waste to Duckweed (Lemna Minor)—Duckweed Growth Inhibition Test*; ISO: Geneva, Switzerland, 2005.
49. Mor, S.; Ravindra, K.; Dahiya, R.P.; Chandra, A. Leachate Characterization and Assessment of Groundwater Pollution Near Municipal Solid Waste Landfill Site. *Environ. Monit. Assess.* **2006**, *118*, 435–456. [CrossRef]
50. Vriens, B.; Voegelin, A.; Hug, S.J.; Kaegi, R.; Winkel, L.H.E.; Buser, A.M.; Berg, M. Quantification of Element Fluxes in Wastewaters: A Nationwide Survey in Switzerland. *Environ. Sci. Technol.* **2017**, *51*, 10943–10953. [CrossRef]
51. Rengasamy, P. World salinization with emphasis on Australia. *J. Exp. Bot.* **2006**, *57*, 1017–1023. [CrossRef]
52. Cañedo-Argüelles, M.; Kefford, B.J.; Piscart, C.; Prat, N.; Schäfer, R.B.; Schulz, C.-J. Salinisation of rivers: An urgent ecological issue. *Environ. Pollut.* **2013**, *173*, 157–167. [CrossRef] [PubMed]
53. Schuler, M.S.; Cañedo-Argüelles, M.; Hintz, W.D.; Dyack, B.; Birk, S.; Relyea, R.A. Regulations are needed to protect freshwater ecosystems from salinization. *Philos. Trans. R. Soc. B Biol. Sci.* **2018**, *374*, 20180019. [CrossRef] [PubMed]
54. Baberschke, N.; Irob, K.; Preuer, T.; Meinelt, T.; Kloas, W. Potash mining effluents and ion imbalances cause transient osmoregulatory stress, affect gill integrity and elevate chronically plasma sulfate levels in adult common roach, *Rutilus rutilus*. *Environ. Pollut.* **2019**, *249*, 181–190. [CrossRef] [PubMed]
55. Herrmann, M.C.; Entrekin, S.A.; Evans-White, M.A.; Clay, N.A. Salty water and salty leaf litter alters riparian detrital processes: Evidence from sodium-addition laboratory mesocosm experiments. *Sci. Total Environ.* **2021**, *806*, 151392. [CrossRef]
56. Bandow, N.; Gartiser, S.; Ilvonen, O.; Schoknecht, U. Evaluation of the impact of construction products on the environment by leaching of possibly hazardous substances. *Environ. Sci. Eur.* **2018**, *30*, 14. [CrossRef]
57. Heisterkamp, I.; Ratte, M.; Schoknecht, U.; Gartiser, S.; Kalbe, U.; Ilvonen, O. Ecotoxicological evaluation of construction products: Inter-laboratory test with DSLT and percolation test eluates in an aquatic biotest battery. *Environ. Sci. Eur.* **2021**, *33*, 1–14. [CrossRef]
58. Choi, J.B.; Bae, S.M.; Shin, T.Y.; Ahn, K.Y.; Woo, S.D. Evaluation of Daphniamagna for the Ecotoxicity Assessment of Alkali Leachate from Concrete. *Int. J. Ind. Èntomol.* **2013**, *26*, 41–46. [CrossRef]

59. Strähle, U.; Scholz, S.; Geisler, R.; Greiner, P.; Hollert, H.; Rastegar, S.; Schumacher, A.; Selderslaghs, I.; Weiss, C.; Witters, H.; et al. Zebrafish embryos as an alternative to animal experiments—A commentary on the definition of the onset of protected life stages in animal welfare regulations. *Reprod. Toxicol.* **2012**, *33*, 128–132. [CrossRef]
60. Jalal, K.; Shamsuddin, A.; Rahman, M.; Nurzatul, N.; Rozihan, M. Growth and Total Carotenoid, Chlorophyll a and Chlorophyll b of Tropical Microalgae (*Isochrysis* sp.) in Laboratory Cultured Conditions. *J. Biol. Sci.* **2012**, *13*, 10–17. [CrossRef]
61. ISO 11348 (2007–12). *Water Quality—Determination of the Inhibitory Efect of Water Samples on the Light Emission of Vibrio Fscheri (Luminescent Bacteria Test)—Part 1: Method Using Freshly Prepared Bacteria, Part 2: Method Using Liquid-Dried Bacteria, Part 3: Method Using Freeze-Dried Bacteria*; ISO: Geneva, Switzerland, 2007.
62. Ladislas, S.; El-Mufleh, A.; Gérente, C.; Chazarenc, F.; Andrès, Y.; Béchet, B. Potential of Aquatic Macrophytes as Bioindicators of Heavy Metal Pollution in Urban Stormwater Runoff. *Water Air Soil Pollut.* **2011**, *223*, 877–888. [CrossRef]
63. Leblebici, Z.; Aksoy, A.; Duman, F. Influence of nutrient addition on growth and accumulation of cadmium and copper in Lemna gibba. *Chem. Speciat. Bioavailab.* **2010**, *22*, 157–164. [CrossRef]
64. Varga, M.; Horvatić, J.; Čelić, A. Short term exposure of Lemna minor and Lemna gibba to mercury, cadmium and chromium. *Open Life Sci.* **2013**, *8*, 1083–1093. [CrossRef]
65. Hegazy, A.K.; Kabiel, H.F.; Fawzy, M. Duckweed as heavy metal accumulator and pollution indicator in industrial wastewater ponds. *Desalination Water Treat.* **2009**, *12*, 400–406. [CrossRef]
66. Thomas, G.; Stärk, H.-J.; Wellenreuther, G.; Dickinson, B.C.; Küpper, H. Effects of nanomolar copper on water plants—Comparison of biochemical and biophysical mechanisms of deficiency and sublethal toxicity under environmentally relevant conditions. *Aquat. Toxicol.* **2013**, *140-141*, 27–36. [CrossRef] [PubMed]
67. ISO. *ISO In Water Quality—Fresh Water Algal Growth Inhibition Test with Unicellular Green Algae*; ISO: Geneva, Switzerland, 2012; p. 21.
68. OECD. *OECD Test No. 201: Alga, Growth Inhibition Test*; OECD Guidelines for the Testing of Chemicals, Section 2: Effects on Biotic Systems; OECD Publishing: Paris, France, 2006; ISBN 978-92-64-06992-3.
69. ISO 10260:1992. *Water Quality—Measurement of Biochemical Parameters—Spectrometric Determination of the Chlorophyll-a Concentration*; ISO: Geneva, Switzerland, 1992; p. 6.
70. Poojary, M.M.; Barba, F.J.; Aliakbarian, B.; Donsì, F.; Pataro, G.; Dias, D.A.; Juliano, P. Innovative Alternative Technologies to Extract Carotenoids from Microalgae and Seaweeds. *Mar. Drugs* **2016**, *14*, 214. [CrossRef]
71. Osório, C.; Machado, S.; Peixoto, J.; Bessada, S.; Pimentel, F.B.; Alves, R.C.; Oliveira, M.B.P.P. Pigments Content (Chlorophylls, Fucoxanthin and Phycobiliproteins) of Different Commercial Dried Algae. *Separations* **2020**, *7*, 33. [CrossRef]
72. Ananyeva, N.D.; Ivashchenko, K.V.; Sushko, S.V. Microbial Indicators of Urban Soils and Their Role in the Assessment of Ecosystem Services: A Review. *Eurasian Soil Sci.* **2021**, *54*, 1517–1531. [CrossRef]
73. Goodwin, D. Urban soils and functional trees. In *The Urban Tree*; Taylor and Francis Group: London, UK, 2017; pp. 56–89. [CrossRef]
74. Qi, D.; Wieneke, X.; Tao, J.; Zhou, X.; DeSilva, U. Soil pH Is the Primary Factor Correlating with Soil Microbiome in Karst Rocky Desertification Regions in the Wushan County, Chongqing, China. *Front. Microbiol.* **2018**, *9*, 1027. [CrossRef]
75. Kuwano, B.H.; Nogueira, M.A.; Santos, C.A.; Fagotti, D.S.; Santos, M.B.; Lescano, L.E.; Andrade, D.S.; Barbosa, G.M.; Tavares-Filho, J. Application of Landfill Leachate Improves Wheat Nutrition and Yield but Has Minor Effects on Soil Properties. *J. Environ. Qual.* **2017**, *46*, 153–159. [CrossRef]
76. Shailaja, G.S.J.; Srinivas, N.; Rao, P.P.V.V. Effect of Municipal Solid Waste Leachate on Soil Enzymes. *Nat. Environ. Pollut. Technol.* **2021**, *20*. [CrossRef]
77. Kazlauskaitė-Jadzevičė, A.; Volungevičius, J.; Gregorauskienė, V.; Marcinkonis, S. The role of PH in heavy metal contamination of urban soil. *J. Environ. Eng. Landsc. Manag.* **2014**, *22*, 311–318. [CrossRef]
78. Gholizadeh-Vayghan, A.; Bellinkx, A.; Snellings, R.; Vandoren, B.; Quaghebeur, M. The effects of carbonation conditions on the physical and microstructural properties of recycled concrete coarse aggregates. *Constr. Build. Mater.* **2020**, *257*, 119486. [CrossRef]
79. Ho, D.W.S.; Lewis, R.K. Carbonation of concrete and its prediction. *Cem. Concr. Res.* **1987**, *17*, 489–504. [CrossRef]
80. Malati, M.A. *Experimental Inorganic/Physical Chemistry*; Woodhead Publishing Limited, Abington Hall: Cambridge, UK, 1999; p. 338.
81. Bernal, S.A.; San Nicolas, R.; Myers, R.J.; Mejía de Gutiérrez, R.; Puertas, F.; van Deventer, J.S.J.; Provis, J.L. MgO content of slag controls phase evolution and structural changes induced by accelerated carbonation in alkali-activated binders. *Cem. Concr. Res.* **2014**, *57*, 33–43. [CrossRef]
82. Xiao, R.; Jiang, X.; Zhang, M.; Polaczyk, P.; Huang, B. Analytical investigation of phase assemblages of alkali-activated materials in CaO-SiO_2-Al_2O_3 systems: The management of reaction products and designing of precursors. *Mater. Des.* **2020**, *194*, 108975. [CrossRef]
83. Kikuchi, T.; Kuroda, Y. Carbon Dioxide Uptake in Demolished and Crushed Concrete. *J. Adv. Concr. Technol.* **2011**, *9*, 115–124. [CrossRef]
84. Dodoo, A.; Gustavsson, L.; Sathre, R. Carbon implications of end-of-life management of building materials. *Resour. Conserv. Recycl.* **2009**, *53*, 276–286. [CrossRef]

85. Yang, K.-H.; Seo, E.-A.; Tae, S.-H. Carbonation and CO_2 uptake of concrete. *Environ. Impact Assess. Rev.* **2014**, *46*, 43–52. [CrossRef]
86. Marinković, S.; Dragaš, J.; Ignjatović, I.; Tošić, N. Environmental assessment of green concretes for structural use. *J. Clean. Prod.* **2017**, *154*, 633–649. [CrossRef]
87. Colangelo, F.; Navarro, T.G.; Farina, I.; Petrillo, A. Comparative LCA of concrete with recycled aggregates: A circular economy mindset in Europe. *Int. J. Life Cycle Assess.* **2020**, *25*, 1790–1804. [CrossRef]

MDPI

Article

Circular Economy in the Construction Sector: A Case Study of Santiago de Cali (Colombia)

Aníbal Maury-Ramírez [1,*], Danny Illera-Perozo [2] and Jaime A. Mesa [3]

1 Engineering Faculty, Universidad El Bosque, Bogotá 111711, Colombia
2 Department of Mechanical Engineering, Universidad de la Sabana, Chía 140013, Colombia; danny.illera@unisabana.edu.co
3 Department of Industrial Engineering, Pontificia Universidad Javeriana, Bogotá 110231, Colombia; mesajaime@javeriana.edu.co
* Correspondence: amaury@unbosque.edu.co; Tel.: +57-164-89000 (ext. 1285)

Abstract: The circular economy, a new paradigm of technological and economic development, is of great importance in developing countries, particularly in the construction sector, one of the most relevant in Colombia. In the Latin American context, Colombia has one of the most important construction industries, contributing to the social and productive development of the country. However, this sector is also responsible for serious environmental problems and social conflicts. Therefore, it is imperative to work with all actors of the value chain to transform the construction sector from a linear economy to a circular economy model. Therefore, this article describes the circular economy model proposed for Santiago de Cali, which is mainly oriented to the analysis and efficient use of construction materials, mostly taking into account the recovery of ecosystems and the circular flow of rocky materials. This model includes an analysis of the production of construction materials, construction process, use and operation, and completion of the life cycle of buildings and infrastructure. In particular, the model proposes an innovative product portfolio for the use of construction and demolition waste (C&DW) supported in applied research (case studies). The portfolio consists of family products, such as recycled aggregates or eco-aggregates, eco-concretes and mortars, eco-prefabricated products and modules, and smart construction materials. In addition, this model describes the C&DW management system and some characteristics of the Technological and Environmental Park (TEP), the main strategy for C&DW valorization in the city.

Keywords: construction and demolition waste (C&DW); circular economy; construction; concrete; recycling

Citation: Maury-Ramírez, A.; Illera-Perozo, D.; Mesa, J.A. Circular Economy in the Construction Sector: A Case Study of Santiago de Cali (Colombia). *Sustainability* **2022**, *14*, 1923. https://doi.org/10.3390/su14031923

Academic Editor: Castorina Silva Vieira

Received: 11 December 2021
Accepted: 3 February 2022
Published: 8 February 2022

Publisher's Note: MDPI stays neutral with regard to jurisdictional claims in published maps and institutional affiliations.

1. Introduction

The rapid growth of the Colombian population and its displacement to urban centers in the last 50 years has indirectly stimulated the development of the national construction sector, which represents one of the most relevant in the economy of Valle del Cauca and Santiago de Cali (approx. 3 million habitants), promoting a large number of jobs (direct and indirect) and energizing other subsectors of the economy of the Colombian southwest. According to the Colombian Chamber of Construction (CAMACOL), the national construction sector currently generates annual investments of 77 billion pesos (USD 19.5 million), contributes 46 billion pesos (11.7 million USD) to the economy, demands inputs for COP 34 billion (USD 8.6 million) annually and, together with real estate activities, this generates 1.8 million jobs [1]. In addition, this industry has been relevant in improving the infrastructure for land and air connectivity, the development of the housing sector, the provision of public, hospital, educational services, and the growth of business, tourism, and commercial activity in the Colombian southwest region (Pacific coast). However, mainly due to the segmentation and disconnection of all stakeholders within the value chain, lack of control, and the ignorance of the environmental impacts by all the actors, serious problems have been generated in the Valle del Cauca and, particularly, in the city of Santiago de Cali, which has a rich environment and cultural background. The environmental deterioration of the seven hydrological basins of the rivers across the city is significant, due to the extraction of raw materials caused by handcrafted and illegal mining. Likewise, the proper disposal of construction and demolition waste (C&DW) has become a great challenge for the region. Daily, in the city of Santiago de Cali are generated approximately 2500 m^3 of rubble or C&DW, occupying one of the first places among the Colombian capitals in the generation of this type of waste [2].

Due to the above and in order to improve this situation, the circular economy model for the construction sector of Santiago de Cali seeks to transform the construction sector from a linear model (extraction, production, consumption, and waste disposal) to a more circular model. Since 2012, with the support of the Ellen MacArthur Foundation, the circular economy concept (CE) has gained popularity worldwide. Nevertheless, its development really comes from much earlier times. For example, Agenda 21 of 1992 established, in Principle 8, that "states will have to reduce and eliminate unsustainable patterns of production and consumption". Similarly, the Rome Club in 1972, with a report on "the limits of growth" stated the bases to think about a different economy, one that was no linear [3].

Therefore, considering the sustainable development goals (SDGs) stated by United Nations in 2015, particularly SDG 12 on sustainable consumption and production, the proposed model appropriates the circular economy as a production and consumption system that promotes efficiency in the use of materials, water and energy, taking into account the resilience of ecosystems and the circular use of material flows through the implementation of technological innovations, alliances and collaborations between stakeholders (e.g. raw material producers, building companies, users and final disposal actors), and the promotion of business models that respond to the fundamentals of sustainable development [4]. Therefore, this pioneer model takes into account the following challenges and opportunities to implement circular economy in construction:

(a) Most of the barriers to implementation are related to organisational concerns [5];
(b) There still exists ambiguity and inconsistency in the assessment methods to measure circularity in building projects [6];
(c) Digitalization could be a great help in developing sustainable circular products [7];
(d) The customers' involvement is necessary for creating innovative sustainable circular products using digitalization [7];
(e) The inclusion of people driven factors in the adoption of circular economy practices in the supply chains [8].

The next sections of this article are organized as follows: Section 2 describes the generalities of the circular economy model proposed for Santiago de Cali. Section 3 corresponds to the Technological and Environmental Park (TEP) proposal that integrates C&DW from Santiago de Cali to promote different circular services and products for the construction sector. Case studies related to the potential of the TEP are described in Section 4. Finally, challenges and conclusions are included in Sections 5 and 6, respectively.

2. CE Model Proposal

Although not presented in this article due to the limited space, the circular model for the construction sector of Santiago de Cali required a previous study from the different actors and their environmental and social impacts through the entire construction value chain. In this study, extractors of raw materials (mining), building materials producers, transporters, construction companies, promoters, maintenance, repair and rehabilitation businesses, certifiers and laboratories, construction and demolition waste managers, other economic sectors of interest to symbiosis, the community and the media were included. Thus, this section describes the phases and product portfolio considered around the CE model adapted to the potential and context of Santiago de Cali. The summary of data and analysis of C&DW potential were based on previous research projects that support the need for integrative solutions based on material circularity. Phases and product portfolio are explained in detail as follows. In addition, the references of the previous research projects are included.

2.1. *Phases of the CE Model for C&DW Materials*

Four phases are considered in the CE model for C&DW of Santiago de Cali: (i) extraction of raw materials and building material production, (ii) construction, (iii) use and operation, and (iv) end of the lifecycle. It is possible to find recovery flows in such phases to enable a circular economy model (Figure 1). Each phase is described in detail as follows.

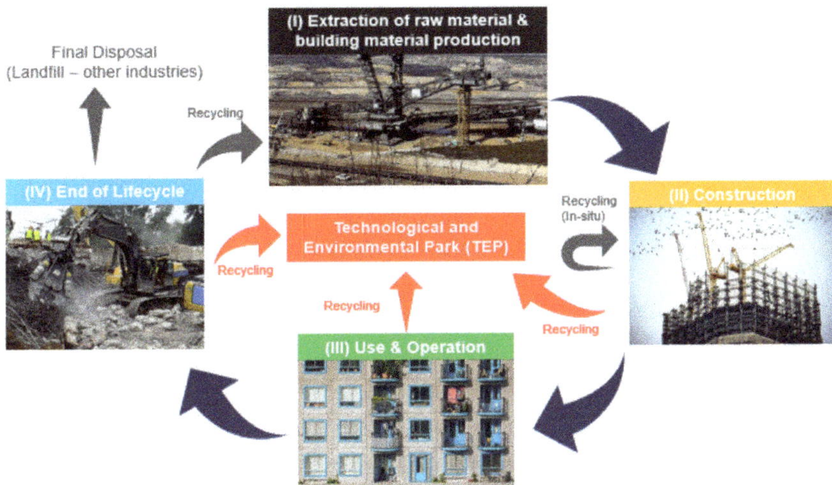

Figure 1. Circular economy model for the construction sector of Santiago de Cali (Colombia) [9].

2.1.1. Extraction of Raw Materials and Building Materials Production

The production of construction materials encompasses two main stages: extraction of raw materials and their processing, which involve severe environmental and social impacts. Therefore, the circular economy model proposes incorporating the principle of environmental assets extraction and clean production to replace the classic view of the production of construction materials. Within the framework of a green economy, environmental assets are associated with ecosystem services, resources, or the processes of natural ecosystems (goods and services) that benefit human beings. They include products such as drinking water or clean air and processes such as waste decomposition. Based on the above premise, the appraisal of environmental assets is directly related to economics, sociology, and biology.

Clean production attempts to preserve raw materials, water, and energy; while reducing toxic raw materials use, emissions and waste, which may be depleted into the water, atmosphere, and soil. Clean production arises from continuous improvement, quality control and the reengineering of process engineering. Its application advocates reviewing operations and unitary processes that are part of a productive or service activity, intending to find the various possibilities for improvement or optimization regarding the use of resources. Clean production is thus defined as a preventive business management strategy applied to products, processes, and work organization, whose objective is to minimize emissions and discharges at the source, reducing risks to human and environmental health while simultaneously increasing competitiveness. This results from the following actions:

- The minimization and efficient consumption of raw materials, water, and energy;
- The minimization of the use of toxic raw materials;
- The minimization of the volume and toxicity of all emissions generated by the production process;

- Recycling the maximum proportion of waste in the processing facility;
- Reduction in the environmental impact of building products during their life cycle;
- The development of end of life strategies for building products defined from early design phases.

To articulate the above mentioned principles, the Portland Cement Association (PCA), one of the oldest and most essential production associations in the United States, has created the Concrete Sustainability Hub, a research center at Massachusetts Institute of Technology (MIT) in collaboration with the Ready Mixed Concrete (RMC) Research and Education Foundation. The Hub was founded with the mission to accelerate emerging advances in concrete science and technology and transfer the best available results to engineering practices. The interdisciplinary team of researchers brings together leaders from academia, industry, and government to facilitate knowledge transfer by aligning world-leading research with end user needs. Meanwhile, Hub researchers are investigating concrete from the nanoscale to address the sustainability and environmental implications of the material's production and use. Their research also aims to refine the composition of concrete, reduce greenhouse gas emissions during its production, and quantify its environmental impact and cost over the lifetime of an infrastructure or construction project [10]. The initiatives mentioned above have significantly impacted the cement industry worldwide. In Colombia, many concrete and cement companies are currently expanding their portfolio of sustainable products based on reducing energy and water consumption during production.

2.1.2. Construction of Buildings and Infrastructure

The construction of a building or infrastructure is the macroprocess of the materialization of architectural and engineering projects (e.g., structural, hydraulic, sanitary, electrical and telecommunications networks) arising as a response to a set of global needs that the project must satisfy during use or operation. However, to achieve that objective, it is necessary to work on three main phases: prior and simultaneous coordination and the follow up of the executed project. In particular, these phases seek the formulation and development of construction techniques to ensure optimal water consumption, energy, and materials and the minimum emission of particulate matter, gases, and noise generation. These techniques allow the development of sustainable construction principles, minimize labor risks, reduce costs, and improve the quality of the works.

Here, it is crucial to define a methodology for developing an integrative project considering the large number of professional specialists involved in the construction phase of buildings and infrastructure. In this sense, the stages proposed by [11] are considered according to the Colombian Regulation NSR-10: Construction, Supervision, and Technical Supervision of Structural Concrete Buildings. Coordination phases are described as follows:

- Preliminary/prior coordination: refers to the understanding of the architectural project when it is not yet defined, and only the criteria indicated by the designer are established. The aim is to establish alternatives to the structural system, type of foundation, the possible location of screens, the magnitude of column spacing and dimensions, height of slabs and characteristics of probable materials, among other aspects.
- Simultaneous coordination: As these projects' architectural and structural definitions progress, knowledge is shared among the design professionals. The impact of the definitions is derived from the dimensioning of the structural elements or aspects that the architecture proposes as the project's conceptual basis is established simultaneously.

To facilitate the development of the proposed methodology under the principles of sustainable construction, technological tools that facilitate interdisciplinary and simultaneous work for the execution of architectural and engineering projects are required. For example, building information modeling (BIM) is a collaborative work methodology applied to the construction sector, a collection of organized data of a building to facilitate the management of engineering, architecture, and construction projects achieving improvements in the result and efficiency in the processes [12,13]. All professionals involved in a construction project can work on a single project in real time, with access to the same information. BIM is associated with geometry, site relationship, geographic information, quantities and the properties of a building or infrastructure components. For example, details of door manufacturers or the energy data of a material can be easily consulted at any stage of the project. In general, BIM can be used to provide information on a building or infrastructure at any time. For example, progress in structural design, maintenance activities and even rehabilitation and demolition processes. Shared material quantities and properties can be easily extracted. In addition, labor aspects, component details and the sequences of construction activities can be isolated and defined. BIM software can achieve such improvements through graphical representations of the parts and components used to construct a building or infrastructure.

Although modularization and prefabrication are technologies that have been used for centuries in the construction industry, their re-emergence as a new trend is associated with the rise of BIM and the requirements of sustainable construction. In particular, modularization refers to the process of manufacturing functional units in controlled off-site environments so that they can be transported in whole pieces from their place of manufacture to their final location. Prefabricated elements include structural elements (e.g., beams, columns, porticos) and nonstructural elements (e.g., pavers, curbs, sidewalks) that make up the functional units (e.g., structural system, foundation). The benefits of using precast and modularization in construction include savings in project schedules, reduction in contingencies, cost optimization for personnel and materials, safety, quality, minimization of C&DW and potential for reuse of modules and precast at the end of the life cycle of buildings and infrastructure [14]. Recently, robotization and 3D printing have emerged as technologies that seek to articulate the construction industry to the demands of environmental, economic, and social sustainability. In particular, 3D printing is the fabrication of structural and nonstructural elements by layering layers of material, just as in paper printing. Even with significant challenges, 3D printing has already been applied to constructing houses, footbridges, and canals in Europe and Asia [15].

2.1.3. Use and Operation of Buildings and Infrastructure

Although it would seem that buildings and infrastructure projects during their operation and use do not have much environmental impact, the truth is that water and energy consumption, as well as the generation and management of solid waste, constitute a severe problem for the sustainability of the planet [16]. This problem has become even more acute due to the increase in the use and operation of residential buildings to prevent the spread of COVID-19, environmental contingencies associated with poor urban air quality, and even social protests that have turned homes into places of work and leisure [17]. Therefore, responsible use and consumption has become a new lifestyle paradigm for society. Responsible consumption is a concept that considers that humanity would change its consumption habits by adjusting them to its real needs and opting in to the market for goods and services that favor environmental conservation, social equality, and the welfare of the less favored classes. Responsible consumption is a way of consuming goods and services considering, in addition to the variables of price and quality, the social and labor characteristics of the production environment and the subsequent environmental consequences. Therefore, this model incorporates technological tools for the management and use of C&DW, efficient use of energy and water in buildings and infrastructure and constitutes a training tool for users of buildings and infrastructure in the search for sustainable lifestyles.

2.1.4. End of Lifecycle

Buildings and infrastructure are exposed to the mechanical action of service loads and other factors, such as environmental factors, which tend to deteriorate and destroy them (i.e., physical, chemical, and biological actions), leading to the termination of their life cycle [18]. Depending on the level of damage and economic resources, maintenance, repair, rehabilitation and repowering activities are prioritized over demolishing a building or infrastructure. These activities are prioritized because, with their relatively low cost and environmental impact, they can significantly increase the useful life of a building or infrastructure. This approach is particularly useful in urban centers, where it is also possible to preserve the architectural and historical heritage of the cities. The above premise raises the need to incorporate architectural flexibility or transformable architecture in the design of buildings and infrastructure. The history of civilizations has shown the importance of rethinking the use of buildings and infrastructure. This principle has been fundamental to solving problems ranging from food and shelter from the elements to battlefield logistics. Flexibility has been used in experimental and lightweight structures for institutions, commerce, and housing. In particular, flexible architectural design aims primarily at adapting and changing to the user's needs and the environment. This position is contrary to traditional architectural design, which, to a greater extent, is fixed or static towards the needs of man. Thus, flexible architecture redesigns itself over time because it allows it to develop, eliminate, and modify parts, keeping the structure in continuous service. The benefits of flexible architectural design include low cost, lower environmental impact, greater creative field, but, above all, it allows the development of a transitional architecture and search for ingenious solutions in times of need and scarcity, such as the one in which humanity is currently living [19].

However, in cases where, after an integrated analysis, it is found that the total or partial demolition of a building or infrastructure is necessary, the process must be carried out considering the environment and societal impacts. Therefore, it is imperative to consider demolition as the opposite process to construction, which includes well designed sequential operations. This conceptualization of the demolition process is also known today as deconstruction (Figure 2). Within the framework of the deconstruction process, selective demolition has become an advantageous technique for the utilization of CDW. The main objective of this demolition technique is to improve the classification and use at the source, extending the life cycle of construction materials, favoring reuse and generating less waste whose final destination will be the landfill. The technique consists of dismantling architectural finishes (e.g., floors, windows, doors), the heating–ventilation systems and, finally, the demolition of the structure, starting with the superstructure and then the foundations. It is important to mention that the foundation waste is challenging to use because of its continuous exposure to groundwater and soil minerals.

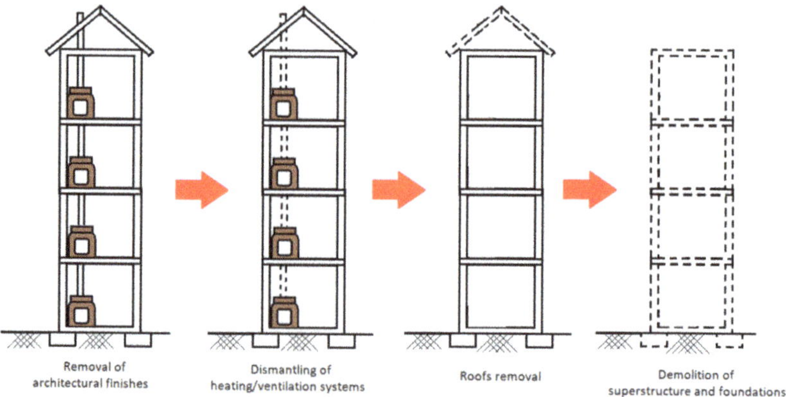

Figure 2. Sequence for the deconstruction process of a building. Adapted from [20].

3. Technological and Environmental Park (TEP) Project

Santiago de Cali's Technological and Environmental Park will be a public–private initiative that integrates highly trained human resources and a modular and flexible platform for researching and developing an innovative C&DW based construction materials portfolio. Therefore, in addition to the technology required for creating the product portfolio, a materials laboratory will be included for the adequate physical, chemical, mechanical, durability, and environmental impact characterization of the products developed. This laboratory will be linked to the region's network of universities with extensive experience in this area.

It is worth noting that the use of C&DW is developed under the principle of the 7Rs (rethink, redesign, reuse, repair, remanufacture, recycle and recover), an initiative on consumption habits, initially promoted by Greenpeace as 3Rs (reduce, reuse and recycle), which seeks to encourage habits such as responsible consumption. This concept refers to waste management strategies that seek to be more environmentally responsible, prioritizing reducing the volume of waste generated over reuse and recycling. These processes generally consume more energy and involve higher environmental and economic impacts. However, given that, in Santiago de Cali, approximately 2600 m^3/day of C&DW is generated with a percentage of 96% usable due to its ceramic nature [21], this model offers specific tools for onsite recycling (in situ) and in the TEP. Table 1 summarizes the consumption of construction materials for Santiago de Cali.

Table 1. Consumption of construction materials according to the constructive system in Santiago de Cali [21].

Materials	Industrialized Prefabricated Products (kg/m^2-%)	Structural Masonry (kg/m^2-%)	Masonry Confined with Porches (kg/m^2-%)
Aggregates	542.19–42.7	404.05–29.3	624.99–26.0
River sand	445.21–35.1	349.09–25.3	733.63–30.5
Grey cement	156.74–12.4	138.78–10.1	306.12–12.7
Rock (excavation soil)	46.6–3.7	152.24–11.0	372.52–15.4
Fired ceramics	39.98–3.2	301.28–21.9	358.08–14.9
Steel	26.68–2.1	20.31–1.5	9.44–0.4
Wood	5.02–0.4	3.58–0.26	0.13–0.0
Roof tiles	3.15–0.25	5.92–0.43	-
PVC	2.35–0.19	2.06–0.15	2.39–0.1
Copper	0.42–0.03	0.13–0.01	-
White cement	0.37–0.03	0.48–0.03	-
Paint	0.32–0.03	0.55–0.04	-

Recycling is conceived as an environmentally friendly process that seeks to convert construction and demolition waste into new products for subsequent use in the construction sector and other sectors of the regional economy. Recycling avoids the disuse of potentially valuable materials that would typically be disposed inappropriately or at a high cost. This technology reduces the consumption of new raw materials to manufacture construction materials which, according to ACODAL (2017), are dominated in Santiago de Cali by masonry confined with frames (46.4%), prefabricated (27.2%), structural masonry (22.4%) and other systems (4%). Therefore, to satisfy the demand for the above construction materials, guaranteeing their mechanical properties and durability together with a significant reduction in environmental impact, the circular economy model, through the TEP, presents the following family of products (Table 2):

(a) Recycled aggregates or eco-aggregates

In addition to the massive and traditional use of aggregates in concrete and mortar, aggregates have recently been used to generate permeable surfaces, vegetated surfaces (green roofs and green walls) and sustainable urban drainage systems (SUDS). Eco-aggregates or recycled aggregates are granular materials with physical, chemical, and mechanical properties that meet the conditions of mechanical strength and durability for the different applications described.

(b) Eco-concretes and mortars

Concrete and mortar are the most widely used building materials in the regional, national and global construction sectors. Eco-concretes seek to replace the most critical components of traditional concrete, such as aggregates (fine and coarse) and portland cement, with recycled aggregates and supplementary cementitious materials such as ashes and slag. The latter being byproducts of other productive sectors such as thermoelectric plants and agribusiness. Eco-concretes and mortars may be used in buildings and infrastructure when they meet Colombian regulations' mechanical strength and durability conditions.

(c) Eco-prefabricated products and modules

These elements belong to construction systems based on the design and production of components and subsystems that are mass produced in a factory away from their final location and brought to their final position to assemble the building or infrastructure after a relatively simple and precise assembly phase. The eco-prefabrications proposed in this manual include structural elements (e.g., beams, columns, slabs, frames) and nonstructural elements (e.g., blocks, pavers, curbs, and sidewalks).

(d) Smart construction materials

These are active and adaptive materials that can respond in an autonomous, functional, and controlled manner to changes in their condition or the environment to which they are exposed to. Self-cleaning concrete and mortars with the capacity to purify the surrounding air stand out. These materials, incorporating titanium dioxide (TiO_2) nanoparticles, when exposed to solar energy (UV-A fraction), can generate a photocatalytic process to degrade organic and inorganic pollutants, both in solid and gas phases. Such materials have been used in buildings and infrastructure worldwide.

Table 2. Materials and infrastructure considered in the product portfolio.

Material/Product	Description	Example
Recycled Aggregates	Employed during construction projects to generate permeable surfaces, green roofs (walls and roofs), and urban drainage systems. Aggregates are granulated materials with physical, chemical, and mechanical properties that fulfill specific mechanical strength and durability requirements.	
Eco-concrete and mortars	The most widely used construction materials in the regional, national, and global construction sectors. Eco-concretes seek to replace the most critical components of traditional concrete, such as aggregates (fine and coarse) and portland cement, with recycled aggregates and supplementary cementing agents such as ash and slag, the latter byproducts from other productive sectors such as thermoelectric plants and agribusiness.	
Eco-prefabricated products and modules	These elements belong to construction systems based on the design and production of components and subsystems produced in series in a factory outside their final location. They are brought into their final position to assemble the building or infrastructure after a relatively simple assembly phase and precise. Prefabricated products include structural (e.g., beams, slab columns, frames) and nonstructural (e.g., blocks, paving stones, gutters, curbs, and platforms) elements.	

Table 2. *Cont.*

Material/Product	Description	Example
Smart construction materials	Active and adaptive materials that can respond autonomously are helpful and controlled to changes in their condition or the environment they are exposed to. Self-cleaning concretes and mortars stand out, with the ability to purify the surrounding air. These materials, incorporating nanoparticles of titanium dioxide (TiO_2), when exposed to solar energy (UV-A fraction), can generate a photocatalytic process of degradation of organic and inorganic pollutants, both in the solid and gas phases.	Self-cleaning building [22]

4. Management and Use of C&DW

It is worth noting that the recycling of C&DW is not a substantially new technique in the construction sector; the first modern scientific reports date back to the 1940s in Central Europe, with a substantial increase in the 1970s due to the research developed by the Building Contractors Association (BCA) and the Ministry of Public Works (Japan) after the oil crisis. Today, this Asian country has a system for the exclusive use of C&DW, mainly for flexible pavements in sub-bases. In Taipei, interest is growing in the use of concrete waste. It is estimated that close to 90% is recovered, with 95% of old concrete being used. In the Netherlands and Finland, robust legislation has been put in place that allows almost all C&DW to be recovered, except at the production stage. More recently, 38 US states have approved recycled aggregates in road subbases, and 11 states allow it in new concrete. Brazil already has C&DW recycling plants, particularly Belo Horizonte and Sao Paulo. Although there are significant differences between the rates of C&DW utilization in northern, central, and southern Europe, due to its legislative and technological progress in the utilization (reuse and recycling) of C&DW, this region is of great interest to the world. Precisely, the European Community, which had a 28% utilization rate, intended to reach an average utilization rate of 70% for the 15 member states by 2020 [23].

In Colombia, although there is an adjustment in progress, it is expected to reach at least 2032 using 30% of usable C&DW in the total weight of the materials used for construction [24]. Considering the above, the European Standard EN 12620 (2002) is proposed to classify construction and demolition waste for the Technological and Environmental Park. With a higher level of detail than the one proposed in Resolution 472 of the Ministry of Environment and Sustainable Development (2017), this standard allows better use of C&DW. This standard is based on eight main waste groups and seven categories indicating the composition of the leading group.

4.1. Technical Assessment for New Materials and Products

The product portfolio is designed and fabricated based on the type of C&DW material, considering ashes and slag and the technical requirements such as mechanical strength and durability. Therefore, due to the mixture design of these new materials it is necessary to perform a rigorous research process that includes the first phase of physical, chemical, and mechanical characterization of each of the mixture components, highlighting the characterization of the aggregates from C&DW, ashes, slags, and new components such as nanomaterials. Using recognized mix design methods, the design and fabrication of the test mix proceed based on this information. Subsequently, the evaluation of properties before hardening for mixtures is carried out, measuring consistency and fluidity. When these mixtures satisfy the conditions for placing the product for concretes and mortars fabricated in situ (or prefabricated products), mechanical and durability evaluation in the hardened state is developed according to the demands of the building or infrastructure. Similarly, autonomic properties such as self cleaning and air purification are evaluated in the case of smart building materials. Figure 3a summarizes the process for the technical assessment of mixtures from C&DW.

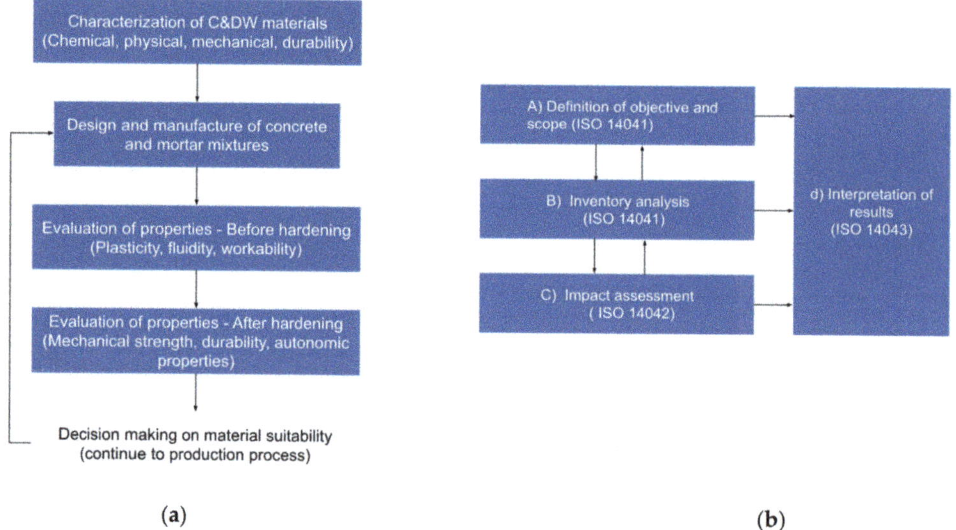

Figure 3. (**a**) Diagram for the technical assessment for developing new materials and products in the TEP of Santiago de Cali. (**b**) LCA process to study the TEP product portfolio Adapted from [23].

Although the mixture design methods serve as a guide for an adequate proportioning of the components, the design process with recycled aggregates, other residues and nanomaterials demands a recurrent adjustment depending on the characteristics in the state before and after hardening until the final mix design and the manufacturing process are obtained. The reliability of the results is based on the experimental design that uses the comparison with concrete and mortar mixtures, which, using natural aggregates and conventional cement, meet the same mechanical and durability requirements.

In addition to the technical assessment, a life cycle analysis (LCA) will be developed to measure and improve the sustainability performance of new products and materials. This tool assesses a product or service's environmental and social impacts during all stages of its existence: extraction, production, distribution, use, and end of the life cycle. LCA involves quantifying resource consumption (energy, water, and materials) and environmental emissions to air, water, and soil associated with the system being evaluated. According to the ISO 14040-14043 standards, the LCA consists of four stages: (a) definition of the objective and scope, (b) creation of the inventory, (c) evaluation of the impact and (d) interpretation of results (Figure 3b).

The first stage of the definition of the objective and the scope must define the application and intended use of the results and users (target audience). The typical objectives of an LCA study are to compare two or more products that fulfill the same function (e.g., concrete with natural aggregates and concrete with recycled aggregates), to identify possibilities for the improvement of existing products, or even the innovation and design of new products. The definition of the scope of an LCA study involves the establishment of the limits of the evaluation. The following elements should be clearly described in the scope definition: the system to be studied and its function, the functional unit, the system boundaries, types of impact and impact assessment methodology, data quality requirements, assumptions and limitations [25].

Inventory analysis is the second stage in an LCA. This involves data collection and calculation procedures to quantify a product system's relevant inputs and outputs. These inputs and outputs include using resources, emissions to air, water, and soil, and generating waste associated with the system. The inventory analysis must be supported by a process tree (process diagram, flow tree) that defines the phases in the life cycle of a product. Each of the different phases can be composed of the different unit processes, for example, production with different types of raw materials to combine in the production phase of the material. Transport processes often connect the different phases. Data on material, water, energy consumption, waste, and emissions must be collected for all process units in a product's life cycle [23].

In the third stage, the potential environmental impacts of the modeled system are evaluated. This stage consists of three mandatory elements: (i) selection of impact categories, category indicators, and characterization models, (ii) classification and (iii) characterization. The impact categories are selected to describe the impacts caused by the analyzed product or product family. This is a follow up to the decisions made in the targeting and scoping phase. Some of the impact categories that are usually considered are the consumption of nonrenewable sources, water consumption, global warming potential, ozone layer depletion potential, eutrophication potential, acidification potential, the potential for the formation of smog, human toxicity (carcinogenic and noncarcinogenic), ecological toxicity, waste generation, land use, air pollution and alteration of habitats.

Finally, the interpretation of results is the fourth phase of the LCA. It includes the following main aspects: identification of significant environmental problems, evaluation of results to establish their reliability (integrity, sensitivity, and coherence), conclusions and recommendations.

4.2. Technology for Obtaining Recycled Aggregates from C&DW

The process that seeks to separate materials from paper, polymers, wood, steel, soil and other contaminants must follow several steps to guarantee the obtention of adequate recycled aggregates. Figure 4 shows the overall process to separate different materials from C&DW. This process should include previous weighing and drying, important information and condition, respectively, to manage the following steps. Particularly, weighing can be carried out by a weighing scale for trucks or a load cell in hoppers. In general, the selection of technologies for the process are recommended using criteria such as initial investment, energy consumption, dust emissions, maintenance cost and useful life.

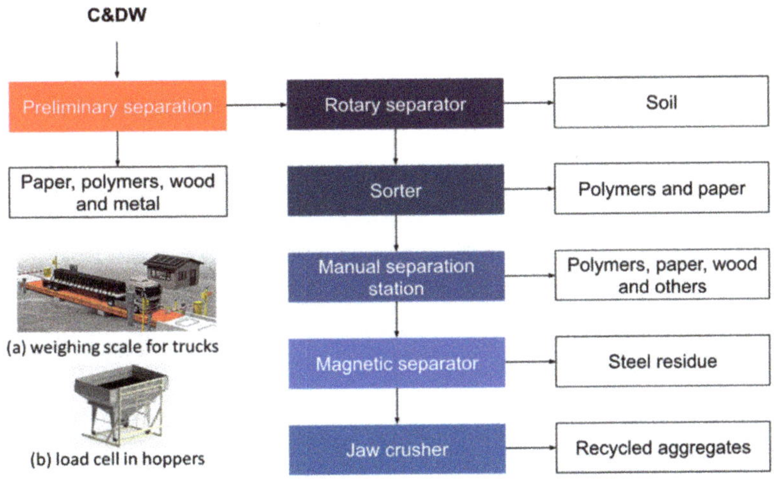

Figure 4. Separation process for the production of recycled aggregates from C&DW. Adapted from [20].

5. Case Studies

The product families and methodologies proposed in the circular economic model for construction are mainly supported by previous research in the Master Program of Civil Engineering at the Pontificia Universidad Javeriana de Cali. Below are some case studies of great value for the proposed circular economic model (Figure 5).

Figure 5. Family of ecoproducts from the TEP of Santiago de Cali.

5.1. Paving Stones

In this research project, the resistance to the bending and compression of nine concrete mixtures (called M1 to M9) with partial and total replacements of the fine and coarse natural aggregate by recycled aggregate from the crushing of concrete paving stones from the Plaza of the Municipality of Almaguer (Cauca, Colombia) were evaluated. The methodology followed included five phases, from sampling the aggregates to selecting the concrete mixes with the best mechanical performance for their use in new paving stones. The results showed that the concrete mixtures replacing 50%, by weight, of the fine natural fraction (called M2) and 50%, by weight, of the natural coarse fraction (called M4) meet the strength required by the NTC-2017 to manufacture new paving stones. Although techniques should be investigated to improve the wear resistance of concrete mixtures when used as paving stones, the percentages of use of recycled aggregates presented in this project concerning those reported in the literature are much higher. Therefore, the solution proposed in this project not only has the potential to significantly reduce the negative environmental impact caused by the improper disposal of construction waste and extraction of nonrenewable resources, but it also has the potential to reduce costs in construction projects [9].

5.2. High Strength Concrete

This research aimed to evaluate the effect of the partial replacement of natural coarse aggregate by recycled concrete aggregate (RCA) on the mechanical properties of high strength concrete (HSC). An experimental methodology composed of five stages was followed: the selection of the concrete residues and subsequent production of the RCA, the characterization of the materials used, the manufacture of concretes with different contents of RCA (0, 10, 20, and 40%), the determination of the properties in the fresh and hardened state of the above mentioned concretes and, finally, the analysis of the results. It was concluded that the developed HSC not only satisfies the specifications of the Colombian regulations, the incorporation of recycled aggregate also had a positive effect on the mechanical and durability properties of the concretes, obtaining the best performance with the mix using 40% RCA. Therefore, using HSC with RCA as a construction material is technically feasible, positively impacting the environment by reducing the exploitation of nonrenewable natural resources and extending useful life spans of buildings and infrastructure [26].

5.3. Prefabricated Products

This project included the diagnosis of construction waste generated in civil projects in the urban area of the city of Pereira (Colombia). Once classified onsite, C&DW materials were incorporated into prefabricated concrete elements in the same place where they are generated, contributing to the reduction in the exploitation of the natural resource, lower emissions of greenhouse gases, and decrease in land occupation for the disposal of unused waste.

During the project's development, the construction waste from three pilot projects was classified and quantified. From this, components from concrete and natural aggregate were recycled, for which the most relevant physical and mechanical properties were determined through laboratory tests. New concrete mixtures were made with these recycled aggregates, whose mechanical behavior was established through compression and flexural strength tests. The production costs of the recycled aggregate obtained within the source where they were generated were reviewed and compared with market prices, involving activities from the initial phase of the classification process. Some prefabricated elements were manufactured with several samples of these mixtures, and their mechanical behavior was analyzed. The degree of technical and economic viability was finally established for using these recycled aggregates within the same construction activities from which they were obtained and circumscribed to the local conditions of the region related to the city of Pereira [27].

5.4. Mortar

Fly ash from the manufacturing process of the paper industry was used for this research because this possesses physical and chemical properties that make them suitable for reuse as substitutes for raw materials for modified mortars, proving the case of the substitution of Portland cement, evaluating the effect that this substitution has on the physical and mechanical properties of the mortar. The research covered the characterization of the raw materials that make up the mortar, the determination of the mechanical properties of the mortar, such as compressive resistance, and the evaluation of the modified mortar as a joining material for structural walls. This study showed that replacing cement in percentages greater than 10% of fly ash yields a mixture with greater workability but less resistance to traditional mortar. It is concluded that using ash from the paper industry to prepare mortars is mechanically feasible since this material is used in conventional block masonry [28].

5.5. Green Roofs

In this research, the hydraulic, thermal and mechanical performance, impact on dead loads, and costs associated to four semi-intensive green roof systems in which recycled (rubber and HDPE plates) and reused materials (PET bottles) were used for the drainage system were evaluated. Then, results were compared with the conventional drainage system that uses the aggregate of natural origin (basalt gravel). For the evaluated environmental conditions, the results showed that some systems (e.g., recycled rubber) can be more useful when the green roof application intends to reduce the temperature, and others (e.g. HDPE plates) when it is the water retention capacity. In addition, the developed green roof systems, using recycled and reused materials, showed the potential to reduce dead loads and costs compared to traditional green roofs [29].

5.6. Air-Purifing Systems

It is known that air pollution has a direct negative effect on the quality of life of people and ecosystems (including infrastructure). Considering that one of the main sources of pollutant at the urban level are vehicles, different transit strategies to reduce the use of vehicles have been implemented without much success in Colombia. For this reason, in this project, a system was designed to generate air purification and self-cleaning in the Colombia Avenue Tunnel in Santiago de Cali (Figure 6). To achieve the last objective, a mortar and a photocatalytic coating were evaluated, technically and financially, using TiO_2 and artificial light with UV-A rays. In this case, to monitor photoactivity, two artificial colorants (methylene blue and rhodamine b) were applied to the surface of the photocatalytic materials, which were exposed in a controlled manner to UV-A rays using a photoreactor. The removal of the mentioned dyes was followed by digital image analysis with ImageJ software. The efficiencies obtained in removing these colorants, which are indicators of the degradation of organic and inorganic pollutants, were simulated in the Street Canyon model (developed by the National Institute for Environmental Research of Denmark), finding promising results for the removal of nitrous oxides, sulfur dioxide, carbon monoxide, and total hydrocarbons [30].

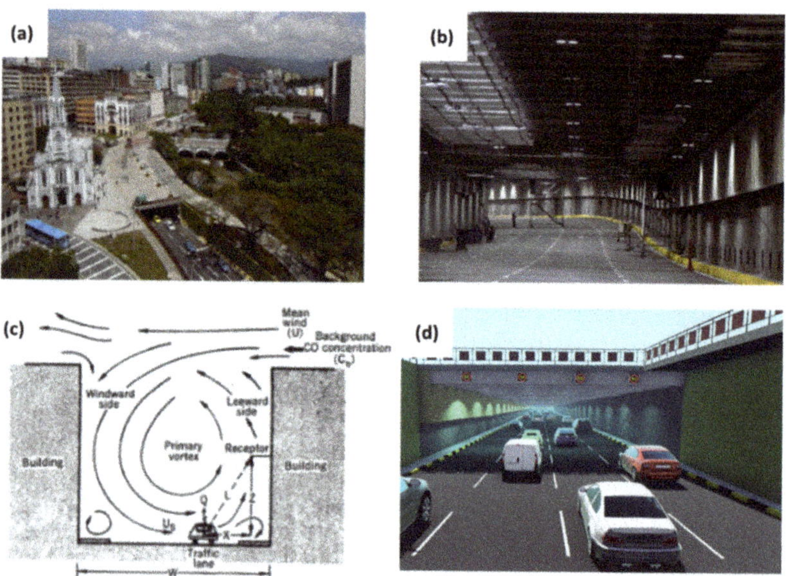

Figure 6. Design of air purification and self-cleaning system in the Colombia Avenue Tunnel in Santiago de Cali. (**a**) Location of the tunnel, (**b**) geometric characteristics of the tunnel, (**c**) model of the air quality used, (**d**) model of the lighting system and photocatalytic coating. Based on [30].

6. Challenges to Face

The proposed circular economy model demands a high degree of innovation for the correct formulation and implementation of concepts and technologies such as the valuation of extraction of environmental assets, the clean production, building information modeling (BIM), robotization, 3D printing, responsible consumption, architectural flexibility, transformable architecture, deconstruction, recycling, and industrial symbiosis. Therefore, the articulated and continuous work between civil society, the government, the business sector, and the academy is proposed to guarantee the proper development of an ecosystem of research, creation, development, innovation, and entrepreneurship.

As a new paradigm of economic development crucial for the postpandemic era, the circular economy generates excellent opportunities for the articulation of the construction sector with vulnerable communities and the environment. In this sense, the proposed model must consider the analysis of energy and water flows in subsequent development and implementation stages, as well as gaseous emissions and particulate matter [31]. Likewise, through tools such as the social life cycle assessment (S-LCA), the assessment of social and sociological aspects are recommended from the production of construction materials, construction processes, use and operation, and completion of the life cycle of buildings and infrastructure [32,33].

In addition, in Colombia, legislation is still under development in C&DW valuation. Therefore, more initiatives are required to accelerate the transition towards a more sustainable ecosystem in the construction sector. Tax benefits and economic support can be implemented to increase the number of companies involved in the construction sector (including logistics, material processing, construction, among others).

7. Conclusions

This article presents a technical proposal to manage and increase the value of C&DW in Santiago de Cali (Colombia) through the development of a Technological and Environmental Park to offer a valuable product portfolio for new construction projects and include benefits in terms of environmental impact more sustainable lifecycle, and cost reduction. Coming from academia, case studies demonstrated that the Colombian southwest region has potential to research and develop new building material products made up of C&DW. This potential can be increased by creating a sustainable production ecosystem of all companies related to the construction sector, together with disposal actors. Similarly, developing a conscious sustainable users' community will drive the product portfolio. In this way, this model becomes a tool for the business sector, academia and the community, sponsored by the government, particularly for this actor the model becomes a tool of performance measurement and benchmarking, consumers want to make the right environmental choices when buying products. Policy makers want to promote sustainable consumption and production to respond to national and international environmental challenges. In addition, businesses want to improve efficiency to boost margins and competitiveness, while contributing to a sustainable society [34,35]. In later stages, the circular economy model proposed for the construction sector of Santiago de Cali must strengthen industrial symbiosis through the valorization of waste of a different nature than ceramic (i.e. C&DW). For example, the environmental use of polymeric and metallic waste from construction is strategic to reduce the impact of the families of products that make up the TEP portfolio. However, in this first stage, the current construction systems demand will potentially guarantee the model's success with the proposed product families based only on C&DW.

Author Contributions: Conceptualization and methodology, A.M.-R.; validation, D.I.-P., and J.A.M.; formal analysis, A.M.-R., D.I.-P. and J.A.M.; investigation and resources, A.M.-R.; data curation, D.I.-P. and J.A.M.; writing—original draft preparation, D.I.-P. and J.A.M.; writing—review and editing, D.I.-P.; visualization, D.I.-P.; supervision, A.M.-R. and J.A.M. All authors have read and agreed to the published version of the manuscript.

Funding: This research was partially funded by Alcaldía de Santiago de Cali.

Institutional Review Board Statement: Not applicable.

Informed Consent Statement: Not applicable.

Data Availability Statement: Not applicable.

Acknowledgments: The authors would like to thank the Engineering Faculties of Universidad El Bosque, Universidad de La Sabana, and Pontificia Universidad Javeriana for their support during the development of this project. Additionally, the authors would like to thank the financial support of Alcaldía de Santiago de Cali.

Conflicts of Interest: The authors declare no conflict of interest.

Abbreviations

ACODAL: Colombian Association of Sanitary and Environmental Engineering, BCA: Building Contractors Association, BIM: building information modeling, CE: circular economy, CAMACOL: Colombian Chamber of Construction, C&DW: construction and demolition waste, HDPE: high density polyethylene, HSC: high strength concrete, LCA: life cycle analysis, MIT: Massachusetts Institute of Technology, NSR-10: Colombian Regulation of Earthquake Resistant Construction, PET: polyethylene terephthalate, PCA: Portland Cement Association, RMC: ready mixed concrete, RCA: recycled concrete aggregate, S-LCA: social life cycle assessment, SDGs: sustainable development goals, SUDS: sustainable urban drainage systems, TEP: Technological and Environmental Park.

References

1. Informe de Gestión 2018–2019—Camacol.co. Available online: https://camacol.co/sites/default/files/descargables/Informe%20de%20Gesti%C3%B3n%202018-2019.pdf (accessed on 10 December 2021).
2. Plan de Ordenamiento Territorial—POT año 2014. Available online: https://www.cali.gov.co/planeacion/publicaciones/106497/pot_2014_idesc/ (accessed on 10 December 2021).
3. La economía Circular y su Papel en el Cambio Climático. Available online: https://www.elespectador.com/actualidad/la-economia-circular-y-su-papel-en-el-cambio-climatico/ (accessed on 10 December 2021).
4. Economía Circular—ANDI. Available online: http://www.andi.com.co/Uploads/economia-circular-1-reporte.pdf (accessed on 10 December 2021).
5. Charef, R.; Morel, J.-C.; Rakhshan, K. Barriers to Implementing the Circular Economy in the Construction Industry: A Critical Review. *Sustainability* **2021**, *13*, 2989. [CrossRef]
6. Zhang, N.; Han, Q.; de Vries, B. Building Circularity Assessment in the Architecture, Engineering, and Construction Industry: A New Framework. *Sustainability* **2021**, *13*, 2466. [CrossRef]
7. Agrawal, R.; Wankhede, V.A.; Kumar, A.; Upadhyay, A.; Garza-Reyes, J.A. Nexus of circular economy and sustainable business performance in the era of digitalization. *Int. J. Product. Perform. Manag.* **2021**. ahead-of-publication. [CrossRef]
8. Sawe, F.B.; Kumar, A.; Garza-Reyes, J.A.; Agrawal, R. Assessing people-driven factors for circular economy practices in small and medium-sized enterprise supply chains: Business strategies and environmental perspectives. *Bus. Strateg. Environ.* **2021**, *30*, 2951–2965. [CrossRef]
9. Bravo-German, A.M.; Bravo-Gómez, I.D.; Mesa, J.A.; Maury-Ramírez, A. Mechanical Properties of Concrete Using Recycled Aggregates Obtained from Old Paving Stones. *Sustainability* **2021**, *13*, 3044. [CrossRef]
10. MIT Concrete Sustainability Hub. Available online: https://www.cement.org/learn/mit (accessed on 10 December 2021).
11. Muñoz Muñoz, H.A. *Construcción, Interventoria y Supervisión Técnica de las Edificaciones de Concreto Estructural: Según el Reglamento Colombiano NSR-10/*; Asociación Colombiana de Productores de Concreto-ASOCRETO: Bogotá, Colombia, 2015; ISBN 9789588564135.
12. Mesároš, P.; Mandičák, T. Exploitation and Benefits of BIM in Construction Project Management. *IOP Conf. Ser. Mater. Sci. Eng.* **2017**, *245*, 62056. [CrossRef]
13. Doumbouya, L.; Gao, G.; Guan, C. Adoption of the Building Information Modeling (BIM) for Construction Project Effectiveness: The Review of BIM Benefits. *Am. J. Civ. Eng. Archit.* **2016**, *4*, 74–79. [CrossRef]
14. Kim, S.; Hong, W.-K.; Kim, J.-H.; Kim, J.T. The development of modularized construction of enhanced precast composite structural systems (Smart Green frame) and its embedded energy efficiency. *Energy Build.* **2013**, *66*, 16–21. [CrossRef]
15. Hager, I.; Golonka, A.; Putanowicz, R. 3D Printing of Buildings and Building Components as the Future of Sustainable Construction? *Procedia Eng.* **2016**, *151*, 292–299. [CrossRef]
16. Hussin, J.M.; Abdul Rahman, I.; Memon, A.H. The Way Forward in Sustainable Construction: Issues and Challenges. *Int. J. Adv. Appl. Sci.* **2013**, *2*, 15–24. [CrossRef]
17. Li, L.; Li, Q.; Huang, L.; Wang, Q.; Zhu, A.; Xu, J.; Liu, Z.; Li, H.; Shi, L.; Li, R.; et al. Air quality changes during the COVID-19 lockdown over the Yangtze River Delta Region: An insight into the impact of human activity pattern changes on air pollution variation. *Sci. Total Environ.* **2020**, *732*, 139282. [CrossRef] [PubMed]
18. de De Freitas, V.P.; Delgado, J.M.P.Q. *Durability of Building Materials and Components*; Springer: Berlin/Heidelberg, Germany, 2013; Volume 3, ISBN 3642374751.
19. Pinto Campos, B.C. Arquitectura y Diseño Flexible Una Revisión Para una Construcción más Sostenible. Ph.D. Thesis, Universitat Politècnica de Catalunya, Barcelona, Spain, 2019; pp. 1–309.
20. Pacheco-Torgal, F.; Cabeza, L.F.; Labrincha, J.; De Magalhaes, A.G. *Eco-Efficient Construction and Building Materials: Life Cycle Assessment (LCA), Eco-Labelling and Case Studies*; Woodhead Publishing: Cambridge, UK, 2014; ISBN 0857097725.

21. Caracterización Exhaustiva de los Residuos de Construcción y Demolición (RCD) que Ingresan a la Estación De Trasferencia de la Carrera 50 en la Ciudad de Cali e Identificación de su Potencial de Transformación en Eco-Productos para la Construcción. Available online: https://www.cali.gov.co/serviciospublicos/publicaciones/146621/caracterizacion-exhaustiva-de-los-residuos-de-construccion-y-demolicion-rcd-que-ingresan-a-la-estacion-de-trasferencia-de-la-carrera-50-en-la-ciudad-de-cali-e-identificacion-de-su-potencial-de-transformacion-en-eco-productos-para-la-construccion/ (accessed on 10 December 2021).
22. Maury Ramirez, A. *Cementitious Materials with Air-Purifying and Self-Cleaning Properties Using Titanium Dioxide Photocatalysis*; Ghent University: Ghent, Belgium, 2011.
23. Reciclaje y Cierre del ciclo de Vida de las Placas de yeso Laminado. Available online: https://www.interempresas.net/Reciclaje/Articulos/109556-Reciclaje-y-cierre-del-ciclo-de-vida-de-las-placas-de-yeso-laminado.html (accessed on 10 December 2021).
24. Ministerio de Ambiente y Desarrollo Sostenible. Resolución No. 0404. Available online: https://www.minambiente.gov.co/wp-content/uploads/2021/08/resolucion-0404-de-2021.pdf (accessed on 10 December 2021).
25. Pacheco-Torgal, F.; Jalali, S.; Labrincha, J.; John, V.M. *Eco-Efficient Concrete*, 1st ed.; Woodhead Publishing: Cambridge, UK, 2013.
26. Diosa, J.S. Experimental Analysis of the Mechanical Properties of High-Strength Concrete Made with Recycled Concrete Aggregates. Pontificia Universidad Javeriana Cali, Cali, Colombia, 2020.
27. Londoño, J. *Mechanical Behavior of Precast Concrete Elements with Recycled Aggregates within the Source that Generates Them*; Pontificia Universidad Javeriana Cali: Cali, Colombia, 2016.
28. Hurtado, I. *Evaluation of the Effect of the Partial Replacement of Portland Cement by Ash from the Paper Industry in Mortars and Walls*; Pontificia Universidad Javeriana Cali: Cali, Colombia, 2019.
29. Naranjo, A.; Colonia, A.; Mesa, J.; Maury-Ramírez, A. Evaluation of Semi-Intensive Green Roofs with Drainage Layers Made Out of Recycled and Reused Materials. *Coatings* **2020**, *10*, 525. [CrossRef]
30. Medina Medina, A.F.; Torres Rojas, D.F.; Meza Girón, G.; Villota Grisales, R.A. *Design of a System to Generate Air Purification and Self-Cleaning in the Surfaces of the Colombia Avenue Tunne*; Pontificia Universidad Javeriana Cali: Cali, Colombia, 2016.
31. Mesa, J.A.; Fúquene-Retamoso, C.; Maury-Ramírez, A. Life Cycle Assessment on Construction and Demolition Waste: A Systematic Literature Review. *Sustainability* **2021**, *13*, 7676. [CrossRef]
32. Life Cycle Initiative. Available online: https://www.lifecycleinitiative.org/wp-content/uploads/2021/01/Guidelines-for-Social-Life-Cycle-Assessment-of-Products-and-Organizations-2020-22.1.21sml.pdf (accessed on 10 December 2021).
33. Dong, Y.H.; Ng, S.T. A social life cycle assessment model for building construction in Hong Kong. *Int. J. Life Cycle Assess.* **2015**, *20*, 1166–1180. [CrossRef]
34. da Cruz, N.F.; Marques, R.C. Scorecards for sustainable local governments. *Cities* **2014**, *39*, 165–170. [CrossRef]
35. European Commission, Joint Research Centre. *Making Sustainable Consumption and Production a Reality. A Guide for Business and Policy Makers on Life Cycle Thinking and Assessment*; European Union: Brussels, Belgium, 2010.

Article

Challenges and Opportunities for Circular Economy Promotion in the Building Sector

Rafaela Tirado [1,2,*], Adélaïde Aublet [1], Sylvain Laurenceau [1] and Guillaume Habert [2]

[1] Scientific and Technical Centre for Buildings (CSTB), University Paris-Est, 24 Rue Joseph Fourier, 38400 Saint-Martin-d'Hères, France; adelaide.aublet@cstb.fr (A.A.); sylvain.laurenceau@cstb.fr (S.L.)
[2] Sustainable Construction Department, IBI, ETH Zurich, Stefano-Franscini-Platz 5, 8093 Zürich, Switzerland; habert@ibi.baug.ethz.ch
* Correspondence: tdeanira@student.ethz.ch

Abstract: The accelerated development of cities involves important inflows and outflows of resources. The construction sector is one of the main consumers of raw materials and producers of waste. Due to its quantity and potential for recovery, waste from the construction sector constitutes significant deposits and requires major action by bringing together different stakeholders to achieve the objectives of a circular economy. Consequently, it is crucial to understand the current knowledge of urban metabolism, deposits, and recovery practices. This article aims to investigate the role of local authorities in the planning of strategies to facilitate a circular economy; in particular, this article aims to answer how local authorities facilitate circular economy initiatives in the building sector and what opportunities and obstacles they encounter in the process. The strategy used for the study was to conduct semistructured interviews with those responsible for circular economy projects within local authorities that were pioneering circular economy projects in metropolitan France. The results highlight the importance of community involvement in the implementation of circular economy principles in the building sector. Thus, it is essential to identify the different stakeholders and their respective challenges to build an operational framework.

Keywords: circular economy; local authorities; urban metabolism; interview; building

Citation: Tirado, R.; Aublet, A.; Laurenceau, S.; Habert, G. Challenges and Opportunities for Circular Economy Promotion in the Building Sector. *Sustainability* **2022**, *14*, 1569. https://doi.org/10.3390/su14031569

Academic Editors: Anibal C. Maury-Ramirez and Jaime A Mesa

Received: 16 December 2021
Accepted: 21 January 2022
Published: 28 January 2022

1. Introduction

The activities of the construction sector, which are still based on a linear economic model [1], are mainly responsible for Greenhouse Gas (GHG) emissions, the depletion of natural resources, and the production of a considerable quantity of waste [2–4]. The ecological context and recent health crisis have highlighted the imperative need for a more sustainable, circular, resilient, and inclusive economy. Circular Economy (CE) theory is based on the efficiency and optimization of the use of resources and the reduction of waste throughout the life cycle of goods and products while creating economic opportunities [5]. Because of its ecological and socioeconomic impact, the construction sector is considered a sector with a high potential to generate value and take advantage of practices at several scales.

Resource use efficiency has traditionally focused on production and consumption [6]; however, territories have the ability to manage and implement larger-scale CE strategies thanks to their roles in, for instance, urban planning and their relationship with economic players and their consequential understanding and mastery of urban metabolism. At the territorial level, the application of CE in the built environment requires collaborative, transdisciplinary work and multiscalar and prospective reflection to develop and apply strategies for better consumption, construction (production), and waste management. Consequently, local authorities can become the main catalysts for the development of economic dynamics, given their role within the territories.

In this context, we interviewed various project managers involved in this transition toward a CE, including local authorities that are pioneering the application and development of CE strategies in the construction sector because they recognize the challenges and opportunities of applying a CE due to the initiatives already undertaken in their jurisdictions. Our main goal was to understand their involvement in CE initiatives, strategies, and projects in their jurisdictions and to identify their main needs to encourage the application of CE in building sector projects. Identifying their current and future challenges and opportunities will allow others to be inspired and, above all, to proactively anticipate the challenges and opportunities offered by the transition to CE. This article focuses on the opportunities and challenges encountered by interviewed CE project managers in applying CE principles in the construction sector.

1.1. Circular Metabolism

Circular metabolism is the result of the fusion of CE and urban metabolism. The CE concept was first mentioned in the report "The potential for substituting manpower for energy" by the European Commission [7], which listed the reduction of energy consumption and job creation as objectives and concluded with a definition of the structure and nature of an "economy in loops". During the 2000s, in the context of the fight against climate change, the notion of CE took on much greater importance, based on the "cradle to cradle" theory ("from cradle to cradle") [8]. Then, the Ellen MacArthur Foundation, created in 2010, conceptualized the notion of CE that was later used in the 2013 reports of the European Commission [9]. In France, the concept of CE was widely disseminated during the Grenelle Environment Forum in 2007 [10]. Since then, the CE has been part of the public policies of state and local authorities as well as all of the city's stakeholders (LTEVEC law [11], EPCi law [12]). On February 10, 2020, the French Senate adopted Law No. 2020-105: the Anti-waste and the circular economy (AGEC) law [13]. The metaphor of urban metabolism has been widely used to describe territories as organisms that require resources to support their activities and generate waste during transformation processes. This interest in the study of urban metabolism has enabled the multiscale analysis of the flow and stock of resources and waste in a territory [14–22]. Circular metabolism, therefore, refers to the circulation of the flow of resources on a territorial scale so that the inputs of external resources are minimized to give rise to internal circular practices, in this way reducing environmental impacts and promoting sustainable and resilient territories.

1.2. Circular Economy for the Construction Sector

The building sector in France is responsible for a quarter of the national GHG emissions and consumes 43% of the total energy [23]. It also produces a large part of the construction waste, 40 million tons of waste per year on average [24]. This waste, which is mainly generated by demolition activities [25], partly feeds illegal dumps that are a real environmental and economic problem for communities [26].

Consequently, the AGEC law requires the construction sector to (1) promote the treatment of construction materials, equipment, and products during rehabilitation and demolition so that they do not become waste; (2) increase the use of reused materials in building construction or renovation projects; (3) plan and manage the deployment and networking of CE equipment (storage platforms, sorting centers, resource centers, and recycling facilities).

In summary, the implementation of the strategies of the AGEC law aims to reduce the flow of materials in the territories and the consequential consumption of resources and waste production through a value-retention process of resources (Figure 1). Estimation of resource requirements for built-up areas of the territory, identification of secondary resource deposits, quantification and identification of flows, as well as the development of valuation channels for secondary materials and land planning that will accommodate these resources, therefore, represent important issues for public authorities and communities to address in order to establish an efficient CE system.

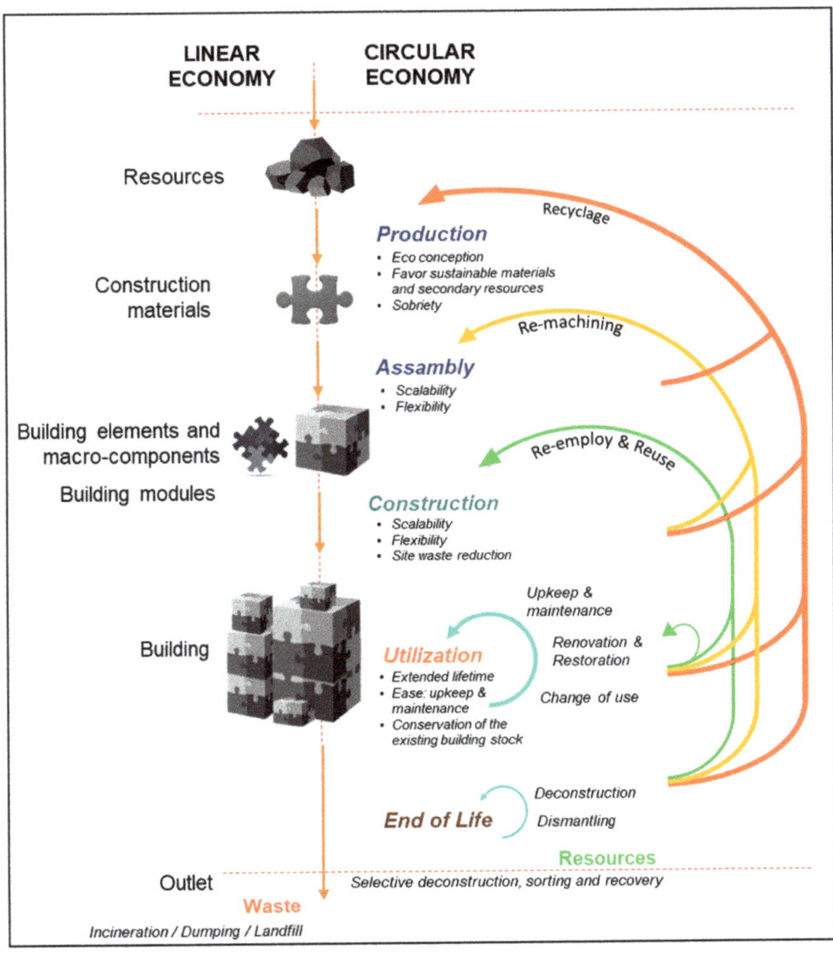

Figure 1. Levels of resource valuation for the construction sector. Inspired from the Ellen MacArthur Foundation, Circular economy diagram [27].

1.3. Role of Territorial Authorities

A local authority maintains various roles within its territory: (1) as the contracting authority, it manages the heritage of the territory; (2) it provides a financial boost in supporting economic activities, for instance, by linking community members and training activities; (3) as social housing financiers, the community finances the building or rehabilitation of social housing. In this context, the community has levers to mobilize landlords to incorporate CE practices; (4) as a planner, since the community manages land rights, it intervenes within the framework of local urban planning, for instance, in building construction or demolition permits. It can therefore facilitate, for instance, land for the establishment of secondary resource storage platforms; (5) as an administrator and pilot, the local authority signs contracts with planners to develop the territory. The town planning department manages these operations, and CE objectives can be included; (6) a local authority can deploy synergies between community members to codevelop strategies to apply a CE to the sector.

Faced with the challenges of transitioning to a CE, communities can promote and apply the 3R (reduce, reuse, and recycle) concept to building materials during their own renovation or deconstruction work, that is, to act as an example. Another advantage is their leadership role, with which they can facilitate partnerships between different stakeholders in the development of their city and various other pioneering cities to promote a CE. They can also support technical developments by implementing tax breaks, for example, and promoting bottom-up initiatives (e.g., innovative ideas, exemplary projects).

1.4. Territorial Diagnosis

CE strategies at the territorial level revolve around several stages: a diagnosis of the territory, a roadmap, public initiation and support policies, and evaluation. To understand circularity, knowledge of a territory's flows is essential because the reuse and recycling channels require securing a supply of resources. A territorial diagnosis is applied to identify and quantify the material stocks and flows in a territory; this diagnosis will answer questions such as: What is the nature and quantity of material stocks and flows? What are the characteristics of waste flows, and how can they be reduced? Spatially and temporally, what are the characteristics and availabilities of resource deposits? What percentage of the flow is, for instance, revalued or recycled? What are the principal recycling and reuse channels in the territory? Which economic actors are involved?

Analysis of urban mines allows the identification of the materials present in the built stock in a territory and the deposits they form. A deposit comprises resources with similar characteristics sharing the same units of the constructive system, time, and space [28]. The material characterization of the stock and flows of buildings and deposits represents one of the main challenges for applying the CE. It represents the basis for the formulation of strategies for employment, opportunities, and the technical conditions for reintegrating resources into the economic loop by preventing them from becoming waste; because the better a material is known, the more efficient is its outlet.

2. Materials and Methods

Information was gathered by following a qualitative semistructured interview format [29,30]. This interview format allows the respondents' perceptions of complex issues such as the challenges and opportunities of the projects they participate in to be captured. The choice to follow a semistructured interview format was also made because it allows the respondents to express themselves freely and thus evoke subjects or problems that would not have been anticipated or identified in the formulation of our questions. Interviews were conducted by telephone or video conference. Representatives of three metropolitan areas or regions in France were interviewed, with 1 to 2 people interviewed at a time; details are given in Table 1. A total of 4 interviews were performed. The interviewees were experts and key people in circular economy projects, mainly managers of circular economy projects in territories that were pioneering CE strategies in their construction sector. Those interviewees were chosen because they presumably had a better viewpoint and understanding of the challenges and opportunities, thanks to the initiatives they had already carried out or that were in progress at the time of the interview. Interviews lasted approximately 45 min to 1.5 h. During the interviews, the interviewer took notes. In addition, most of the interviews were recorded if the interviewee agreed, and some interviews were transcribed if needed.

Table 1. Roles and locations of interviewees.

Reference	Role	Location
I–1	CE and sustainable development project manager	Grenoble Metropolis
I–2	CE and costs studies project manager	Paris City hall
I–3	CE project manager	Lyon metropolis
I–4	CE and energy transition project manager Urban and territorial ecology researcher	Paris Region Institute

The interview guide (Appendix A) details the instructions for the interviewer as well as the semistructured questions related to (1) the interviewees' background in the subject of CE; (2) the projects in which they have participated or are currently participating, project(s) in the city or region where CE strategies are applied, and more specifically, projects in the construction sector; (3) how the projects unfolded, the successes and obstacles encountered, how difficulties were approached, and their recommendations for facing these challenges. Finally, the interviewees were invited to mention their needs to improve the application of a CE in the construction sector at different territorial scales.

Principles of theoretical thematic analysis were used to analyze the information collected from the interviews [31]. The information was grouped into clusters under six key themes. Under these six themes, more specific subcategories were distinguished based on material obtained from the interviews. The analysis was collaboratively and iteratively performed by the coauthors to correctly interpret the information and reach a common understanding.

3. Challenges and Opportunities

3.1. Consideration of a Buildings' Life Cycle

3.1.1. Avoiding Deconstruction

A building is usually constructed for an intended lifespan of 50 years; however, its use will influence its longevity. The lifespan of a residential building is estimated to be between 70 and 100 years, while for a logistics or industrial building, it is only estimated to last 30 or 40 years [32]. Such a difference is not related to the structural resistance of the building but rather its profitability in terms of satisfying the needs of its investors; therefore, these buildings are often demolished or deconstructed before reaching their planned lifespan. As building construction is often inflexible and not adaptable to changes in future uses, buildings that are still in good condition undergo heavy renovations or are deconstructed for new construction. Interviewee I–3 mentioned: "the main issue is not waste, but how can we avoid deconstruction? Because today we sometimes deconstruct buildings when we do not need to". Therefore, today, deconstruction practices only strengthen the linear economy and increase environmental impacts. Consequently, deconstruction is still a sensitive issue because it is currently a common practice and can take yet some time to change if stakeholders are not aware of the subject and if they do not enact strategies to communicate and change this practice. In the context of circularity, buildings are intended to be preserved instead of deconstructed through regular maintenance, restoration, and renovation activities. Deconstructing a building should be the last resort if the building can offer adequate structural and sanitary conditions. Moreover, when this stage has arrived, it is necessary to institute a selective dismantling and deconstruction process because dismantling an element without damaging it allows it to be reinserted in the circular chain, allowing resources to be better separated and their reuse and recycling to be optimized.

3.1.2. Selective Deconstruction

Obtaining secondary materials at the end of a project and preventing them from becoming waste requires direct action at the source of the deconstruction sites by carrying out a selective deconstruction of building components. Selective deconstruction consists of a sequence of activities to help separate and sort building elements and materials [33]. Article 74 of the AGEC law [13] requires that "Any producer or holder of construction and demolition waste sets up sorting of waste at the source and when the waste is not treated on-site, a separate collection of wastes, especially for wood, mineral fractions, metal, glass, plastic, and plaster".

Successful selective deconstruction requires rigorous upstream preparation and the willingness and ability of stakeholders to adapt. In terms of deconstruction, environmental and economic viability must be assessed. According to the authors of [34], environmental sustainability will depend mainly on the characteristics of the building to be deconstructed as well as on local secondary resource markets. Resource diagnoses must also be carried out to dismantle and sort the elements as well as possible; for this, technical experts must be trained and sensibilized.

3.2. Knowledge of Territorial Resources

The application of a CE, given its interdisciplinarity and the large number of constituents it involves, can increase the number of complex practical questions.

At the end of life of a project (for instance, an urban project or a building project), it is essential to have an efficient waste management perspective, but for circularity, it is also essential to prevent outgoing resources from becoming waste. This requires knowledge of the resources (e.g., their nature, quantity, and state) to initiate and apply procedures to give the outgoing materials a second life.

Understanding the spatial and temporal composition and organization of stocks and flows is a major issue both for researchers [17–19,35] and communities: Interviewee I–3 highlights that "There is a whole issue around knowledge, knowledge sharing; therefore, both knowing the quality of the existing building, knowing the building stock material composition". To enhance the value of resources, it is necessary to have a detailed description of the resources and the deposits: I–2, speaking about the deposits, explained that "we would mainly need a characterization of the materials and the state of the materials ... It would be useful to have files on resource inventories, for example". That is, it is important to (1) identify the materials present in the urban mine and the potential receiving sites; (2) identify the economic actors and operators to reinsert the elements into the economic loop; (3) identify the sources offered by the possibility of developing reuse and recycling channels; (4) orchestrate the logistics of resource flows [28]. I–1 confirmed this need for information about deposits and building stock materials: "There is work to structure the visibility of resource diagnoses to know exactly where, when, and in what state they are available; from there, we can know if they can be used on our operations".

The interviewees shared their experience and their need to characterize the deposit and material flow in the building sector. I–1 explained that Paris has some initial elements of a territorial diagnosis that allow understanding of the waste flows of its building sector based on the studies of the Paris Urbanism Agency (APUR). This expert mentioned that: "To calculate the portion of inert, nonhazardous and hazardous waste in Paris, the APUR used a waste ratio production for building typologies and their historical knowledge of certain buildings. They currently have different needs to create reuse platforms according to the target material (for instance, concrete and gypsum). Consequently, we need a more detailed analysis [than inert, hazardous or nonhazardous waste] to identify companies that will agree to treat specific waste and reuse materials. Moreover, concerning the land areas and quantities of materials, we need to know what is relevant and what is anecdotal".

In turn, I–3 explained that for the metropolis of Lyon, "we have the first elements of diagnosis but they are not very precise ... We carried out an urban metabolism study with a technical office; the study was based on standard data and a table of inputs and outputs of economic flow (in euros) in the territory; then, material flows were estimated (in tons) ... The downside of this tool is that if we improve the reuse and recycling of materials, or the reduction of material consumption in the territory, it is unclear how the tool is converting the inputs and outputs in euros into material flows in tons; as long as we spend these euros, it appears that we are continuing to produce and consume ... Therefore, this tool provides an order of magnitude, but it does not reflect reality in the territory, in particular, related to our progress".

For the Grenoble metropolis, I–2 mentioned that "the metropolis has jurisdiction over household waste and wastes of a similar nature and composition (DMA) but not over professional waste ... Within the framework of the CODEC (Contract for Waste Objective Circular Economy), which has been signed, what is asked of us by ADEME is to model and quantify the flows: (1) The flows of DMAs ... For these flows, we have all the elements to be able to quantify them since we are the ones who collect them and we are the ones who process them; (2) they also asked us to quantify and model the flows of construction and public works in our territory and ... of what is called waste from economic activities (DAEE). On the other hand, both regarding the flow in the construction industry and in DAEE, we have no visibility".

3.2.1. Spatial Scale

An assessment scale is defined by geographical, economic, social, and administrative parameters. In a circular economy, the scale is instead defined according to the influences of an urban development project; that is, concerning the geographical perimeter, it is a question of acting rather locally, with an extended network of administrative and socioeconomic actors to promote partnership exchanges [28].

I–3 explained: "If we are working at the building scale, we need to know the quality of the materials and which sector they can go to; can they be reused or recycled, and so on... At the district scale, it is preferable to prepare for mass reuse and recycling. At the metropolis scale, a macro analysis is needed to structure the material sectors". I–2 specified: "We realize that we are all in our territories trying to create material reuse channels and each territory will perhaps have to specialize in one flow or in a main flow and secondary flow. However, there is no territory that will be able to handle all of the buildings' flows because there are too many of them. Today, we are still using newly emerging economic models. There are storage issues that are important; there are also transformational issues. Therefore, we thought that perhaps the best scale was the regional scale so that everyone could see how to process and target one or two flows to have at least a more significant number of flows processed. It appeared to be a relevant scale to us. Therefore, all deposits need to be identified, materials need to be reconditioned, and transformation needs to occur. In addition, after that, outlets need to be identified ... All of that is very difficult to set up in each territory. Suddenly, it occurred to us that maybe the reuse sector could be at a regional scale and the deposits could be at the metropolitan scale".

I–1 indicated: "I think that, in fact, you have different problems depending on your perspective, whether you are in a very dense urban, dense urban, rural or semirural area. The needs are not at all the same because, quite simply, the deposit is not the same, and the possibilities of creating platforms are not the same. Therefore, I think that when you are in a rural or semirural environment, you can plan more efficiently at the department or urban community level. You must have a large enough deposit volume to be able to create platforms and have a large enough territory. When you are in an urban or semiurban environment, it is even a little more problematic than in rural areas. Then, I think that there is one more elements to consider: the need for transit platforms. It is necessary to have an existing offer for these platforms. The supply must be substantial enough to justify these short-term and medium-term investments in terms of economic return. For this reason, it is difficult to say what the scale is because it depends on the territory in which we are located. Then, this answer will vary depending on, once again, whether you are located in a site that is urban, urban-dense, etc. In addition, you also need to find land, which is not easy in urban-dense areas ... Therefore, for me, it is difficult to think along lines of district collectives, intermunicipalities, departments, and regional scales".

3.2.2. Temporal Scale

I–2 mentioned: "Indeed, the modeling of the built stock of territory allows us to know its urban mine, but it does not give the full picture (dynamic) of this urban mine; it can perhaps be built there and never become waste. Therefore, we instead need to set up treatment channels or reuse channels for the quantity of waste in a territory at a given time. Because the problem, particularly in the reuse sector, is that if we do not have these elements, it is difficult for this sector to emerge ... What interests me much more today is to have information on work sites that are being planned ... or are in progress, which effectively gives information on the potential deposits, because in areas where we do not have any deposits, we simply have a snapshot in time of the urban mine".

I–1 explained: "In my opinion, we need foresight. In addition, we know that today, buildings represent an important source and that this is where the significant issue lies in terms of reuse and recycling . . . Everything that is not made of earth, concrete or stone has to be built. Thus, if everything is to be built, the first step is always to know and take stock of both what we have and, at the same time, the prospects to evaluate the issues, and from there, to be able to determine actions".

In Section 3.2, Preliminary results from BTP-Flux project [22] were shown to the interviewees to better frame the study and meet the needs of the CE. BTP-Flux considered various needs that were expressed by the interviewees regarding several points: (1) it allows a detailed description, including the nature and quantity, of the materials present in the building stock and the demolition waste flow (a more refined resolution than inert, hazardous, and nonhazardous waste); (2) the results can be obtained at different territorial scales in France (for instance, departmental, regional, and national–territorial divisions); depending on the availability of data, the study could be applied to finer territorial scales such as neighborhoods or districts; (3) the model also aims to assess the robustness of the results to be able to communicate reliable results to possible users (for instance, local authorities and building stock managers). The interviewees' feedback on the model results was positive and affirmed that it addresses the primary issue: knowledge of the material building stock and waste flow. It will allow them to diagnose their territory and then formulate optimal CE strategies in the construction sector.

In conclusion, the application of the circular economy must consider a multiscalar approach. For example, for the sustainable and circular management of a certain category of waste, the articulation of the different scales for the characterization of deposits, creation of transit platforms, and possible outlets is essential to identify the actors but also to ensure solidarity between territories and avoid pressure between them. Construction or urban development projects are quite long; therefore, the integration of circular economy strategies must consider the evolution of projects at different time scales. For instance, the construction materials present in built stock can be mobilized in the medium term during the maintenance or renovation of projects or, in the long term, at the end of a project. Therefore, it is necessary to have dynamic images of the deposits to plan the development of progressive strategies. This generates a need for data on the material stock composition and deposits and the information should include uncertainties about these availabilities.

3.3. *Census and Synergy of Actors—Organizational Brakes*

The interdisciplinary and dynamic nature of the circular economy involves a diversity of construction actors with various missions and skills. The CE needs the involvement of all the constituents in a city's development to be effective, including those who already work together and those who are not used to working together. That is, they need to think about, articulate, and agree on common environmental, technical, and economic objectives from the design stage, through production, until the end of the building's lifespan, and reinsert secondary materials in the economic chain. Each of these constituents plays a role that will contribute to the success of the transition.

The role of communities is to develop synergies between actors, that is, to manage governance. Governance is characterized by the development of the capacity for promotion and collaboration between actors. The project's governance is based on the sharing and complementarity of the stakeholders' skills in various areas of the city's development (housing, mobility, environment, town planning, etc.). The main goal of this governance is to converge the strategic (coherent and effective policies) and operational (application of the CE in projects) parts of the application of the seven pillars of the circular economy in a sector, taking into account the temporal and spatial scales influencing its projects.

Strengthening the link between all the players is therefore essential to move toward an iterative approach to harmonize working methods and modes of operation. To achieve this, the construction sector players must have a clear vision of the issues; in this sense, public figures can put in place strategies to raise awareness and train their constituents on integrating CE issues into their territory, foster the capacity to work collaboratively, and benefit from the collective intelligence.

Referring to a circular economy experience in their territory, I–1 mentioned: "It showed that to be able to do it, we needed to inventory the structures that were made using eco-design and development, structures that transformed, ensured, restored, and were verified in their adaptability, and were compliant with technical performance in relation to use, etc. Therefore, we needed this collaboration of construction sector players. I think that this needs to be known and that all the actors are kept informed".

3.3.1. Training and Awareness

The training and awareness of stakeholders throughout the life cycle of a project is essential to encourage them to consider the challenges and levers raised by the CE and thus evolve current linear practices. It is about changing mentalities and practices both in the short term and in the long term; for example, on how to design a project, carry out diagnosis, or articulate the networks and sectors, as well as on enhancing and integrating reuse in construction projects. Today, few professions are entirely dedicated to the circular economy; currently, it is more a question of adapting the skills of professionals in other professions.

To act at a building or regional scale, we must train the technicians performing the diagnoses so that they can correctly diagnose the materials and identify those that can be reused, re-employed, or recycled and how it will be done. The feedback from local authorities was clear. I–1 explained: "In deconstruction, several issues need to be resolved. The first is the qualification and training of the actors, particularly the diagnosticians, so that they integrate the notion of reuse and have knowledge of the materials ... For me, one of the challenges is in this qualification and training, because to diagnose, one must reconcile both the knowledge of materials in the building sector and the knowledge of waste and potential sectors for reuse and recycling. Moreover, there is no interaction today between these two components, and the objective is that these diagnosticians will speak to each other; or better yet, that the same person can carry out both functions of this diagnosis". Therefore, training and requalification of professionals will allow them to adapt to this evolution in the sector and respond to new demands. "The second issue is, of course, whether there will be training and qualification work for architects, project managers and companies on this new [reuse] technique, including inspection offices, so that authorities know exactly how to deal with unusual techniques related to reused materials".

3.3.2. Control and Monitoring of Demolition Crews

I–1 quoted: "The other issue is the control and monitoring of demolition crews. Today, they know how to deconstruct ... it is not a question of knowledge, but there are two issues: for them, deconstructing can generate a hardship for their staff and a loss of productivity. Therefore, for these two reasons, and given their economic model, deconstruction requests are complicated. In addition, in any case, if we ask them to do something, whatever it is, there is work to be done in order to be able to monitor and control what they actually do because even if we apply regulations, enforcing them can be difficult. In fact, what would be interesting is if we did not assume a policing role, but instead, the process would come naturally for these crews... that they would not only be sensitized to the issues, but that they suddenly find technical and economic means to deconstruct in a more natural way, with cost sharing between the client and the demolition company. Therefore, there is something to be done ... I do not think we have a choice but to go through control and monitoring at the beginning, but it is exhausting and it is not satisfying to play this role of police officer; and I think that there is an issue, in any case, with that approach. I think that if several clients demand it, it will be done, but in any case, it will be done with reluctance. Thus, I think the best approach would be for the trade chambers to take interest in this subject and understand that it is also in their interest to find the levers for a new technical and economic model".

3.4. Traceability and Profitability

3.4.1. Waste Traceability and Digitization

Traceability is defined according to ISO 9000: 2015 as "the implementation or location of an object (3.6.1)", and it also states that "in the case of a product, it can be linked to the origin of materials and components, production history, distribution and location of the product after delivery". In summary, the main objectives in the building-waste sector are (1) to compile the information to follow the flows and share them with all the players involved. This information may relate, for instance, to the identification of waste streams and their collection, preparation, treatment, transport, and performance checks, and (2) to understand the technical performance issues of a resource to ensure its quality and be able to reinsert it more quickly into the economic loop, thereby increasing its value.

The development of traceability models is, therefore, a significant challenge. I–1 explained that "if we truly want to recycle materials, there is a need to get more information from a company than what it receives from the manager of its processing company. Upstream of the deconstruction, we chose to examine traceability; in fact, we did not settle for the information we received from the massification platform; we asked where it was going beyond that . . . We also asked for documents attesting that the companies' treatment facilities are ICPE (Installations classified environmental protection) certified, and that they had the proper authorizations or declarations for their activities. This control was extremely laborious and complicated to deliver, but it made it possible to ensure that we were working with a third party that respects the rules, which is not a given, even today. Downstream, we systematically asked for the company to follow up with a photo of their dumpster. From there, we asked if it could be posted online on a document management platform, and that was truly useful in allowing us actually to invoice what was removed and follow up. This is how we discovered, in particular, that inert waste ultimately did not go to recycling but to landfills. So that is fundamental for me. In addition, I think that on this point, there is upstream traceability work to be done so that it is not up to each client to re-ask every one of their service providers for authorizations as well as their secondary outlets".

The traceability of the waste will allow us to adequately follow up on the waste throughout its treatment and ensure the compliance of the sectors in terms of ICPE regulations, for example, as mentioned by I–1. It is, therefore, necessary to have reliable tracking slips. In addition, article L541-2 of the Environmental Code mentions that "Any producer or holder of waste is responsible for the management of this waste until its elimination or final recovery, even when the waste is transferred for third party processing". Article 36 of the CCAG (General and technical administrative clauses of works) considers the producer as the owner and the holder as the contractor.

With respect to this point, article 106 of the AGEC law specifies that "the person in charge of the waste collection facility must deliver free of charge to the company having carried out the work a deposit slip specifying the origin, nature, and the quantity of waste collected". In addition, "the company that carried out the work mentioned must be able to prove the traceability of the waste from the sites for which it is responsible by keeping the slips issued by the waste collection facility. The company that has carried out the work sends the slips to the commissioner of the work or the competent authority mentioned in article L.541-3 at their request".

The development of waste treatment platforms is evident. It is clear today, after feedback, that there is a significant lack of traceability tools and that the sector must adopt new digital tools to promote traceability and therefore optimal flow management. I–1 mentioned, "today, it is a paper document which is distributed, passing from hand to hand between different actors . . . Therefore, the issue is digitization, so that suddenly it is not a paper document, but slightly more computerized, with a tablet or other device, at the moment the work begins until it arrives at its final outlet. With digitization, it would be possible to link actors, create networks for accessing data and help decision-making. Having the information in real-time will also make it possible to observe any slippages and react quickly; for example, an operator who, despite alerts, sent the waste to an unauthorized service provider". Certain waste traceability tools, such as "Trackdéchets" [36], a digital version of the monitoring slip for hazardous waste, are being developed by the public authorities in France. This type of device could be extended to other types of waste and could give project owners better visibility of what happens to their waste once a platform takes it.

The traceability of site waste coupled with digital tools can meet several needs, including the quantification of flows and real-time monitoring of resources, the increase and control of their quality and possibly a reduction of the costs of waste treatment due to the better overview of their origin, and the ability to share information so that each stakeholder can be made aware of their responsibility throughout the treatment chain, from the site to the outlet; this would make the system more reliable and encourage stakeholders to improve their waste recovery.

3.4.2. Performance of Secondary and Regulatory Brakes

Speaking on secondary resources, I–1 explained: "we still have the whole problem of testing to verify the performance of these materials in the context of their use". Moreover, indeed, the use of secondary materials has slowed down because of (1) the properties and quality of the secondary resources recovered and (2) the regulations, standards, or insurance not yet being ready because there are still problems related to testing the performance of these materials for their intended use because they are not homogeneous. Therefore, the owners and insurers, who may not be confident about the technical and sanitary quality of the resources (for instance, stability, flexibility, resistance to deformation, strength), may hesitate to use them in their projects.

The main challenge is the coordination between all the stakeholders in the project. Contracting authorities can stipulate the use of reused materials in their contracts. Then, the intervention step will be for architects to identify in their first sketches the potential deposits or macroelements of the project that are likely to use reused elements by relying on specialized resellers. One of the sites that censuses these suppliers is Opalis [37]. Then, building element and material quality checks can be supported by experienced craftsmen, so that if the materials meet expectations (for instance, technical and sanitary expectations), they can be used, and insurers can include them. In addition, even if it can be challenging to know the performance of reused and recycled materials, if not impossible, the construction industry should make its best effort to maintain the materials' value as much as possible by addressing the 3R principle.

Another obstacle to using secondary resources is that the stakeholders do not yet have feedback on the cost-effectiveness of applying CE throughout the material's value chain. I–3 explained, "I think that feedback is still rare, but from what I understand, reuse or recycling requires more upstream engineering. However, afterwards, it saves costs downstream. Therefore, basically, you have to anticipate all the flows. Furthermore, I do not know that the additional engineering costs compensate for the avoided costs downstream; that is always the big question. In any case, there is a displacement of costs". At the territorial level, actions to massify waste flows are not yet common; nevertheless, they would be necessary to optimize profit from infrastructure investments required to manage secondary resources. At the material or product level, recycling costs remain mostly high compared to the prices of certain raw materials, so producers will favor virgin materials over recycled materials given their price. In addition, current product prices do not incorporate the price of their environmental impact, and the environmental quality of a product is not yet a major asset for its marketing.

3.5. Massification and Storage Platforms and Land

3.5.1. Platforms

Resource collection and recovery companies are essential to better manage waste. In France, site waste can be transported to a network of community or professional waste collection centers. Companies specializing in waste management carry out waste collection at construction sites, incorporating several steps dedicated to different waste categories; then, all of the waste flow will converge on aggregation, sorting, or processing platforms or go directly to recovery or disposal outlets [38].

The regrouping, sorting, and pretreatment platforms are established with multiple objectives: (1) collect waste, (2) offer local solutions, (3) sort waste off-site when sorting on-site is not possible, and (4) optimize the costs of disposal (by negotiating with the disposal and final recovery channels). Currently, for the treatment of inert waste, there are specific platforms for this type of waste; however, most nonhazardous waste in the sector is managed by multiactivity platforms that can both receive industrial and/or household waste, which makes listing them more complex compared to platforms that are dedicated only to construction waste [38].

The sector needs to develop new recycling channels, in particular for the treatment of insulation and joinery. Some producers may decide to invest directly in the development of these sectors since the law makes it possible to reduce their environmental impacts when they participate directly in achieving the collection objectives in article 72 of the AGEC law. It is clear that the players wish to avoid being confronted with economic difficulties at all costs. By articulating public contracts, ecomodulations, and incorporated regulatory rates, extended producer responsibility (EPR) must "strengthen the links between demolishers, recyclers, producers of materials and construction companies, in order to find effective solutions together".

3.5.2. Storage and Land

The availability of land is also necessary to store the materials to stabilize deposits. One of the main advantages of communities is their administration of public land. Dedicated spaces or areas for the storage, sorting, and treatment of fixed and temporary materials should be zoned and included in town planning documents, or communities must provide land to store material according to the sites scheduled in the territory. I–1 explained, "the problem is also to find storage to be able to ensure the stability of, or guarantee, the deposit... Between the deposit and the transformation, there is a need for a space for storage ... At the same time, it is necessary to know whether, in the public domain, there is land that could be vacant temporarily or for longer term, on which we could set up massification facilities. This subject is complicated because those in charge of operations in town planning or elsewhere are subject to an eventual work plan that evolves with politics ... An inventory of the installations must then be performed. The objective in theory is to identify the residual capacities of the installations to know if they can accept this increase in work linked to urban changes ... The final consideration is to know if there is any potentially land in the public domain ... The challenge now is to find available land to continue to work on this issue and the need to communicate with and educate elected officials on these issues to try to study the prefeasibility of the platform on this land. Then, we will try to move forward gradually so that we can eventually land on reuse and recycling platforms".

3.6. *Toward a New Territorial Cohesion*

The transition to a circular economy in the construction sector will involve demonstrating opportunities to develop new economic perspectives so that the sector helps the territory become attractive. Attractiveness requires a revitalization of the territory by creating jobs and economic activities resulting from the initiatives taken by political and socioeconomic figures. These new economic activities can be the development of waste treatment and management, reuse, and recycling channels. The emergence of projects with a high locally added value will increase a territory's resilience because it will secure resources and optimize territorial ecosystems. Collective intelligence, territorial cooperation, innovation, and awareness-raising among stakeholders can become the new development engine. The local authorities' unique roles as facilitators, guides, and catalysts of new, practical circular initiatives can accelerate changes in current practices and the transition to new ones.

For the territories, more comprehensive questions arise, including the articulation of other regional plans such as the PCE (Climate Energy Plan), the biodiversity plan, urban travel plans, and the circular economy plan. In addition, the interaction of geographical and temporal scales of the circular economy, its multidisciplinary nature, and its governance requires the decompartmentalization of activities and establishment of hybrid thinking; that is, making mixed action systems worthwhile because today's silo-type organization is no longer tenable. Moreover, this transversality goes further, for instance, when sharing data, planning, and implementing actions and strategies. Currently, it is still difficult to access what is now considered confidential information to identify, for example, whether one project's waste can become another's resource; this requires transparency in details that are sometimes kept private; therefore, cooperation is required.

4. Discussion

The circular economy is presented today as a tool that can energize our territories by linking stakeholders in a process that incorporates solidarity, proximity, and applicability at all scales, and transverses all levels of the territorial structure. The implementation of the circular economy is essential to the development of a circular metabolism and is the key to helping cities reduce their ecological footprint and orient them toward resilience and sustainability. The main levers for the territories are their reductions of quantities of waste to be treated and imported raw materials, the relocation of their supply of materials by including secondary materials in the loop, and their creation of local jobs and development of a social, unified economy.

The building sector is one of the priority sectors for achieving the CE objectives and will involve many economic players in the production of buildings at several scales and beyond the buildings' life cycles. Deploying the circular economy at the stock level is essential because it will provide a better understanding of the flows of materials, energy, water, and goods. This makes it possible to target the priority sectors and resources where interventions will be necessary. In addition, it will allow identification of the players and the missing pieces needed to create valuation loops. Finally, innovative systems can be created by setting up scenarios and strategies to improve circularity and link specific complementary actors and organizations and diminishing traditional systems by creating hybrid development ecosystems.

To succeed in this transition, local authorities acting as facilitators, catalysts, and regulators can promote the development of CE by raising the identified challenges and taking advantage of the opportunities they offer.

One of the main opportunities is to prevent outgoing materials during building renovation or deconstruction, stopping them from becoming waste. Therefore, the main challenge is to increase the recovery rate; for this, it is necessary to know the nature, quantity, and temporal and spatial organization of the material flows and deposits. However, there are still a large number of technical challenges to address, mainly the integration of circularity clauses such as selective deconstruction and reuse in the contracts before deconstruction. Moreover, it is important to diagnose structures because there are specificities for new and existing buildings. Existing buildings are not easily removable, and in addition, the buildings may have undergone renovations or maintenance, and the materials may therefore be of different ages and conditions; in addition, they are not standard. Thus, the person doing the diagnostic evaluation must be made aware and trained to correctly identify all the materials and technical points and, with his/her diagnosis, to increase the degree of recovery of materials and products during deconstruction. At the territorial level, upward flow models can be potential tools in formulating CE strategies, particularly for establishing territorial logistics for managing secondary resource processing infrastructures and ensuring transparency on local channels.

The building sector also has challenges related to the lack of communication between actors and a broad short- and long-term vision of the impacts of the sector's activities throughout and beyond the life cycle of its projects. In addition, given the relatively long lifespan of projects, the complexity of the fleet, and the number of stakeholders involved, it can be challenging to assign responsibilities to stakeholders at a specific stage of the entire life cycle of projects, buildings, and/or products. Nevertheless, the sharing of collaborative strategies and tools can offer a better overview of the built stock and deposits, tools such as shared databases, and common and appropriate methods to carry out deconstruction and manage secondary resources. To achieve this, stakeholders can communicate on certain cases and implement research and innovation within their structures to develop resources to facilitate CE deployment.

Technological factors can play an essential role in the lack of tools and digital logistics systems. The development of applications or platforms to stimulate this market will make it possible to promote materials better; in addition, digital tools and the sharing of information between stakeholders can help with the monitoring of resources from their diagnosis, highlighting their potential for development until they arrive at the appropriate outlets.

The sector must face resistance to change that is rooted in certain actors; awareness-raising and training of stakeholders is key not only for the deployment of actions but also for the generation of jobs, the creation of new professions, and the development of economic opportunities within the territories. From an organizational point of view, the main challenges and opportunities are to map the stakeholders and the territorial issues, then activate a territorial coordinator to facilitate the development of strategies and public policies to promote cooperation between stakeholders. Public authorities have the ability to find consensus for the application of the CE at different scales and levels, to provide normative and regulatory support for the management of resources, and to help stakeholders by reducing administrative procedures; they can also, for instance, give tax incentives to exemplary actors.

Social and cultural challenges are essentially linked to the lack of interest, knowledge, and/or commitment to applying CE strategies; in this context, communities have an opportunity to lead by example with their projects, by promoting collaboration between actors and by participating in the education, training, and qualification of the actors. Communities also have the opportunity to develop short- and long-term partnerships to promote the deployment of the CE. Social challenges also provide the opportunity to create social ties and consciousness to shape resilient, inclusive, circular, and sustainable territories.

5. Conclusions

The reduction of resource consumption, as well as the optimization of material use, prevention and improvement of waste management, and reduction in environmental impacts over the entire life cycle (and beyond) of building and building stock have become priority issues in the construction sector, which is faced with alarming signs of climate change and resource scarcities. These issues encompass actions relating to design, engineering, and management of materials, buildings, the building stock, and their cycles of use, maintenance, renewal, and deconstruction. All these actions mobilize many construction players who will contribute to their applicability, the consolidation or adaptation of existing sectors, and the development of new EC sectors in their territories while generating local jobs.

This article presents the results of semistructured interviews carried out with managers of CE projects in cities that are pioneering CE strategies in their construction sectors. The study highlights some of the main challenges and opportunities that interviewees raised within their territories, based on their experience. First, a life-cycle, multiscale, and multicriteria approach is required to meet the challenges of circularity in the construction sector. Then, the key to knowing the territory involves knowledge of its resources (for instance, materials, wastes, construction players, material sectors) to ensure the materials' circulation and traceability while relying on collaborative tools and technologies. All this will be possible only if there is increased awareness among stakeholders, allowing them to collaborate to pool resources and tools and develop synergies. Pioneer communities and actors are called upon to strengthen support and training to increase the skills of stakeholders and future professionals. The approaches also need to consider applicability to other subjects and sectors such as environmental, economic, sanitary, and social sectors. The results of this study can help construction stakeholders develop action plans in favor of a CE that can share visions, objectives, timeframes, and even common scopes that can take different forms, such as roadmaps and strategies.

Author Contributions: Conceptualization, methodology, formal analysis, investigation, resources, and data curation: R.T., S.L., A.A. and G.H.; writing—original draft preparation: R.T.; writing—reviewing and editing: S.L., A.A. and G.H.; visualization: R.T.; supervision: S.L., A.A. and G.H. All authors have read and agreed to the published version of the manuscript.

Funding: The research presented in this article is part of the research work of the Ph.D. thesis of the first author planned from 2018–2022; this research is funded by the CSTB—funding number: 18.000835.01.01.

Institutional Review Board Statement: Not applicable.

Informed Consent Statement: Not applicable.

Data Availability Statement: Not applicable.

Acknowledgments: The authors would like to thank all the interviewees.

Conflicts of Interest: The authors declare no conflict of interest.

Appendix A

Interview guide

1. Instructions for the Interview

The research was done by interviewing the managers of circular economy projects in the municipalities of France. During the interviews, experiences, challenges, needs, and opportunities to promote the circular economy project in the building sector were discussed.

The participants were first approached by email, where they received a general overview of the interview subject and context. Interviews were conducted by telephone or video conference and recorded if the interviewees agreed. Notes were taken in every interview and were later used in the analysis and refined based on the recordings. Interview results were anonymous.

The main questions addressed concerned the application of circular economy strategies in the construction sector within their jurisdictional boundaries:

- What strategies are adopted in the territory in terms of CE in the construction sector?
- What is the scale of the projects' scope, and what is the projects' nature according to the action scale?
- In projects in which CE strategies have been implemented, what were and/or are still the main challenges and opportunities to CE strategy implementation?
- In a potential model of material flow assessment, what kind of information—qualitative or quantitative—will be needed to facilitate the implementation of CE strategies?
- Based on the interviewee's experiences, what scale would be relevant for the analyses of deposits?
- In your projects, what are the main data or information needs to facilitate or improve their CE strategy application?

References

1. Anastasiades, K.; Blom, J.; Buyle, M.; Audenaert, A. Translating the circular economy to bridge construction: Lessons learnt from a critical literature review. *Renew. Sustain. Energy Rev.* **2020**, *117*, 109522. [CrossRef]
2. Krausmann, F.; Schandl, H.; Eisenmenger, N.; Giljum, S.; Jackson, T. Material Flow Accounting: Measuring Global Material Use for Sustainable Development. *Annu. Rev. Environ. Resour.* **2017**, *42*, 647–675. [CrossRef]
3. Ghisellini, P.; Ripa, M.; Ulgiati, S. Exploring environmental and economic costs and benefits of a circular economy approach to the construction and demolition sector. A literature review. *J. Clean. Prod.* **2018**, *178*, 618–643. [CrossRef]
4. Pomponi, F.; Moncaster, A. Circular economy for the built environment: A research framework. *J. Clean. Prod.* **2017**, *143*, 710–718. [CrossRef]
5. Witjes, S.; Lozano, R. Towards a more Circular Economy: Proposing a framework linking sustainable public procurement and sustainable business models. *Resour. Conserv. Recycl.* **2016**, *112*, 37–44. [CrossRef]
6. Weterings, R.; Bastein, T.; Tukker, A.; Rademaker, M.; de Ridder, M. *Resources for Our Future: Key Issues and Best Practices in Ressource Efficiency*; Amsterdam University Press: Amsterdam, The Netherland, 2013.
7. Stahel, W.; Reday-Mulvey, G. Jobs for Tomorrow: The Potential for Substituting Manpower for Energy. Available online: https://www.researchgate.net/publication/40935606_Jobs_for_tomorrow_the_potential_for_substituting_manpower_for_energy (accessed on 28 April 2021).
8. McDonough, W.; Braungart, M. *Cradle to Cradle—Remaking the Way We Make Things*; North Point Press: New York, NY, USA, 2010.
9. Ellen MacArthur Foundation. Towards a Circular Economy: Business Rationale for an Accelerated Transition. Available online: https://www.ellenmacarthurfoundation.org/publications/towards-a-circular-economy-business-rationale-for-an-accelerated-transition (accessed on 28 April 2021).
10. Ministère de la Transition Ecologique. LOI n° 2009-967 du 3 août 2009 de Programmation Relative à la Mise en Œuvre du Grenelle de L'environnement (1). Available online: https://www.legifrance.gouv.fr/loda/id/JORFTEXT000020949548/ (accessed on 6 May 2021).
11. Ministère de la Transition Ecologique. LOI n° 2015-992 du 17 Août 2015 Relative à la Transition Énergétique Pour la Croissance Verte. Available online: https://www.legifrance.gouv.fr/dossierlegislatif/JORFDOLE000029310724/ (accessed on 6 May 2021).
12. Légifrance LOI n° 2015-991 du 7 Août 2015 Portant Nouvelle Organisation Territoriale de la République (1). Available online: https://www.legifrance.gouv.fr/loda/id/JORFTEXT000030985460/ (accessed on 6 May 2021).
13. Ministère de la Transition Ecologique. LOI n° 2020-105 du 10 Février 2020 Relative à la Lutte Contre Le Gaspillage et à L'économie Circulaire (1)-Légifrance. Available online: https://www.legifrance.gouv.fr/jorf/id/JORFTEXT000041553759/ (accessed on 6 April 2021).
14. Augiseau, V.; Barles, S. Studying construction materials flows and stock: A review. *Resour. Conserv. Recycl.* **2017**, *123*, 153–164. [CrossRef]
15. Augiseau, V.; Kim, E. Spatial characterization of construction material stocks: The case of the Paris region. *Resour. Conserv. Recycl.* **2021**, *170*, 105512. [CrossRef]
16. Lanau, M.; Liu, G.; Kral, U.; Wiedenhofer, D.; Keijzer, E.; Yu, C.; Ehlert, C. Taking Stock of Built Environment Stock Studies: Progress and Prospects. *Environ. Sci. Technol.* **2019**, *53*, 8499–8515. [CrossRef] [PubMed]
17. Kleemann, F.; Lederer, J.; Rechberger, H.; Fellner, J. GIS-based Analysis of Vienna's Material Stock in Buildings. *J. Ind. Ecol.* **2017**, *21*, 368–380. [CrossRef]
18. Stephan, A.; Athanassiadis, A. Towards a more circular construction sector: Estimating and spatialising current and future non-structural material replacement flows to maintain urban building stocks. *Resour. Conserv. Recycl.* **2018**, *129*, 248–262. [CrossRef]
19. Heeren, N.; Hellweg, S. Tracking Construction Material over Space and Time: Prospective and Geo-referenced Modeling of Building Stocks and Construction Material Flows. *J. Ind. Ecol.* **2019**, *23*, 253–267. [CrossRef]
20. Schiller, G. Urban infrastructure: Challenges for resource efficiency in the building stock. *Build. Res. Inf.* **2007**, *35*, 399–411. [CrossRef]
21. Barles, S. Society, energy and materials: The contribution of urban metabolism studies to sustainable urban development issues. *J. Environ. Plan. Manag.* **2010**, *53*, 439–455. [CrossRef]
22. Tirado, R.; Aublet, A.; Laurenceau, S.; Thorel, M.; Louërat, M.; Habert, G. Component-Based Model for Building Material Stock and Waste-Flow Characterization: A Case in the Île-de-France Region. *Sustainability* **2021**, *13*, 13159. [CrossRef]
23. Ministère de la Transition Écologie et Solidaire Chiffres clés du Climat—France, Europe et Monde. Available online: https://www.statistiques.developpement-durable.gouv.fr/chiffres-cles-du-climat-france-europe-et-monde-edition-2019 (accessed on 28 April 2021).
24. European-Union Eurostat by Statistical Office of the European Union. Available online: https://ec.europa.eu/eurostat/data/database (accessed on 28 February 2020).
25. Lützkendorf, T. Sustainability in Building Construction—A Multilevel Approach. In *Proceedings of the IOP Conference Series: Earth and Environmental Science*; Institute of Physics Publishing: Prague, Czech Republic, 2019; Volume 290. [CrossRef]
26. Adème; Ecogeos. Caractérisation de la Problématique des Déchets Sauvages; France; 2019. Available online: https://www.actu-environnement.com/media/pdf/news-33455-etude.pdf (accessed on 12 September 2021).

27. Ellen Macarthur Foundation Circular Economy Diagram. Available online: https://ellenmacarthurfoundation.org/circular-economy-diagram (accessed on 14 February 2021).
28. Bellastock REPAR #2; 2018. Available online: https://www.bellastock.com/projets/repar-2/ (accessed on 24 April 2021).
29. Miles, J.; Gilbert, P. *A Handbook of Research Methods for Clinical and Health Psychology*; Oxford University Press on Demand: New York, NY, USA, 2005.
30. Gill, P.; Stewart, K.; Treasure, E.; Chadwick, B. Methods of data collection in qualitative research: Interviews and focus groups. *Br. Dent. J.* **2008**, *204*, 291–295. [CrossRef] [PubMed]
31. Braun, V.; Clarke, V. Using thematic analysis in psychology. *Qual. Res. Psychol.* **2006**, *3*, 77–101. [CrossRef]
32. Swiss Life Group What is the Lifespan of a House? Available online: https://www.swisslife.com/en/home/hub/what-is-the-lifespan-of-a-house.html (accessed on 20 April 2021).
33. European Commission. EU Construction and Demolition Waste Protocol and Guidelines. Available online: https://ec.europa.eu/growth/content/eu-construction-and-demolition-waste-protocol-0_en (accessed on 12 May 2021).
34. Pantini, S.; Rigamonti, L. Is Selective Demolition Always a Sustainable Choice? *Waste Manag.* **2020**, *103*, 169–176. [CrossRef]
35. Augiseau, V. Using construction materials as secondary resources: Constraints and perspectives for territorial policies. *Flux* **2020**, *116–117*, 26–41. [CrossRef]
36. Ministère de la Transition Écologique et Solidaire Trackdéchets | La Traçabilité des Déchets en Toute Sécurité. Available online: https://trackdechets.beta.gouv.fr/ (accessed on 5 May 2021).
37. FCRBE Opalis. Available online: https://opalis.eu/fr (accessed on 5 May 2021).
38. FFB; Adème. *Guide de Conception et de Fonctionnement des Intallations de Traitement des Déchets du BTP*; Paris, France, 2014. Available online: https://www.seddre.fr/media/guide-de-conception-et-de-fonctionnement-des-installations-de-traitement-des-dechets-du-btp.pdf (accessed on 26 February 2020).

Article

Circular Economy in Construction and Demolition Waste Management in the Western Balkans: A Sustainability Assessment Framework

Ana Nadazdi *, Zorana Naunovic and Nenad Ivanisevic

Faculty of Civil Engineering, University of Belgrade, 11000 Belgrade, Serbia; znaunovic@grf.bg.ac.rs (Z.N.); nesa@grf.bg.ac.rs (N.I.)
* Correspondence: anikolic@grf.bg.ac.rs

Abstract: Population growth, consumerism and linear (take-make-dispose) economy models have been piling up waste for decades. The construction industry is also based primarily on linear economy models, but the good news is that most of the waste can be re-used or recycled. So far, numerous models for managing construction and demolition waste in a sustainable way have been developed, but only a few models have included circular economy approaches. The main objective of this study is to propose an integrated framework for the sustainability assessment of CDW management. Apart from the economic, environmental and social aspects of sustainability, this model also includes circular economy principles. The proposed framework is based on the integration of existing methods: bottom-up materials stock approximation; cost–benefit analysis for criteria calculation; and scenario and multi-criteria decision-making analysis for sustainability. It is suggested that the European average recovery rates should be used for future scenario development. With higher re-use and recycling rates, the potential for the circularity of the recovered waste grows. In an effort to increase circularity in the region, particular attention was devoted to customize the framework and examine its potential for use in the Western Balkan countries. The framework may also be useful in countries with immature construction and demolition waste management.

Keywords: circular economy; green deal; construction and demolition waste; quantification; waste management; recycling; re-use; material stock analysis; multi-criteria decision-making

Citation: Nadazdi, A.; Naunovic, Z.; Ivanisevic, N. Circular Economy in Construction and Demolition Waste Management in the Western Balkans: A Sustainability Assessment Framework. *Sustainability* **2022**, *14*, 871. https://doi.org/10.3390/su14020871

Academic Editors: Anibal C. Maury-Ramirez and Jaime A Mesa

Received: 8 December 2021
Accepted: 10 January 2022
Published: 13 January 2022

1. Introduction

More than a third of the waste in Europe comes from construction and demolition activities [1]. The amounts of this type of waste generated worldwide reached 3 billion tons in 2012, with China, India and the United States as the main contributors [2]. In the same year, Europe generated 0.85 billion tons of construction and demolition waste [3]. In 2018, the data for Europe shows a steady increase in the amount of construction and demolition waste (CDW), amounting to almost 1 billion tons [3].

The typical composition of CDW highly depends on several factors. The key factors are the type of activity that generates the waste and the type of structure constructed or demolished. Depending on the type of activity, waste may come from construction activities, i.e., construction waste and from demolition activities, i.e., demolition waste. Researchers mainly agree that demolition waste constitutes a larger portion of CDW [4]. When it comes to different structures, buildings contribute the most to waste generation due to their mass construction and frequent demolition, while infrastructures are sporadically constructed and rarely demolished. Therefore, depending on the location and the common construction practice in a specific area, CDW consists of mineral waste from construction and demolition (brick, concrete, insulation, etc.), other mineral waste (gravel, rock, sand, etc.), glass and wood waste, metallic waste (ferrous and non-ferrous), soils and dredging spoils [5]. More importantly, almost all streams have a great potential for further processing (treatment).

In the following order of priority, CDW may be re-used, recycled and incinerated [6]. However, CDW is mostly disposed in landfills or, in some cases, illegally dumped [7,8]. CDW is mainly considered as inert [9], although there can be a small percentage of toxic substances coming from asbestos, gypsum, coal tar and heavy metals [10]. Exposure to asbestos can occur during demolition, when asbestos fibers that are released into the air are breathed in and can cause scarring and inflammation of the lungs; this can affect breathing and lead to serious health problems [11]. Asbestos has also been classified as a known human carcinogen [12–15]. Drywall can leach toxins and release hydrogen sulfide gas in landfills while occupational exposure to coal tar increases the risk of skin cancer [16].

Processing of CDW may decrease the consumption of primary raw materials and the emission of greenhouse gases (GHG) [17]. These characteristics, alongside with the large amounts of CDW generated worldwide, were recognized by both scientists and governments. They turned this potential into strategies, action plans and legislations, based on sustainable development goals (SDG).

The most recent effort of the European Union resulted in the Circular Economy Action Plan. The first Action Plan that was published in 2015 revolved around the transition from linear to circular economy business models [18]. The transition stipulates that instead of taking primary raw materials from the environment to make different products that will be disposed as waste at the end of their use, circular economy should be oriented on the prevention of products becoming waste or prolongation of their use by employing re-use or recycling [18]. The Plan identified 54 actions designed to support SDG; goal 12.5 was defined as the "reduction of waste through prevention, reduction, recycling and reuse" and named five priority waste streams, including CDW [19]. However, apart from publication of a non-binding Protocol on how to properly manage CDW and pre-demolishing assessment guidelines, not much else was achieved in the area of CDW management [20].

The New Circular Economy Action Plan was published in 2020 and will hopefully amend this problem [21]. This new plan was designed as a part of a Green Deal initiative to make Europe the first climate neutral continent by 2050 [22]. The EU will devote at least EUR 1 trillion for this goal, out of which EUR 100 billion will go to the most affected regions [23]. As for the construction industry, the highest expectations are laid upon the specific waste reduction targets which were missing in the previous Waste Framework Directive (WFD). Although the Waste Framework Directive (WFD) set a 2020 objective for re-using, recycling or other recovery of non-hazardous fractions of CDW to a minimum of 70% by weight [6], many practitioners and researchers highlighted the need to make this target treatment specific (setting a specific re-using target, a recycling target, etc.) [23]. Apart from the specific targets and changing of product regulations to include recycled materials, the New Plan also included two new strategies: the Renovation Wave and Sustainable Built Environment Strategy [21]. While the former aims to double the energy efficient renovation rate (currently at 1%) by 2030 [24], the Sustainable Built Environment Strategy is designed to revise construction product regulations, reconsider waste regulations and promote circular economy principles.

So, what has Europe done so far to increase circularity in the construction and demolition sector? For instance, the latest data for the year 2018 shows that in the EU, mineral and non-hazardous CDW is very much recovered (in many counties is around 90%), and that the WDF target is already achieved. In the Western Balkans, four out of eight countries have reported that they achieved the 70% recovery target from the WFD. Two countries, Slovenia and Croatia, are EU members and as such are guided by the WFD; they reported a 98% and a 78% CDW recovery rate, respectively. Serbia and North Macedonia, as EU candidate countries, reported a CDW recovery of 81% and 100%, respectively [25]. However, these high percentages may be misleading as no information is provided on whether the recovered materials were further used in high-grade or low-grade applications such as backfilling. This is due to different interpretations of the terms recycling and backfilling in EU Member States as backfilled CDW is often reported as recycled [1]. For example, the Netherlands reported an almost 100% recovery rate for CDW in 2018; the fact is that only around 3% of the CDW was used in new concrete production [26] and not all CDW was recovered for high-grade applications. The lack of confidence in the quality of the recycled materials, due to a lack of quality standards, is the most cited reason for low levels of high-grade applications [27,28]. The other barriers for adopting the circular economy approach in CDW management practices identified by the academia are undeveloped markets for the recovered materials and the low prices of raw materials [29].

Therefore, what are the exact elements that CDW management practices should include to facilitate the future needs of circular economy in construction? What type of a sustainability assessment framework may be implemented to evaluate different management scenarios and find the optimal one? Why should the Western Balkans be in the focus and what is the circular economy perspective in the construction and demolition sector in these countries? This paper will suggest a framework that may answer these questions and facilitate these needs. It will take into account the particularities of a country and its economies and the best practices currently available in Europe to design a methodology for the assessment of CDW management options from the economic, environmental and social perspective.

To address the above questions, the current state-of-the art in the CDW management domain had to be analyzed. Desktop research was conducted in two directions. This included a review of CDW management publications in the most significant scientific journals, relevant EU policies and technical papers published in the last 15 years (from 2005). The first direction was to identify and analyze the existing sustainability assessment studies in order to identify their main elements, properties and techniques used for the assessments. This scientific review also helped to highlight the research gap that the proposed framework needed to fill. The second direction was to search for current best management practices and relevant policies that promote sustainable waste management and circular economy principles in the built environment in order to design better CDW management scenarios.

2. Previous Studies on Sustainability Assessments of CDW Management

The scientific community has been focused on CDW management for decades. The most recent studies (from 2011 to 2021) observed and analyzed CDW management from different aspects. The most analyzed were policies and circular economy strategies [30,31], opportunities and barriers in adopting circular economy in the built environment [32,33], stakeholders' awareness [34,35], comparison of different management practices [9,36,37] and application of information technologies, such as GIS and big data in CDW management [38–41].

When it comes to the sustainability assessment, the studies mostly focused on environmental and economic effects and to a lesser extent on the social effects of different treatment options and scenarios for different waste streams. A large number of studies, as expected, covered the environmental performance of recycling [42–44], recycling and landfilling [45,46] and in some cases the reduction of CDW [47]. The second most investigated aspect of CDW management that was analyzed either as a stand-alone element [48,49] or in combination with the environmental [50–52] or social aspects [53] was the economic aspect.

To the best of the authors' knowledge, the first paper that suggested an integration of all three sustainability pillars in waste management systems was written by Taelman et al. in 2020 [54]. In addition to an assessment framework, they developed a set of impact categories and indicators to address both the global and local impacts of a waste management system in five areas (prosperity, human well-being, human and ecosystem health and national resources).

The first to apply this framework in the CDW management sector was Iodice et al. in 2021 [55]. They analyzed and assessed the sustainability of three CDW management scenarios (baseline, linear and best practice) in the Campania region in Italy. Sustainability was assessed from the economic, environmental and social aspects and treatment options that were considered were mobile and stationary recycling and landfilling. The results showed that the implementation of the best practice scenario that includes selective demolition and increased recycling may benefit both the environment and society.

Although rich in sustainability factors, as 20 factors from all three domains were observed, the study by Iodice et al. was limited just to two treatment options [55]. The other limitation concerned the data on CDW generation and composition, particularly the CDW material breakdown. Namely, the data for the waste from traditional demolition were taken from the available studies and local environmental agencies while the waste data from the selective demolition were based on assumptions and best-guess estimates.

It appears that the majority of the above-mentioned sustainability assessment studies based their calculations on the statistical records of CDW quantities or estimations from practitioners and academia. However, CDW statistics in most countries are underdeveloped or vague. Additionally, academia and practitioners base their estimations on the amount of total construction and demolition waste per GDP or capita (population). In both cases there are no reliable data on the quantities of particular waste streams such as the mineral, metallic, glass or wood waste streams. This data would facilitate the development of more appropriate treatment strategies, designed and developed for each stream that could finally lead to better and more informed decision-making.

When it comes to different CDW streams, recycled concrete (alone or as a part of mixed CDW) is the most analyzed material, owing it to the fact that concrete is the largest contributor in the overall quantity of CDW: however, one must not overlook the treatment potential of other materials such as brick, wood and steel.

Finally, the circular economy approach is yet to be fully adopted in these assessments as most of the studies were focused mainly on recycling and landfilling, while the preparation for re-use and energy recovery as treatment options were rarely analyzed.

The proposed framework will address these limitations. Firstly, rather than using statistical data on the total generation of CDW, or CDW generation rates obtained from previous studies or practitioners, the framework will calculate and forecast the quantities of each particular CDW stream. Secondly, these quantities will be calculated on the basis of construction material quantities built into residential buildings and they will form a unique database of material stock. Finally, the CDW management scenarios will encompass all possible treatments of CDW, from preparing for re-use to energy recovery and disposal. Additionally, the CDW management scenarios will include high re-use and recycling rates to investigate the full circular economy potential of the proposed CDW management option.

3. Proposed Integrated Framework for Sustainability Assessment of the CDW Management

3.1. Sustainability Assessment Framework

A three-stage multidisciplinary methodology for the evaluation of CDW management options' sustainability is proposed. In general, the framework is designed to transform corresponding input data that may be related to the construction and demolition practice into qualitative and quantitative output data. The transformation process shown in Figure 1 uses several analyses and methods from different scientific domains, mainly waste and project management, that are integrated into one overarching methodology for the sustainability assessment.

As stated previously, this assessment should enable more informative decisions in CDW management. To facilitate this, specific goals of this framework were to form a database of materials (i.e., Material stock database) used for the construction of buildings and consequently to estimate the potential quantity of construction and demolition waste. The former presents stage one, while the latter presents stage two of the proposed framework.

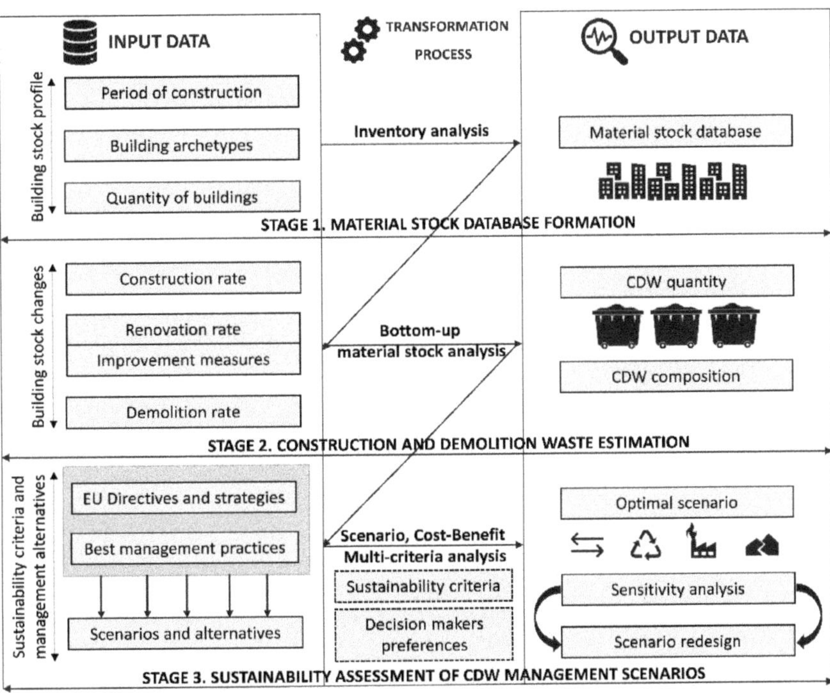

Figure 1. The framework for the sustainability assessment of CDW management scenarios.

The formation of the Material stock database requires a comprehensive set of building information, such as the type of materials built into construction elements and the physical characteristics of these elements for the entire building stock of a country. This information is rarely available. To overcome this, a bottom-up material stock analysis is suggested. The bottom-up material stock analysis includes the aggregation of the entire building environment stock into groups of objects with similar age and physical characteristics. Each group has its typical representative (archetype), which is then used further in the analysis. The transformation process in this stage of the methodology involves the segregation of archetypes into construction elements made of different materials and the calculation of their quantities. The materials are then grouped into different types (concrete, bricks, wood, metal, etc.) to facilitate the easier division of waste streams and their further treatment.

The second stage of the framework starts with the definition of the construction and demolition profile of the building stock. The building stock is a dynamic system, which means that it changes over time. Therefore, a simulation of future stock development is necessary before the quantity of CDW may be estimated. Apart from construction and demolition rates, this simulation encompasses a list of improvement measures that will be performed in the corresponding renovation cycle (usually 20 to 50 years depending on the construction element). If the projected timeframe is 50 years or more, this framework stage should also include the occurrence of natural disasters that cause significant demolition quantities, especially earthquakes in this region. Consecutively, the framework should use some method for earthquake loss estimation (including demolition quantities), similar to the one proposed in Stojadinović et al. [56]. And finally, renovation cycles serve for the definition of the building stock's renovation rates. These rates present the share of building stock from different groups of objects that will be constructed, renovated or demolished in a given year. Coupled with the material composition of the corresponding archetypes, the quantity of waste generated from these activities may be estimated.

The final stage of the integrated methodology is to choose the optimal strategy and make the right decision with respect to different CDW management scenarios. The first step in this stage is to define the treatment rates for different waste streams and different scenarios. The sustainability of each scenario is then assessed through a set of criteria and sub-criteria that may have a positive or negative effect on the society. The authors propose a set of 16 sub-criteria grouped into three criteria (Table 1) to grasp the entire sustainability domain.

Table 1. An overview of the sustainability assessment criteria and sub-criteria.

Economic (e)	Environmental (en)	Social (s)
1. Capital expenditures 2. Operational expenditures 3. Replacement works 4. Clearance works 5. Sale of recovered materials 6. Sale of recovered energy (heat and electricity) 7. Landfill taxes and gate fees 8. Residual values	1. Avoided GHG emissions through: recovered materials energy recovery improved WMS [1]	1. Social capital expenditures 2. Social operational expenditures 3. Social replacement works 4. Social clearance works 5. Public discomfort due to landfill presence 6. Arable land consumption 7. Social residual values

[1] Waste Management System.

The first group of criteria are costs and revenues and they belong to the economic field (labeled e in Table 1). These are capital, operational, replacement and clearance expenditures (costs) of different treatment options. The revenues from these options may come from sales of recovered product including heat and electricity and the savings from different taxes and fees, such as landfilling taxes and gate fees.

The environmental set of criteria (labeled en in Table 1) considered the positive and negative effects of different scenarios. The positive effects are reflected in decreased greenhouse gas (GHG) emissions and the reduced use of primary raw materials. There are even some treatment options that have a negative effect on society, such as landfilling and illegal dumping. Even recycling and energy recovery emit GHGs, mostly from transport and operation, but the positive effects of these waste treatment operations prevail over the negative effects.

And finally, the social set of criteria (labelled s in Table 1) includes the social adjustment of the cost and revenues to illustrate the local market imbalance with respects to taxation, customs and unemployment rates, as well as land degradation and discomfort caused by different treatment options.

To compare the scenarios with respect to these criteria and sub-criteria several multi-criteria decision-making analyses are available. An application of several Multi-Criteria Decision-Making (MCDM) techniques (AHP, VIKOR, TOPSIS, etc.) can be applied as in Tirth et al. [57]. Authors propose the Analytical Hierarchy Process (AHP) to be used in this framework, due to its simplicity and flexibility. AHP was developed by Saaty in 1980 and it uses pairwise comparisons of criteria and sub-criteria and alternatives with respect to each criterion and sub-criterion in the form of matrixes of judgment. The Saaty scale [58] was used to describe the comparison: 1, 3, 5 and 9 were used to grade equal (1) to extreme (9) importance of one sub-criteria over another; 2, 4, 6 and 8 were used when compromise was needed. The matrixes of judgments are symmetrical with reciprocal values, meaning that "when activity *i* has one value assigned when compared with activity *j*, then *j* has the reciprocal value when compared to *i*".

An example of the sub-criteria comparison matrix is given in Figure 2a,b. Two Serbian waste management experts were interviewed and asked to compare pairs of sub-criteria. The experts were selected on the basis of their preferences: one that was mainly economically oriented and one that was mainly environmentally oriented. The economically oriented expert gave higher importance to the economic criteria when compared to two other criteria. For instance, a very strong (7) or even extreme importance (9) was assigned to capital expenditures (sub criteria e_1) over environmental (en_1) and social criteria (s_1–s_7).

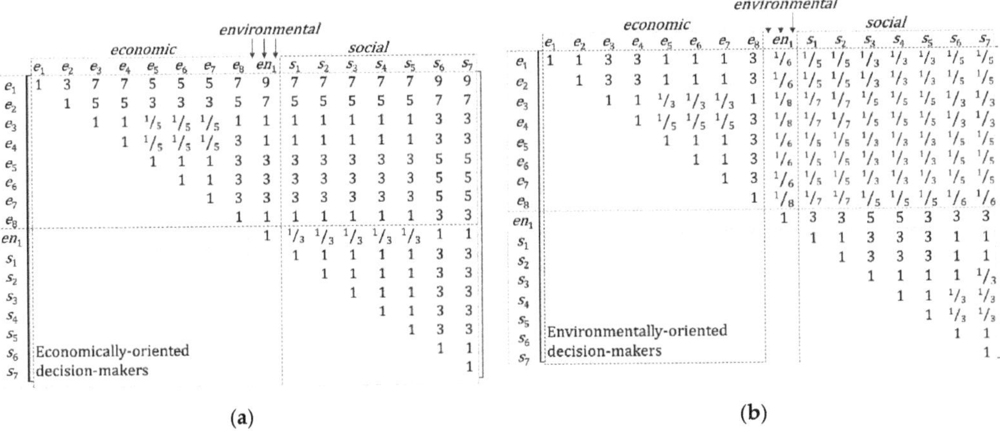

Figure 2. An example of a sub-criteria judgment matrix from Serbia: (**a**) Economically oriented decision-maker; (**b**) Environmentally oriented decision-maker.

As expected, the environmentally oriented expert preferred environmental and social criteria over economic (e_1–e_8); thus, environmental criterion en_1 had moderate importance (3) over social criteria s_5 (public discomfort due to landfill presence) and s_6 (arable land consumption), but a very strong importance (6, i.e., 1/6) over capital (e_1) and operational expenditures (e_2).

The eigenvector of each matrix is the vector of criteria priorities [58]. The CDW management strategy scenarios (alternatives) are compared with respect to each criterion and form the alternative judgment matrices. The eigen vector of these matrices is the vector of alternative priorities meaning that the alternative with the highest priority represents the optimal scenario for managing CDW.

Considering that the proposed framework serves to build a model that will illustrate a real-life CDW management system, verification and validation of whether the model is a good representation of the real-life system is advised. Both of these processes should be conducted in all three stages of the framework. The verification of a model should be carried out during the implementation process to search for errors that may hinder the implementation. The authors propose a three-step validation: the extreme conditions test; a sensitivity analysis; and the comparison of results with the real or analytical data, if real data are not available.

3.2. CDW Management Scenarios

To reach appropriate levels of circularity in the Western Balkan countries, one must start from different positions in each of the particular countries. The current state of CDW management differs from country to country and, for that reason, a baseline scenario for each country should be established and described before developing possible management scenarios. The future practice (or scenarios) should be established by investigating the principal policy elements that support the European best practice, mainly the economic instruments and legal requirements. Aside from the policy elements, possible scenarios should be designed on the basis of the composition of CDW and the maximum share of particular waste streams and the appropriate and available technology for their treatment.

3.2.1. Current CDW Management Scenarios

As shown in Figure 3, countries with larger populations and a higher GDP per capita generate more CDW, which is in line with claims from several authors [1]. Official statistical quantities of CDW for each country are given in Figure 3, with assumptions made for Albania which had no data on CDW [3]. As shown, the current generation of CDW in the

Western Balkans is below the European average and reportedly ranges from 35.6 (North Macedonia) to 1260 kilotons (Croatia) of CDW per year. Almost 30% of the total CDW generated belongs to the mineral fraction of CDW that is most suitable for further treatment and high-grade applications. However, these amounts are mostly underestimated as there is a significant amount of CDW waste generated and disposed of in landfills without reporting or even illegally dumped (ranging from 2.4% to 24%) [59].

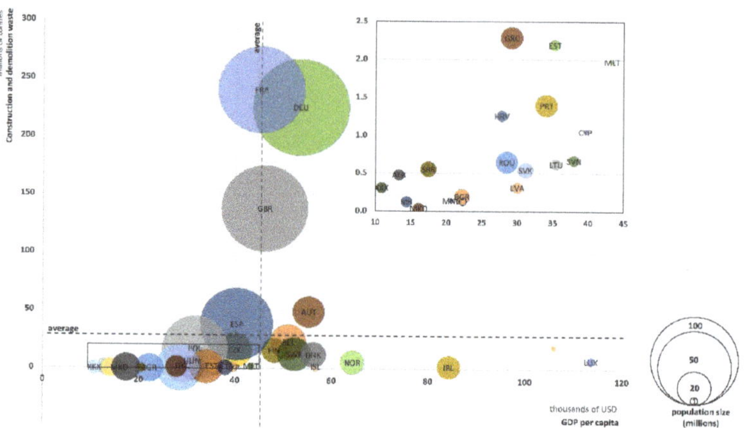

Figure 3. European CDW generation rates in relation to GPD per capita (data for 2018). (ISO-3166-Aplha 3-digit codes were used for countries abbreviations).

The current treatment rates alongside with the principal elements of current CDW management are shown in Table 2. The rates are based on the national statistics published by Eurostat [25]. These numbers are most likely the result of poor management practices in the non-EU Western Balkans countries.

Table 2. Principal elements of CDW management and treatment rates in Western Balkans countries.

Country	Elements of CDW Management						Treatment Rates		
	End-of-Waste Criteria	Quality Standards	Recovered Mat. Market	Incentives	Bans or Taxes	Recovery Targets	RC	BF	D
ALB	no	no	immature	no	no	no	n/a	n/a	n/a
BIH	no				no	no	n/a	n/a	n/a
HRV	yes				foreseen	WFD	70%	8%	22%
XKX *	no				no	no	n/a	n/a	n/a
MNE	no				no	no	0%	0%	100%
MKD	no				no	no	100%	0%	0%
SRB	no				no	no	0%	81%	19%
SVN	yes				11(22) €	WFD	98%	0%	2%

ALB—Albania; BIH—Bosnia and Herzegovina; HRV—Croatia; XKX—Kosovo; *—under UN Resolution 1244; MNE—Montenegro; MKD—North Macedonia; SRB—Serbia; SVN—Slovenia; RC—recycling; BF—Backfilling; D—disposal (landfilling).

In most of the countries here, there is a certain legal waste framework primarily based on the Waste Directive Framework [6], however the implementation is partial. Albania, Bosnia and Hercegovina and Kosovo (under UN Resolution 1244) are yet to adopt the Circular Economy approach [59].

The EU countries, Slovenia and Croatia, have high recycling rates due to the recovery targets from the WFD. The Slovenian landfilling tax of EUR 11 and 22 per ton of non-hazardous and hazardous waste, respectively [60], may show the efficiency of tax as an economic instrument for increased recycling and recovery (98%). However, the share of high-quality application of these recycled materials is still in question.

With an immature market in the rest of the countries and no quality standards that would increase confidence in the recycled materials, the recycling rates are zero. North Macedonia, with a reported recycling rate of 100% is an exception; however, as no recycling facilities were reported in North Macedonia this may be attributed to statistical misinterpretation. Clearly, there is an enormous potential for change in these rates.

3.2.2. Future CDW Management Scenarios

A starting point when considering the possible CDW management scenarios for the Western Balkan countries may be the average European treatment rates shown in Figure 4. The other point may be the management drivers from the countries with high recovery rates and similar GDP per capita. The data for 2018 shows that the EU average recycling and backfilling rate of the CDW mineral fraction are 83% and 7%, respectively, while only 10% of the waste is disposed. This means that, on average, the EU has reached the WFD target even before 2020. When it comes to CDW management drivers, almost all EU countries have implemented either a landfill tax (ranging from EUR 5 to more than 100 per ton) or a landfill ban [60].

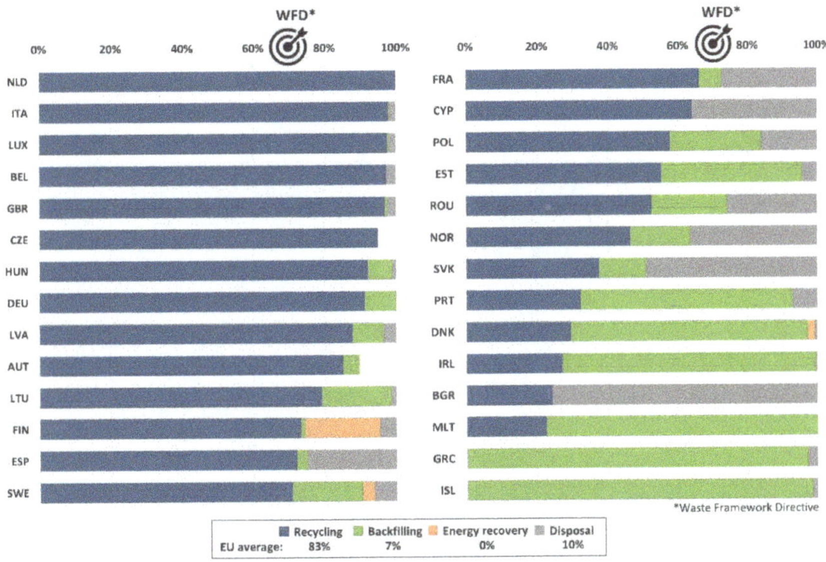

Figure 4. CDW management treatment rates in Europe (data for 2018). (ISO-3166-Aplha 3-digit codes were used for countries' abbreviations) AUT—Austria; BEL—Belgium; BGR—Bulgaria; CYP—Cyprus; CZE—Czech Republic; DEU—Germany; DNK—Denmark; ESP—Spain; EST—Estonia; FIN—Finland; FRA—France; GBR—Great Britain; GRC—Greece; HUN—Hungary; IRL—Ireland; ISL—Iceland; ITA—Italy; LUX—Luxemburg; LVA—Latvia; LTU—Lithuania; MLT—Malta; NLD—Netherlands; NOR—Norway; POL—Poland; PRT—Portugal; ROM—Romania; SVK—Slovakia; SWE—Sweden.

Apart from the introduction of landfill taxes, recycling in Slovenia and Croatia is further encouraged by the end-of-waste criteria developed for iron, steel, aluminum, copper and glass [61–63]. These criteria are oriented to increase the confidence of practitioners in recovered material use and increase the maturity of the recovered material market development. However, there are currently no quality standards developed specifically for the recovered materials and their use in new products.

One of the main drivers for high recovery rates are well established and mature markets of recovered materials. For instance, valuable lessons for secondary material market development may be acquired from the Netherlands, Germany and Denmark [9] for Europe, or the United States, Australia and South Korea [64] for the rest of the world. Additionally, aside from taxes and bans, market development barriers may be successfully and more efficiently overcome by governmental subsidies or environmental credits for green building certified by BREEAM or LEAD.

Evidently, one of the possible scenarios may be to reach the European average by a prescribed period of time. On the other hand, countries that signed the Paris Agreement [65] and that are devoted to the reduction of global GHG emission should not stop there. The transition to a circular economy directly supports this goal, so scenarios based on this approach should be included in future CDW management. Apart from quality standards and end-of-waste criteria, which have been developed by the EU, the elements of these scenarios should include high re-use and high-quality recycling rates achieved through recovery thresholds at least on public projects [35] or economic instruments that would subsidize markets of recovered materials.

3.3. Spatial and Temporal Scope of the Framework

The authors suggest that the proposed framework should be applied to the territory of the Western Balkans for several reasons. To begin with, in addition to the EU Member States Slovenia and Croatia, these countries share a combined territory of 292.3 thousand km^2 (7% of Europe) and have almost 23.7 million inhabitants [66,67]. They also share similar architectural characteristics and construction practices, especially for the buildings built in the aftermath of World War II, when most of these countries were part of one federation (Yugoslavia). As the economies still very much lean one on another, there is also a large potential for the expansion of the recovered material market to nearby countries. Finally, most of these countries are yet to adopt circular economy principles and develop new business models to support these.

The other suggestion concerns the types of construction. Taking into account the share of residential buildings in the building stock in one country and the circumstances where infrastructure objects (bridges, tunnels, dams, etc.) are rarely renovated or demolished, the authors recommend that the framework focus should be on residential buildings.

When it comes to the time limits of the study, the authors propose that residential buildings with a higher probability that either renovation or demolition activity will occur before 2050 be used in the future case studies. These are the buildings built from 1946 to 1990, which will be aged between 50 and 94 years by 2050 and will undergo renovation in order to increase their energy efficiency. Apart from demolition, two other circumstances that may generate large amounts of CDW in the future may be frequent earthquakes or major development projects. The period prior to World War II was not considered, as it was assumed that most of the building stock in these countries was built after the war. For instance, in the case of Serbia and Bosnia and Herzegovina more than 84% and 98% of the buildings, respectively, were built after 1946 [68].

3.4. Data Availability

Optimal and informed decision making requires quality and reliable data. The proposed framework requires different categories of data for all three stages. These data may come from various sources (Table 3), however the authors suggest that the assessment should utilize official sets of data whenever possible. Data from the national statistic offices are easily available and the most used data in literature. This framework uses data on the quantity and type of residential buildings in a given country and their period of construction. Most of the counties gather these types of data from their censuses of population, households and dwellings, which are organized every 10 years (i.e., 2001, 2011, 2021). Considering that most of the countries delayed their censuses for a few months or even a year due to the COVID-19 outbreak, results from the previous censuses, held in 2011 (or 2013 in case of Bosnia and Herzegovina), are used.

Table 3. A review of the framework input data and data availability.

Country	Formation of MS Database (Based on Archetypes)		Estimation of CDW Quantity			Sustainability Assessment of CDW Management Options		
	Source for Quantity Estimation and Period of Construction	Architectural Plans	CR	RR	DR	Criteria	Alternatives	Decision-Makers Preferences
Albania	instat.gov.al accessed on 1 November 2021	to be adapted	to be calculated			ECO ENVI SOC as in Table 2	as described in Section 3.2	based on local stakeholders' preferences
BIH	bhas.gov.ba accessed on 1 November 2021	TABULA	to be calculated					
Croatia	dzs.hr accessed on 1 November 2021	to be adapted	to be calculated					
Kosovo *	ask.rks-gov.net accessed on 1 November 2021	TABULA	Sandberg et al.					
MNE	monstat.org accessed on 1 November 2021	to be adapted	to be calculated					
MKD	stat.gov.mk accessed on 1 November 2021	to be adapted	to be calculated					
Serbia	stat.gov.rs accessed on 1 November 2021	TABULA	to be calculated					
Slovenia	stat.si accessed on 1 November 2021	TABULA	Sandberg et al.					

BIH—Bosnia and Herzegovina; Kosovo *—under UN Resolution 1244; MNE—Montenegro; MKD—North Macedonia; MS—Material stock; CR—construction rate; RR—renovation rate; DR—demolition rate; ECO—Economic; ENVI—Environmental; SOC—Social.

The other set of data, related to the archetypes of building, the physical characteristics of their elements and the type of materials from which these elements were made, can be obtained from the direct surveying of the residential building stocks. Conveniently, 13 European countries participated in the TABULA project, which was performed from 2009 to 2013, including Bosnia and Herzegovina, Serbia and Slovenia. This project was cofounded by the European Union with the aim of developing building typologies and calculating possible energy savings by the implementing improvement measures [68]. Buildings were classified on the basis of their type and period of construction; namely, single house and multi-house buildings (including high-rise buildings) were further divided into six groups depending on the period in which they were built (for example: before 1945, 1946–1960, 1961–1970, 1971–1980, 1981–1990, 1991–2011).

In these periods, 29 archetypes were developed for Bosnia and Herzegovina, 39 for Serbia and 24 for Slovenia [68]. In the absence of better results for the other Western Balkans countries, the authors suggest that these archetypes may also serve as the starting points for Croatia, Kosovo, Montenegro and North Macedonia, especially if we analyze the building stock for the period of 1946–1990, when all of these countries were part of one federation. One may choose to use building archetypes from the adjacent regions of neighboring countries with similar economies when deciding which available archetype to use. For instance, certain archetypes from Slovenia may be used for north Croatia, or archetypes from south Serbia may be used for Kosovo, etc.

Contrary to static building stock modelling, such as a time snapshot of building stock in 2013, the second stage of the proposed framework involves dynamic stock modelling, i.e., future changes of building stocks as developed by Sandberg et al. [69]. They used information from national statistics and demolition and renovation probability to perform the simulation of the building stock in 11 European countries by 2050. The results, based on data from 11 countries with different construction practices, showed similar results (the renovation rate ranged from 0.6% to 1.6% while the demolition rate ranged from 0.4% to 1.2%). For Serbia and Slovenia, who participated in the Sandberg et al. study [69], the future CDW management sustainability assessments may use the direct results from this study, while other Western Balkan countries should develop their own simulations on the basis of this model.

4. Discussion

4.1. Comparison with Other Sustainability Assessment Framework

The assessment framework proposed in this study is compared with other sustainability assessments studies in the CDW management domain. Details of this comparison are provided in Table 4. As shown in Table 4, other studies most often include the comparison of two CDW treatment scenarios (mostly recycling and disposal) end evaluate only the environmental and economical aspect of sustainability, while the proposed framework encompasses all CDW treatment options (from re-use to disposal) and all sustainability aspects.

Table 4. Comparison of the proposed sustainability framework and literature data.

Reference	Scope			Treatment Options	Sustainability		
	Level	Forecast	Waste Stream		ECO	ENVI	SOC
Proposed framework	NAT	√	bricks, concrete, stone, metal, wood, plastic, gypsum, glass	RU, RC, ER, D	√	√	√
[55]	REG	-	mixed CDW	RC, D	√	√	√
[44]	PRO	-	mixed CDW, wood, plastic, plasterboards, glass, metal	n/a		√	
[43]	NAT	-	concrete	RC, D	√	√	
[46]	CIT	-	mineral CDW	RC, D		√	
[53]	PRO	√	concrete	RC	√	√	√
[42]	PRO	-	concrete	RC		√	
[52]	REG	-	gypsum, cement, concrete	RC	√	√	
[45]	NAT	√	mineral CDW, metal, wood, mixed CDW	RU, RC, ER, D	√	√	
[48,49]	CIT	√	mixed CDW	RC	√		
[51]	NAT	√	mixed CDW	RC, D	√	√	
[50]	REG	-	mixed CDW	RC, D	√	√	√

NAT—national level; REG—regional level; CIT—city level; PRO—project level; RU—re-use; RC—recycling; ER—energy recovery; D—disposal; ECO—Economic; ENVI—Environmental; SOC—Social.

The proposed framework also aims to include more waste streams then other studies that were mainly focused on concrete or mixed CDW stream. The information on CDW streams quantities may lead to better forecasting and more effective CDW management decisions. To facilitate this, the proposed framework integrates a bottom-up material stock analysis, rather than using practitioners' estimations or statistical data on the CDW quantities.

When it comes to the comparison of the scope of the studies, the presented framework is designed for sustainability assessments on a national level, but can easily be applied on a regional or a city level. The principles of the framework proposed here are not country specific. The framework may be used worldwide in countries that are early adopters of the circular economy approach and with no reliable statistics related to CDW management. However certain preconditions are needed. The framework relies on the (arche)type of residential buildings to determine the material stock and this has to exist in some form in the particular country. Additionally, the renovation and demolition profile and the CDW management scenarios are also country specific and are subject to the level of economic development and the legal framework within the country.

4.2. Limitations of the Framework

The proposed framework has several main limitations. One of the most important is related to the material and building stock analysis. Namely, only residential buildings are included on the basis of the assumption that as a result of the Renovation Wave strategy, future activities will predominantly be the renovation of the residential building stock. For the same reason and due to their age, only typical residential buildings that will have existed for 50 or more years in 2040 are included.

Another limitation concerns the generalization of the entire residential building stock and its approximation with typical building representatives. As much as these representatives are rich in details regarding their physical and geometrical characteristics, they still cannot represent the entire stock. However, with no other reliable data on the stock, the authors believe that this is a good representation especially when it comes to construction practices, element types and materials used in a certain period. In order to cover the uncertainties in the quantity of materials and consequently the waste generated, the authors suggest that a sensitivity analysis should be conducted in each stage of the framework.

Finally, the environmental aspect of the sustainability assessment is focused around GHG emissions, CO_2 emissions in particular. This was suggested in order to simplify the analysis. CO_2 emissions are easily calculated and monetized. On the other hand, other emission to air such as dust, dioxins, etc. were excluded.

Furthermore, emissions to water (both surface and underground) and soil (leachate, heavy metals) were not included in the study as they are considered minimal due to the inert nature of the construction and demolition waste. However, future holistic assessments should include them as well.

4.3. Implications for CDW Management

The primary focus of the presented work was to create a framework that is based on sound research and that can be applied in practice and be as user friendly as possible. Further research will focus on framework validation through case studies, as a platform for benchmarking. Additional framework development implies the inclusion of more evaluation criteria, non-residential buildings, a wider temporal scope and different methods for MCDM analysis. The validation of the calculated waste quantities is expected to be done with Building Information Modeling (BIM) and Geographic Information System (GIS) modeling.

Decision-makers and practitioners can use the material stock database when planning renovation and demolition strategies in certain locations as well as when designing their business models. Both national and local governments may formulate more informative CDW management strategies based on the proposed framework that would benefit the environment and society at large.

5. Conclusions

The circular economy is gaining momentum worldwide, requiring that more complex and more sustainable CDW management strategies are being developed. The existing models for sustainability assessment rarely include all three aspects of sustainability (economic, environmental, social). On the other hand, almost all of the models are based on statistical records on CDW quantities that often underestimate the real CDW generation rates.

The integrated framework proposed in this work uses scenario and multi-criteria analysis that are frequently used in the decision-making process and couples them with a bottom-up material stock analysis. The framework is designed to yield three sets of results: a material stock database, the quantity of CDW and the optimal scenario for managing this CDW. Aside from the obvious use in CDW management decision-making, the authors believe that the material stock database may be a valuable resource for circular economy planning. This database scales the type and the quantity of materials built into the residential building stock. At some point this material will become waste due to renovation and demolition activity. By knowing its quantity and composition, different

CDW management strategies with higher re-use and recycle rates may be developed that support the Circular economy. The authors propose the use of AHP for the selection of the optimal CDW management scenario. This is completed in two steps: (1) through economic, environmental and social criteria comparison on the basis of experts' preferences; and (2) scenarios (alternatives) comparison on the basis of these criteria.

Future frameworks may include non-residential buildings and cover the entire building stock as well as infrastructure objects. As well as this, a temporal boundary of future research may be extended to cover buildings built after 1990. Considering that circular economy-oriented legislation is expected in the years to come, future scenarios should analyze its impact on the society in these countries.

Considering that the sustainability assessment suggested in the framework will be the first of its kind in the Western Balkans as well as in Croatia and Slovenia, the authors believe it would be very beneficial both to governments and practitioners. Governments may use it when analyzing and deciding on the circular economy policies and strategies for CDW management at both the national and local scale, while practitioners may use it to create more sustainable and more circular business models.

Author Contributions: A.N.—conceptualization, investigation, methodology, visualization and writing—original draft preparation; Z.N.—investigation, methodology and writing—review and editing; N.I.—writing—review and editing and supervision. All authors have read and agreed to the published version of the manuscript.

Funding: This research received no external funding.

Institutional Review Board Statement: Not applicable.

Informed Consent Statement: Not applicable.

Data Availability Statement: Not applicable.

Conflicts of Interest: The authors declare no conflict of interest.

References

1. Sáez, P.V.; Osmani, M. A diagnosis of construction and demolition waste generation and recovery practice in the European Union. *J. Clean. Prod.* **2019**, *241*, 118400. [CrossRef]
2. Akhtar, A.; Sarmah, A.K. Construction and demolition waste generation and properties of recycled aggregate concrete: A global perspective. *J. Clean. Prod.* **2018**, *186*, 262–281. [CrossRef]
3. Eurostat. Generation of Waste by Waste Category, Hazardousness and NACE Rev.2 Activity. Available online: https://ec.europa.eu/eurostat/databrowser/view/ENV_WASGEN$DEFAULTVIEW/default/table (accessed on 18 October 2021).
4. Bossink, B.; Brouwers, H.J.H. Construction Waste: Quantification and Source Evaluation. *J. Constr. Eng. Manag.* **1996**, *122*, 55–60. [CrossRef]
5. Eurostat European Commission. *Manual on Waste Statistics. A Handbook for Data Collection on Waste Generation and Treatment*; Publication Office of the European Union: Luxembourg, 2013.
6. The European Parliament and The Council of the European Union. *Directive 2008/98/EC on Waste and Repealing Certain Directives*; Publication Office of the European Union: Luxembourg, 2008; pp. 312/3–312/30.
7. Wu, H.; Wang, J.; Duan, H.; Ouyang, L.; Huang, W.; Zuo, J. An innovative approach to managing demolition waste via GIS (geographic information system): A case study in Shenzhen city, China. *J. Clean. Prod.* **2016**, *112*, 494–503. [CrossRef]
8. Mihai, F.-C. Construction and Demolition Waste in Romania: The Route from Illegal Dumping to Building Materials. *Sustainability* **2019**, *11*, 3179. [CrossRef]
9. Gálvez-Martos, J.-L.; Styles, D.; Schoenberger, H.; Zeschmar-Lahl, B. Construction and demolition waste best management practice in Europe. *Resour. Conserv. Recycl.* **2018**, *136*, 166–178. [CrossRef]
10. European Commission. *European Waste Catalogue*; Publication Office of the European Union: Luxembourg, 2000; pp. 1–46.
11. Agency for Toxic Substances and Disease Registry. Health Effects of Asbestos. Available online: https://www.atsdr.cdc.gov/asbestos/health_effects_asbestos.html (accessed on 1 December 2021).
12. Agency for Toxic Substances and Disease Registry. Toxicological Profile for Asbestos. National Toxicology Program. Asbestos. 2001. Available online: https://www.atsdr.cdc.gov/toxprofiles/tp61.pdf (accessed on 1 December 2021).
13. U.S. Department of Health and Human Services. Report on Carcinogens. Fourteenth Edition. National Toxicology Program. 2016. Available online: https://hero.epa.gov/hero/index.cfm/reference/details/reference_id/3827262 (accessed on 1 December 2021).
14. U.S. Environmental Protection Agency. Health Effects Assessment for Asbestos. EPA/540/1-86/049 (NTIS PB86134608). 1984. Available online: https://bit.ly/3rchVpI (accessed on 1 December 2021).

15. International Agency for Research on Cancer (IARC). Evaluation of Carcinogenic Risk to Humans. Arsenic, Metals, Fibres and Dusts. (IARC Monographs on the Evaluation of Carcinogenic Risks to Humans, No. 100C.). 2012. Available online: https://publications.iarc.fr/Book-And-Report-Series/Iarc-Monographs-On-The-Identification-Of-Carcinogenic-Hazards-To-Humans/Arsenic-Metals-Fibres-And-Dusts-2012 (accessed on 1 December 2021).
16. National Cancer Institute. Coal Tar and Coal-Tar Pitch—Cancer-Causing Substances—National Cancer Institute. Available online: https://www.cancer.gov/about-cancer/causes-prevention/risk/substances/coal-tar (accessed on 26 December 2021).
17. de Wit, M.; Ramkumar, J.H.S.; Douma, H.F.A. The Circularity Gap Report. An Analysis of the Circular State of the Global Economy. 2018. Available online: https://www.circle-economy.com/news/the-circularity-gap-report-our-world-is-only-9-circular#.W09Rh9IzZPY (accessed on 1 November 2021).
18. European Commision. Clossing the Loop—An EU Action Plan for the Circular Economy, COM(2015). Brussels, 2015. Available online: https://eur-lex.europa.eu/legal-content/EN/TXT/?uri=CELEX:52015DC0614 (accessed on 1 November 2021).
19. United Nations General Assembly. *Transforming Our World: The 2030 Agenda for Sustainable Development*; United Nations: New York, NY, USA, 2015.
20. European Commisssion. Report on the Implementation of the Circular Economy Action Plan. Brussels, 2019. Available online: https://ec.europa.eu/info/publications/report-implementation-circular-economy-action-plan-1_en (accessed on 1 November 2021).
21. European Commisssion. A New Circular Economy Action Plan. For a Cleaner and More Competitive Europe. Brussels, 2020. Available online: https://eur-lex.europa.eu/legal-content/EN/TXT/?qid=1583933814386&uri=COM:2020:98:FIN (accessed on 1 November 2021).
22. European Commisssion. The European Green Deal. 2019. Available online: https://eur-lex.europa.eu/legal-content/EN/TXT/?uri=COM%3A2019%3A640%3AFIN (accessed on 1 November 2021).
23. European Commisssion. Sustainable Europe Investment Plan. European Green Deal Investment Plan. 2020. Available online: https://eur-lex.europa.eu/legal-content/EN/TXT/?uri=CELEX%3A52020DC0021 (accessed on 1 November 2021).
24. European Commisssion. A Renovation Wave for Europe—Greening Our Buildings, Creating Jobs, Improving Lives. 2020. Available online: https://eur-lex.europa.eu/legal-content/EN/TXT/?uri=CELEX%3A52020DC0662 (accessed on 1 November 2021).
25. Eurostat. Treatment of Waste by Waste Category, Hazardousness and Waste Management Operations. Available online: https://ec.europa.eu/eurostat/databrowser/view/ENV_WASTRT__custom_1426044/default/table (accessed on 18 October 2021).
26. Zhang, C.; Hu, M.; Yang, X.; Miranda-Xicotencatl, B.; Sprecher, B.; Di Maio, F.; Zhong, X.; Tukker, A. Upgrading construction and demolition waste management from downcycling to recycling in the Netherlands. *J. Clean. Prod.* **2020**, *266*, 121718. [CrossRef]
27. Di Maria, A.; Eyckmans, J.; Van Acker, K. Downcycling versus recycling of construction and demolition waste: Combining LCA and LCC to support sustainable policy making. *Waste Manag.* **2018**, *75*, 3–21. [CrossRef]
28. Badraddin, A.K.; Rahman, R.A.; Almutairi, S.; Esa, M. Main Challenges to Concrete Recycling in Practice. *Sustainability* **2021**, *13*, 11077. [CrossRef]
29. European Commisssion. EU Construction Sector: In Transition towards a Circular Economy. 2019. Available online: https://ec.europa.eu/docsroom/documents/34904 (accessed on 1 November 2021).
30. Bao, Z.; Lu, W. Developing efficient circularity for construction and demolition waste management in fast emerging economies: Lessons learned from Shenzhen, China. *Sci. Total Environ.* **2020**, *724*, 138264. [CrossRef]
31. Ruiz, L.A.L.; Ramón, X.R.; Domingo, S.G. The circular economy in the construction and demolition waste sector—A review and an integrative model approach. *J. Clean. Prod.* **2019**, *248*, 119238. [CrossRef]
32. Liu, J.; Wu, P.; Jiang, Y.; Wang, X. Explore potential barriers of applying circular economy in construction and demolition waste recycling. *J. Clean. Prod.* **2021**, *326*, 129400. [CrossRef]
33. Ratnasabapathy, S.; Alashwal, A.; Perera, S. Exploring the barriers for implementing waste trading practices in the construction industry in Australia. *Built Environ. Proj. Asset Manag.* **2021**, *11*, 559–576. [CrossRef]
34. Chen, J.; Hua, C.; Liu, C. Considerations for better construction and demolition waste management: Identifying the decision behaviors of contractors and government departments through a game theory decision-making model. *J. Clean. Prod.* **2018**, *212*, 190–199. [CrossRef]
35. Ghaffar, S.H.; Burman, M.; Braimah, N. Pathways to circular construction: An integrated management of construction and demolition waste for resource recovery. *J. Clean. Prod.* **2020**, *244*, 118710. [CrossRef]
36. Tam, V.W.-Y.; Lu, W. Construction Waste Management Profiles, Practices, and Performance: A Cross-Jurisdictional Analysis in Four Countries. *Sustainability* **2016**, *8*, 190. [CrossRef]
37. Nunes, K.; Mahler, C.F. Comparison of construction and demolition waste management between Brazil, European Union and USA. *Waste Manag. Res.* **2020**, *38*, 415–422. [CrossRef]
38. Lu, W.; Chen, X.; Peng, Y.; Shen, L. Benchmarking construction waste management performance using big data. *Resour. Conserv. Recycl.* **2015**, *105*, 49–58. [CrossRef]
39. Lu, W.; Webster, C.; Chen, K.; Zhang, X.; Chen, X. Computational Building Information Modelling for construction waste management: Moving from rhetoric to reality. *Renew. Sustain. Energy Rev.* **2017**, *68*, 587–595. [CrossRef]
40. Li, C.Z.; Zhao, Y.; Xiao, B.; Yu, B.; Tam, V.W.; Chen, Z.; Ya, Y. Research trend of the application of information technologies in construction and demolition waste management. *J. Clean. Prod.* **2020**, *263*, 121458. [CrossRef]

41. Han, D.; Kalantari, M.; Rajabifard, A. Building Information Modeling (BIM) for Construction and Demolition Waste Management in Australia: A Research Agenda. *Sustainability* **2021**, *13*, 12983. [CrossRef]
42. Marinković, S.; Dragaš, J.; Ignjatović, I.; Tošić, N. Environmental assessment of green concretes for structural use. *J. Clean. Prod.* **2017**, *154*, 633–649. [CrossRef]
43. Mah, C.M.; Fujiwara, T.; Ho, C.S. Life cycle assessment and life cycle costing toward eco-efficiency concrete waste management in Malaysia. *J. Clean. Prod.* **2018**, *172*, 3415–3427. [CrossRef]
44. Pantini, S.; Rigamonti, L. Is selective demolition always a sustainable choice? *Waste Manag.* **2019**, *103*, 169–176. [CrossRef]
45. Dahlbo, H.; Bachér, J.; Lähtinen, K.; Jouttijärvi, T.; Suoheimo, P.; Mattila, T.; Sironen, S.; Myllymaa, T.; Saramäki, K. Construction and demolition waste management—A holistic evaluation of environmental performance. *J. Clean. Prod.* **2015**, *107*, 333–341. [CrossRef]
46. Yazdanbakhsh, A. A bi-level environmental impact assessment framework for comparing construction and demolition waste management strategies. *Waste Manag.* **2018**, *77*, 401–412. [CrossRef]
47. Ding, Z.; Zhu, M.; Tam, V.W.; Yi, G.; Tran, C. A system dynamics-based environmental benefit assessment model of construction waste reduction management at the design and construction stages. *J. Clean. Prod.* **2018**, *176*, 676–692. [CrossRef]
48. Coelho, A.; de Brito, J. Economic viability analysis of a construction and demolition waste recycling plant in Portugal—part I: Location, materials, technology and economic analysis. *J. Clean. Prod.* **2013**, *39*, 338–352. [CrossRef]
49. Coelho, A.; de Brito, J. Economic viability analysis of a construction and demolition waste recycling plant in Portugal—Part II: Economic sensitivity analysis. *J. Clean. Prod.* **2013**, *39*, 329–337. [CrossRef]
50. Coronado, M.; Dosal, E.; Coz, A.; Viguri, J.R.; Andrés, A. Estimation of Construction and Demolition Waste (C&DW) Generation and Multicriteria Analysis of C&DW Management Alternatives: A Case Study in Spain. *Waste Biomass Valorization* **2011**, *2*, 209–225. [CrossRef]
51. Marzouk, M.; Azab, S. Environmental and economic impact assessment of construction and demolition waste disposal using system dynamics. *Resour. Conserv. Recycl.* **2014**, *82*, 41–49. [CrossRef]
52. Silgado, S.S.; Valdiviezo, L.C.; Domingo, S.G.; Roca, X. Multi-criteria decision analysis to assess the environmental and economic performance of using recycled gypsum cement and recycled aggregate to produce concrete: The case of Catalonia (Spain). *Resour. Conserv. Recycl.* **2018**, *133*, 120–131. [CrossRef]
53. Wijayasundara, M.; Mendis, P.; Crawford, R.H. Integrated assessment of the use of recycled concrete aggregate replacing natural aggregate in structural concrete. *J. Clean. Prod.* **2018**, *174*, 591–604. [CrossRef]
54. Taelman, S.; Sanjuan-Delmás, D.; Tonini, D.; Dewulf, J. An operational framework for sustainability assessment including local to global impacts: Focus on waste management systems. *Resour. Conserv. Recycl. X* **2019**, *2*, 100005. [CrossRef]
55. Iodice, S.; Garbarino, E.; Cerreta, M.; Tonini, D. Sustainability assessment of Construction and Demolition Waste management applied to an Italian case. *Waste Manag.* **2021**, *128*, 83–98. [CrossRef]
56. Stojadinović, Z.; Kovačević, M.; Marinković, D.; Stojadinović, B. Rapid earthquake loss assessment based on machine learning and representative sampling. *Earthq. Spectra* **2021**, 1–26. [CrossRef]
57. Tirth, V.; Singh, R.K.; Islam, S.; Badruddin, I.A.; Abdullah, R.A.B.; AlGahtani, A.; Mahmoud, E.R.; Arabi, A.; Shukla, N.K.; Gupta, P. Kharif Crops Selection for Sustainable Farming Practices in the Rajasthan-India Using Multiple Attribute-Based Decision-Making. *Agronomy* **2020**, *10*, 536. [CrossRef]
58. Saaty, T.L. How to make a decission: The Analytical Hierarchy Process. *Eur. J. Oper. Res.* **1990**, *48*, 9–26. [CrossRef]
59. Maleš, I. Guidelines on Circular Economy for the Countries of the Western Balkans and Turkey. Brussels, 2020. Available online: https://eeb.org/library/guidelines-on-the-circular-economy-for-western-balkan-countries-and-turkey/ (accessed on 1 December 2021).
60. The Confederation of European Waste-to-Energy (CEWEP). Landfill Taxes and Bans. Overvew. 2021. Available online: https://www.cewep.eu/landfill-taxes-and-bans/ (accessed on 6 September 2021).
61. Muchová, L.; Eder, P. *End-of-Waste Criteria for Iron and Steel Scrap: Technical Proposals. EUR 24397 EN*; Publications Office of the European Union: Luxembourg, 2010.
62. Villanueva, A.; Eder, P. *End-of-Waste Criteria for Glass Cullet: Technical Proposals. EUR 25220 EN*; Publications Office of the European Union: Luxembourg, 2011.
63. Muchova, L.; Eder, P.; Villanueva, A. *End-of-Waste Criteria for Aluminium and Aluminium Alloy Scrap. Technical Proposals. EUR 24396*; Publications Office of the European Union: Luxembourg, 2011.
64. Wu, Z.; Yu, A.T.; Poon, C.S. Promoting effective construction and demolition waste management towards sustainable development: A case study of Hong Kong. *Sustain. Dev.* **2020**, *28*, 1713–1724. [CrossRef]
65. United Nations Climate Change. Paris Agreement—Status of Ratification | UNFCCC. Available online: https://unfccc.int/process/the-paris-agreement/status-of-ratification (accessed on 2 November 2021).
66. The World Bank. Land Area (sq. km) | Data. Available online: https://data.worldbank.org/indicator/AG.LND.TOTL.K2 (accessed on 2 November 2021).
67. The World Bank. Population, Total. Available online: https://data.worldbank.org/indicator/SP.POP.TOTL (accessed on 1 November 2021).

68. IEE Project TABULA. Typology Approach for Building Stock Energy Assessment. Available online: https://episcope.eu/iee-project/tabula/ (accessed on 1 November 2021).
69. Sandberg, N.H.; Sartori, I.; Heidrich, O.; Dawson, R.; Dascalaki, E.; Dimitriou, S.; Vimm-R, T.; Filippidou, F.; Stegnar, G.; Zavrl, M.Š.; et al. Dynamic building stock modelling: Application to 11 European countries to support the energy efficiency and retrofit ambitions of the EU. *Energy Build.* **2016**, *132*, 26–38. [CrossRef]

Review

Life Cycle Assessment on Construction and Demolition Waste: A Systematic Literature Review

Jaime A. Mesa [1], Carlos Fúquene-Retamoso [1] and Aníbal Maury-Ramírez [2,*]

1 Departamento de Ingeniería Industrial, Facultad de Ingeniería, Pontificia Universidad Javeriana, Bogotá 110231, Colombia; mesajaime@javeriana.edu.co (J.A.M.); cfuquene@javeriana.edu.co (C.F.-R.)
2 Engineering Faculty, Universidad El Bosque, Bogotá 111711, Colombia
* Correspondence: amaury@unbosque.edu.co; Tel.: +57-1-6489000 (ext. 1285)

Abstract: Life Cycle Assessment (LCA) is considered an innovative tool to analyze environmental impacts to make decisions aimed at improving the environmental performance of building materials and construction processes throughout different life cycle stages, including design, construction, use, operation, and end-of-life (EOL). Therefore, during the last two decades, interest in applying this tool in the construction field has increased, and the number of articles and studies has risen exponentially. However, there is a lack of consolidated studies that provide insights into the implementation of LCA on construction and demolition waste (C&DW). To fill this research gap, this study presents a literature review analysis to consolidate the most relevant topics and issues in the research field of C&DW materials and how LCA has been implemented during the last two decades. A systematic literature search was performed following the PRISMA method: analysis of selected works is based on bibliometric and content-based approaches. As a result, the study characterized 150 selected works in terms of the evolution of articles per year, geographical distribution, most relevant research centers, and featured sources. In addition, this study highlights research gaps in terms of methodological and design tools to improve LCA analysis, indicators, and connection to new trending concepts, such as circular economy and industry 4.0.

Keywords: demolition waste; management; life cycle assessment; circular economy; sustainability; concrete; recycling

Citation: Mesa, J.A.; Fúquene-Retamoso, C.; Maury-Ramírez, A. Life Cycle Assessment on Construction and Demolition Waste: A Systematic Literature Review. *Sustainability* **2021**, *13*, 7676. https://doi.org/10.3390/su13147676

Academic Editor: Václav Nežerka

Received: 13 June 2021
Accepted: 7 July 2021
Published: 9 July 2021

1. Introduction

The need for more buildings and infrastructure parallels the population growth and natural expansion of cities and urban projects. In this context, the construction industry has an important role in increasing greenhouse gas emissions and global warming [1]. It is estimated that approximately 40% of all raw materials obtained from the lithosphere are consumed by the construction sector, representing almost 50% of global greenhouse gas emissions [2]. In addition, construction involves high consumption of building materials, water use, and improper waste management during the EOL phase [3].

The sustainability issues previously mentioned are contrasted with new policies to promote the reduction, reuse, and recycling of construction and demolition waste (C&DW), which provide essential reductions in the consumption of virgin materials and impacts associated with primary production and transportation. Research efforts during the last two decades have been oriented to include C&DW materials as an aggregate of new concrete mixtures, for example, [4–8]; new design approaches include the use of building information modeling (BIM), e.g., [9,10]; and others works, for instance, [11–13] have evaluated environmental benefits of using C&DW as replacement of virgin materials without affecting mechanical properties significantly.

All of these research approaches have some methodological tools in common. The one most employed is the LCA, which allows consolidating, comparing, and assessing sustainability impacts through environmental, economic, and social indicators. On a broader scope, sustainability of infrastructure and buildings projects are studied from a LCA perspective, since it can highlight environmental and economic drivers in C&DW management [14]; LCA is also employed to make decisions in early design phases, such as the selection of materials or in EOL scenarios to define best suitable options among recycling, reuse, or disposal of materials [15,16]. The usefulness of LCA lies in the determination and comparison of impacts to make decisions that will affect the entire life cycle of any building or infrastructure project.

In the construction field and related explicitly to the C&DW, LCA is commonly used to select the most suitable scenario between landfill, recycling, and incineration [3] but is not commonly considered from a design perspective. A well-defined challenge in this research field is to include life cycle stages of buildings and infrastructures, such as the design, construction, use, operation, and EOL, including environmental interventions related to construction materials (primary production or extraction, transportation, and fabrication of construction materials), construction and maintenance activities, and not only to dismantling operations during the EOL phase [17]

Some literature reviews related to the topic of LCA for C&DW are available; however, they do not cover the same observation window and imply different research purposes. Some examples of the most relevant reviews include Laurent et al., 2014 [18], who presented a comprehensive revision of LCA research works on solid waste management systems to consolidate their main findings and learnings. In addition, they also summarize main findings and research gaps to address new research opportunities. Bovea and Powell 2016 [19] developed a review of the literature related to the application of LCA methodology for assessing the environmental performance of C&DW systems. This work aimed to create a general mapping of existing literature and summarizes the best practices in compliance with the conventional LCA framework.

Similarly, Rodrigues Vieira et al., 2016 [20] presented a review oriented to analyze the use of LCA methodologies during the manufacturing of ecological concrete from C&DW materials. Later, Wu et al., 2019 [21] developed a study of the performance assessment methods (including LCA) for C&DW management. In addition, they proposed a framework for improving the assessment of waste management systems.

Thus, this article presents a systematic and broader literature review of works related to LCA applied to materials derived from C&DW, considering the flow of materials toward the development of circular economy in the construction sector. A bibliometric and content-based analysis highlights common findings, research gaps, and future research trends. On the one hand, bibliometric analysis comprised the evolution of the number of articles during the last two decades, the geographical distribution of works, featured researchers, most relevant research centers and journals, and most cited articles. Furthermore, content-based analysis studied the objective and methodology followed by authors, practical applications, C&DW materials analyzed, and the most common LCA parameters employed in previous studies.

2. Methods

This section describes the methodological steps to collect, screen, and analyze selected works from the existing literature. There is a methodology for the literature review and another for analyzing selected works, which is performed following bibliometric and content-based analysis.

2.1. Literature Review

Search and collection of research articles was performed using the SCOPUS database and complementary searches in Google Scholar. To facilitate the search procedure, we used a search query (Appendix A) that includes keywords such as "LCA", "construction waste", "demolition waste", among others. Furthermore, the search query included additional filters to limit the type of documents (articles from only journals, English language, and matching of extracted keywords).

The PRISMA approach was used to systematically identify the most relevant works regarding the topic of LCA and C&DW (Figure 1). First, 209 records were collected from the SCOPUS database and secondary searches after eliminating duplicates. Later, 173 articles were classified as related works after revising title, abstract, methods, and conclusions. Finally, 150 entries were classified as selected works after performing a detailed and complete revision of articles.

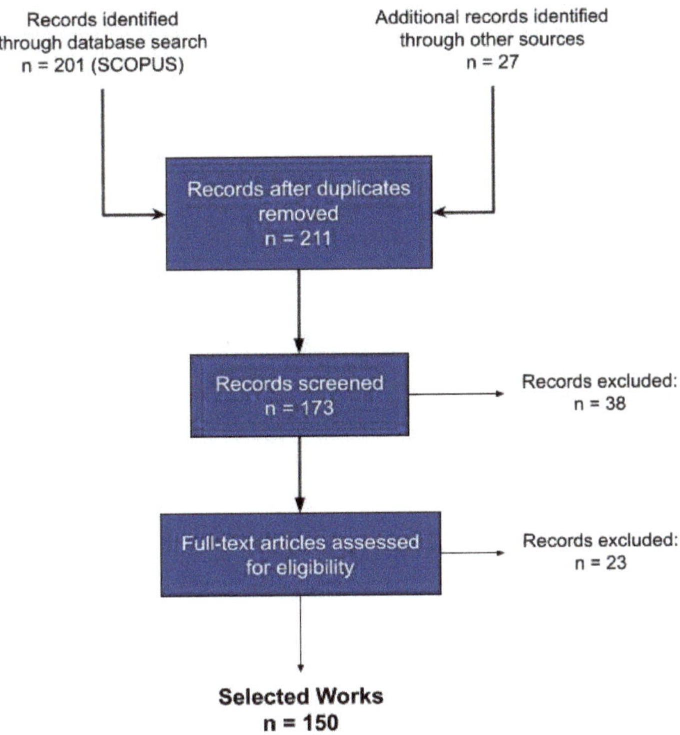

Figure 1. PRISMA literature review methodology to identify selected works.

2.2. *Bibliometric and Based-Content Analysis*

The analysis of the 150 selected works was divided into two main subsections: the first dedicated to summarizing bibliometric attributes, and the second focused on the content-based study of selected works. The bibliometric analysis provides useful information about the overall attributes of articles, authors, journals, among others. For this analysis, bibliometric software was not used for analyzing data.

On the other hand, the content-based analysis aims to study in detail the objectives, methodologies, and applicability of selected works and future research directions. Table 1 summarizes the investigated aspects covered in this study and the possible categories considered for each one. In addition to the research aspects shown in Table 1, this study analyzed the use of LCA indicators, the most common C&DW materials studied in the selected works, and main research topics and trends related to LCA applications on C&DW.

Table 1. Detail of investigated aspects in the content-based analysis.

Investigated Aspect	Categories	Category Definition
Objective	Evaluate, assess, or quantify	The article aims to measure life cycle impacts and compare data
	Propose methodology, guideline, or indicator	The article proposes a new methodological approach or indicator
	Explore or analyze	The article analyzes relationships, data, and results from previous research
Methodology	Analytical	The article is based on analytical models, experimentation, or mixed approaches to calculate and determine key indicators and compare results
	Framework	The article proposed a framework to analyze and evaluate EOL scenarios or life cycle implications of using C&DW materials
	Review	The article employed a literature review approach
	Survey	The article gathers information and research gaps from surveys
Applicability	EOL decision making	The article provides helpful information to compare and select the most suitable EOL scenario
	Material selection	The article provides helpful information to select materials in the fabrication of new products
	Measurement of sustainability impacts	The article provides helpful information about how to measure sustainability impacts in a nonconventional way

3. Results

3.1. Bibliometric Analysis

This subsection includes the evolution of research articles, contributions by region, most prominent research clusters, most relevant journals, and top-cited articles during the last two decades, among others. Each one of such aspects is described in detail below. Figure 2 shows the evolution of the number of articles during the last two decades. It demonstrates that LCA and C&DW are growing topics, especially during the last five years (Table 2). Regarding research articles, an increase from one or two works up to 37 works in less than 15 years is remarkable. This reveals an expanding research interest motivated for worldwide initiatives such as sustainable development goals, policies, and legislation related to sustainability and a more environmentally conscious society.

Table 2. Consolidated summary of selected works obtained from the systematic literature search.

	Time Interval			
	2015 to 2021	**2010 to 2014**	**2005 to 2009**	**2000 to 2004**
References	[1,5–14,19–126]	[3,4,15,18,127–146]	[17,147–149]	[150]

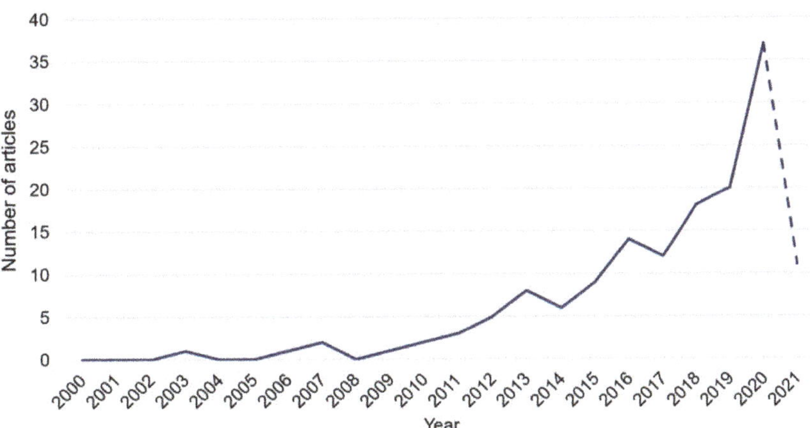

Figure 2. Evolution of selected works regarding LCA + C&DW during the last two decades. (The number of articles for 2021 includes only January through April).

In terms of contributions per region, Europe and Asia have a vast advantage over the rest of the world. Just a few selected works were developed in North America, South America, and Oceania. Africa has minimum participation (one contribution) in the sample of selected works. Figure 3a summarizes the distribution of selected works by region. Figure 3b, on the other hand, shows a detailed distribution of selected works per country. Countries with a higher number of contributions based on author affiliations are China (30), Spain (23), Italy (22), Australia (12), and the USA (12).

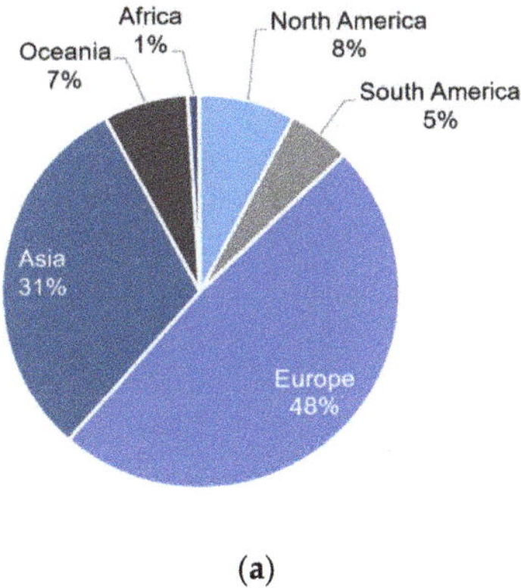

(a)

Figure 3. *Cont.*

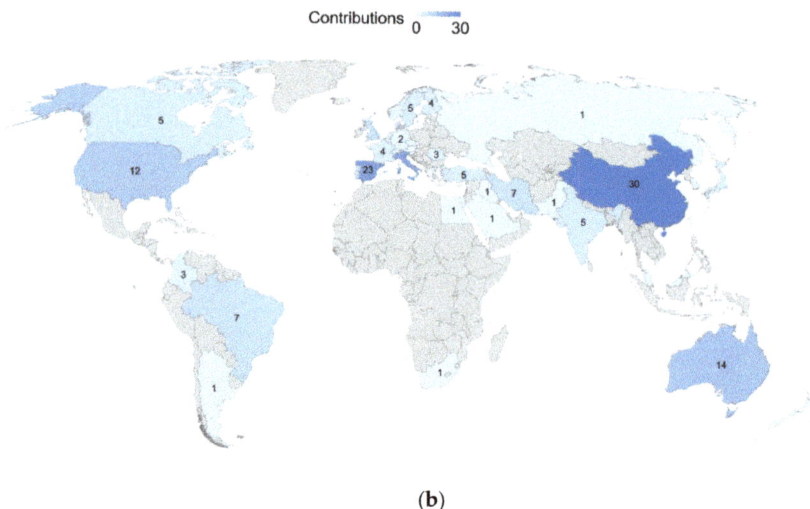

(**b**)

Figure 3. Geographical distribution of selected works: (**a**) distribution per region and (**b**) distribution per country.

After summarizing the analysis of selected works regarding affiliation of authors, three leading universities are identified (Figure 4a): (i) The Hong Kong Polytechnic University (China), (ii) Shenzen University (China), and (iii) The University of Adelaide (Australia). In addition, four universities from Italy and three from Spain were identified and important actors in the topic of LCA and C&DW. Finally, from Latin America, just one university is positioned in this ranking, the University of Campinas (Brazil).

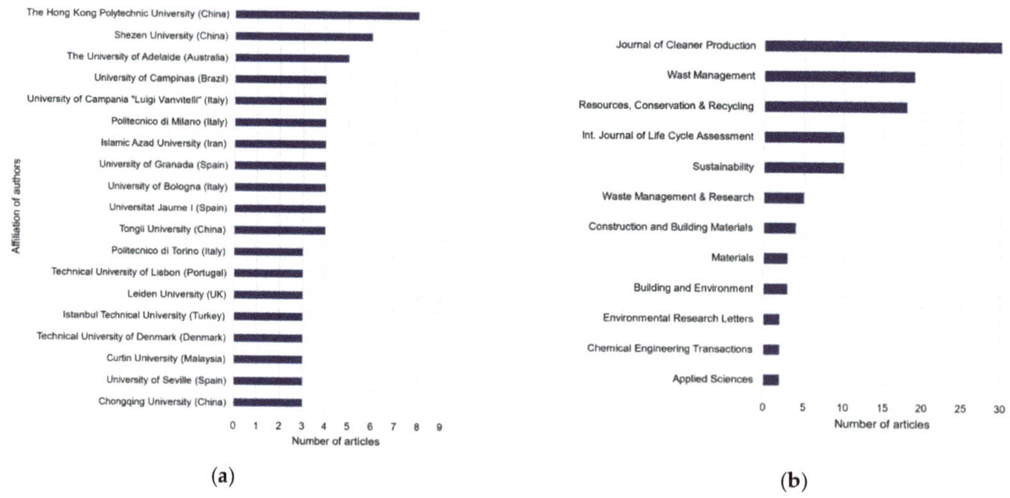

(**a**) (**b**)

Figure 4. Most relevant journals and affiliation of authors. (**a**) Most relevant affiliations of authors (includes affiliations with at least three articles). (**b**) Detail of most representative journals for selected works (includes journals with more than three research articles during the last two decades).

The most relevant journals in the topic of LCA and C&DW are *Journal of Cleaner Production* with 30 selected articles, followed by *Waste Management* (19), *Resources, Conserva-*

tion & Recycling (18), and the *International Journal of Life Cycle Assessment* (10). Figure 4b summarizes the contributions per journal with at least two records. The other journals not mentioned represent individual contributions.

From the analysis of authors' affiliation, it was possible to identify four main research clusters among authors with several contributions in the LCA application on C&DW management and decision-making. Table 3 includes the most relevant research clusters.

Table 3. Main research clusters of researchers (collaborative research works) worldwide during the last two decades.

Research Works	Most Relevant Authors	Topics Studied
[22,60,68,71,72,86,99]	LP Rosado; P Vitale; C Penteado	LCA of natural and mixed recycled aggregate; Waste management of C&DW; Influence of disposal fees on municipal waste management of C&DW; LCA of EOL of residential buildings; Attributional LCA of Italian residential multifamily building.
[56,57,70,85]	MD Uzzal Hossain; CS Poon	Comparative environmental evaluation of construction waste management; Comparative environmental evaluation of aggregate production from C&DW; Upcycling wood waste into cement-bonded particleboard
[31,62,76,91,151]	S Zanni; A Bonoli; GM Cuenca-Moyano	Environmental assessment and life cycle inventory of mortars made of natural and recycled aggregates; Framework for circular economy in buildings; Environmental impact of natural inert and recycled C&DW processing using LCA.
[74,82,96,124]	S Pantini; L Rigamonti; G Borghi	Selective demolition; resource-efficient management of asphalt waste; resource-efficient strategies for managing post-consumer gypsum; LCA of nonhazardous C&DW

From the sample of 150 selected works, 11 articles were identified as key contributions based on their cumulated citations until the literature search date (21 April 2021). Table 4 comprises a brief description of the 11 articles.

3.2. Content-Based Analysis

Selected works were classified according to their objective and methodology, which were defined previously in Section 2.2. Figure 5 shows a graphical summary of the objective vs. methodology for the 150 selected works. As the main result, it is remarkable that 72% of selected works are oriented to evaluate, assess, or quantify sustainability impacts from an analytical perspective. In the second place, articles proposing methodologies, guidelines, or indicators represent 18%, where most works followed an analytical approach, and just one article is categorized as a framework. In third place, articles aiming to explore or analyze represent 10%. Finally, in this last category, most works are review-type, and just a few are based on other methodologies such as surveys or frameworks.

Table 4. List of most cited selected works. Articles with at least 100 citations (until April 2021).

Reference	Citations	Brief Description
Laurent 2014 [18]	323	Review article that analyzes the knowledge from 222 published LCA studies of solid waste management systems from 1995 to 2012.
Blenglini 2009 [17]	286	Study about LCA of residential buildings in an urban area under demolition and redesign.
Yeheyis 2013 [15]	179	Framework to maximize the implementation of 3R strategies (reduce, reuse, and recycling) and minimize the disposal of construction and demolition waste through the use of LCA.
Blengini 2010 [141]	155	Analysis of environmental implications of C&DW recycling chain in Torino, Italy, using a geographical information system and LCA.
Bianchini 2012 [127]	129	Review article that analyzes the social cost–benefit of green roofs. A case study involves the use of C&DW for constructing roof layers.
Coelho 2012 [145]	119	Study of environmental impacts of buildings through their whole life cycle considering different waste/material management options.
Hossain 2016 [56]	119	Comparative LCA of aggregate production using recycled waste materials and virgin sources from first-hand data.
Ortiz 2010 [3]	112	Evaluation of environmental impacts of construction waste in terms of the Life Program Environment Directive of the European Commission.
Cao 2015 [42]	110	Comparison between prefabrication technology and cast in situ technology. Environmental performance is evaluated through LCA.
Knoeri 2013 [132]	108	Analysis of life cycle impacts of recycled concrete mixtures with different cement types and compared to conventional concrete.
Dahlbo 2015 [43]	106	Assessment of the performance of C&DW management systems from an environmental and economic perspective.

The practical application of selected works is summarized in Figure 6. most of the works (40%) are focused on the decision-making of the most suitable EOL strategy (recycling offsite, onsite, and landfill). EOL strategies are commonly contrasted using LCA indicators and transportation or reprocessing costs. Another relevant identified application (38%) compares the technical performance of C&DW materials versus virgin materials. Such analysis provides helpful information and methodological approaches for selecting materials and their fractions to fabricate new products. Lastly, the measurement of environmental or mechanical properties of construction materials using fractions of C&DW is also identified (6%), comprising methods, indicators, and assessment frameworks. Sixteen percent of selected works represent very particular applications or individual contributions, including analyzing waste recovery goals in specific regions, evaluating land consumption, and evaluating conventional vs. selective demolition.

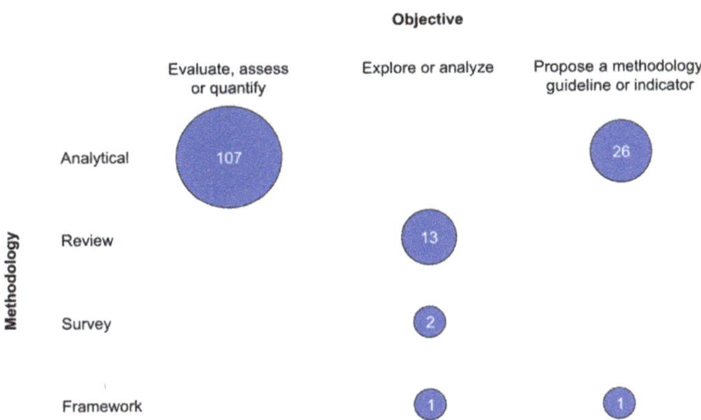

Figure 5. Classification of selected works in terms of objectives vs. methodologies applied.

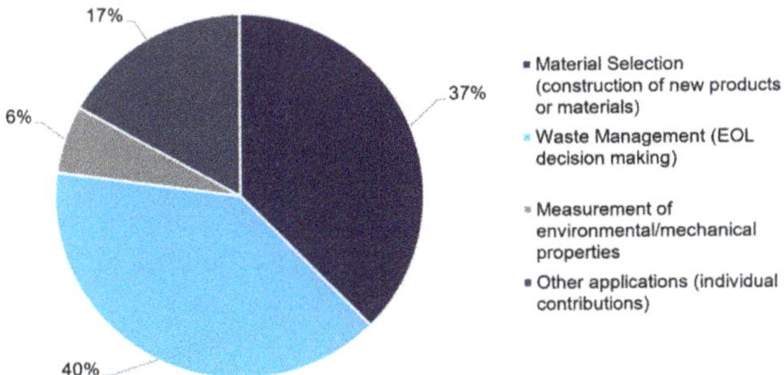

Figure 6. Distribution of selected works based on their practical application.

Regarding the construction materials analyzed in the selected works, it was found that 49% of them (72 works) considered a very diverse group of materials that included ceramics (concrete, cement, bricks, gypsum, drywall, glass), metals (steel, iron, aluminum, copper), and polymers (insulation, rubber). Another important group of works was focused solely on aggregates from the diverse origin (19%), concrete aggregates (10%), and just nine works (6%) were oriented to a unique material (gypsum, asphalt, bricks, polymers). Fifteen percent of selected works (22) did not specify the construction material. Figure 7 shows the distribution of selected works for the categories in terms of absolute values.

LCA parameters most employed are global warming potential, acidification, energy consumption, eutrophication, and photochemical ozone creation potential, which are conventional parameters included in LCA approaches. Some interesting and nonconventional parameters such as cost, person-year equivalent, land occupation, and Eco-Indicator 99 were also found less frequently in the selected works. Figure 8 summarizes and lists the LCA parameters most used in the selected works.

A detailed analysis of data shown in this section regarding bibliometric and content-based analysis is performed in Section 4 (Discussion).

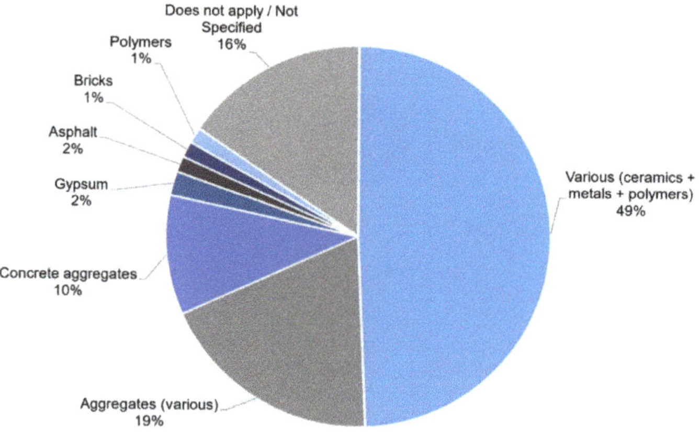

Figure 7. Distribution of materials covered in the selected works.

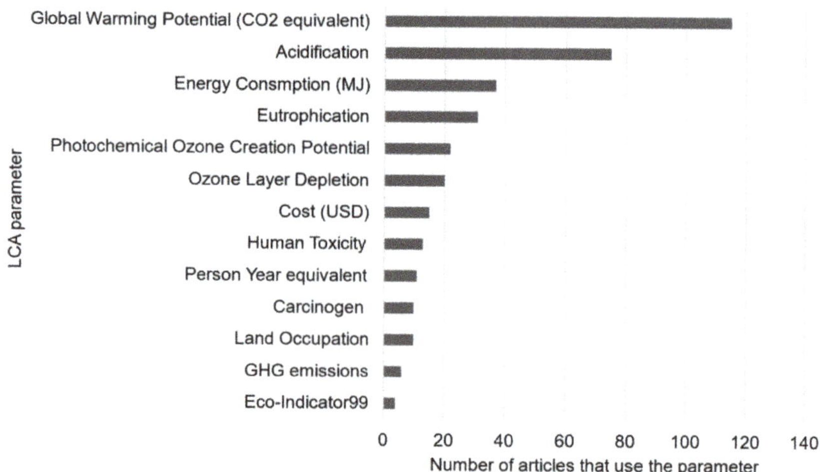

Figure 8. Most used LCA parameters in selected works.

3.2.1. Research Topics

This subsection describes four main topics identified in the literature analyzed. Such topics summarize the most common and relevant themes covered in LCA and C&DW during the last two decades. Each topic is described in detail as follows:

- Material Selection/Definition of C&DW Fraction for Construction Applications

The most common topic related to LCA use is the decision-making on choosing construction materials and defining the range of C&DW fractions in new products or building projects. In this topic, concrete was the material most studied, commonly analyzing the carbon footprint, e.g., [128,152]; carbon footprint and embodied energy [134], greenhouse gas emissions and land-use change [7], or analyzing contamination issues such as content and leaching of sulfates in recycled aggregates [146]. Other works were focused on developing new materials or construction products, and using fractions of C&DW as aggregate or filling material. Typical applications of previous works were focused on concrete, asphalt, and bricks.

Related to concrete, [98] compared two use-cycles of natural aggregate concrete and recycled aggregate concrete to analyze the environmental impacts of recycled materials. Results demonstrated the benefits of using recycled materials, such as environmental savings, primary resource use, embodied energy, embodied emissions, and reduced pressure on landfill sites. Reference [52] evaluated the energy consumption of all processes to manufacture concrete curbs using recycled aggregates for replacing natural sand aggregates.

Other works are related to brick manufacturing with some fraction of C&DW, and acceptable results concerning mechanical and durability properties and significant environmental impact reduction. For example, [83] performed an analysis of concrete and ceramic remains used to partially substitute clay soil to produce unfired bricks. Reference [126] compared reusable blocks with recycled brick aggregate, reusable blocks with recycled concrete, reusable blocks with natural aggregate, and regular concrete wall in mechanical properties and environmental impacts. Reference [33] employed LCA to demonstrate significant environmental benefits with the use of demolished concrete blocks over conventional concrete. Similarly, [57] assessed and compared the environmental impacts and sustainability associated with natural blocks manufactured with virgin materials and three generations of ecoblocks manufactured with C&DW. Other authors, e.g., [62,76], employed LCA to quantify and compare the environmental impacts associated with the production of masonry mortar manufactured with different amounts of fine natural and recycled aggregate from C&DW.

Other less common materials were studied in the selected works analyzed; for instance, [143] performed a study to identify the stages that produce the most significant impact on the environment (materials and processes) in the use of ceramic tiles. Likewise, [137] developed a comparative LCA on the energy requirements and implications of greenhouse emissions of recycling construction and demolition rubble and container glass. Reference [55] performed an LCA approach to evaluate the performance of recovered gypsum waste to manufacture ordinary Portland cement. Lastly, [103] demonstrated the calculation of the economic potential and environmental impact of reused steel building elements.

Regarding nonconventional materials, it was found that some authors worked on possible composite materials not directly employed in construction applications, for example, [37] who compared the use of waste materials derived from C&DW as alternative filers in the production of thermoplastic composites using recycled high-density polyethylene as a matrix material. Likewise, [94] assessed the environmental impact of wood polymer composite production using specific C&DW fractions (wood, plastic, plasterboard, and mineral wool) compared to conventional waste treatment scenarios such as landfilling and incineration. C&DW has been used as complementary material in some cases, e.g., [144], who developed a bamboo and C&DW residential building prototype and its LCA compared to a typical brick–concrete building, demonstrating benefits in embodied energy from the recycling of demolition waste.

Other studies were oriented to analyze different life cycle phases of construction materials, as in the case of [77], who assessed the environmental impacts on the production of aggregates via each scenario using life cycle assessment (virgin vs. recycled aggregates), including energy consumption and CO_2 emissions as the comparative indicator.

- EOL Decision-Making

The selection of the most suitable EOL strategy to manage C&DW is one of the most studied topics in the literature. Commonly, it is used to compare decision scenarios such as landfilling materials, recycling, and analysis of C&DW use in different fractions and incineration to obtain thermal energy. Commonly, the LCA approach is used to determine what alternative is better, according to the case study. Some relevant examples of this LCA application include [3], who evaluated environmental impacts considering landfilling, recycling, and incineration of demolition waste. Other works, such as that proposed by [139], assessed the EOL of a building to identify the demolition process variables that affect energy consumption and GHG emissions. Other authors, e.g., [119], combined LCA and LCC (life cycle cost) to analyze the environmental and economic drivers of three different waste disposal scenarios (landfilling, recycled aggregate, and recycled powder). Likewise, [115] compared the life cycle environmental implications of two C&DW management alternatives (inert landfilling and integrated wet recycling). Reference [60] presents analysis of six different scenarios for C&DW that included a combination of landfill, sorting, and recycling, and the use of material for landfill roads. Reference [41] analyzed the potential environmental impacts associated with C&DW utilization in road construction compared to landfilling, including analysis of transportation distances and the entire life cycle of construction products. In terms of industrialized treatment processes for C&DW, [29] analyzed mixes of recycled aggregates from C&DW treatment plants to evaluate the viability of their use in the construction of road layers. It was determined that these materials, when they come from C&DW with selective collection at origin, cause less environmental impact than the impact caused by the use of natural aggregates to build road layers.

- Waste Management Systems

Several studies were focused on or related to strategies around the waste management systems for C&DW. Such works commonly have a managerial approach and included broader analysis on how technical factors influence the sustainability performance of the waste management process. We can here highlight some interesting approaches. For example, [150] proposed a model to evaluate waste management systems, including environmental, economic, and social aspects of C&DW. References [56,145] quantified environmental impacts within an LCA for buildings in which life cycle stages were adjusted to several waste/material management options. Overall analysis, such as the one proposed by [141], consisted of analyzing the energy and environmental implications of the C&DW recycling chain in a particular region of Italy. It included land use, transportation, and avoiding landfills. Moreover, data related to generate life cycle inventories, such as that described in [129], were used to develop and analyze a life cycle inventory of C&DW management systems based on primary data. More recent studies, as in [15], were oriented to develop conceptual C&DW management frameworks to maximize reduction, reuse, and recycling of materials. Lastly, some interesting recent works [34,73] focused on waste prevention scenarios from early design phases. These studies aimed to predict or simulate waste generation before the construction phase and mitigate future environmental impacts derived from C&DW management.

- Other Trends

Some interesting and novelty approaches were identified in the research articles selected. In the case of sustainability indicators dedicated to C&DW, [15] proposed a cycle-based C&DW sustainability index to assist designers during the selection of material, sorting, recycle/reuse, and treatment or disposal options for C&DW. In the field of methodological approaches, LCA has been combined with different tools to improve results and robustness of the life cycle analysis. Several variations of LCA and combinations can be noted. For instance, [4] proposed LCSA (life cycle sustainability analysis) applied to concrete recycling. Approaches based on building information modeling (BIM) were also identified, e.g., [9,10,39], which analyzed the environmental impact of materials converted into waste by evaluating with a BIM tool.

3.2.2. Common Findings

This subsection summarizes the most representative results from selected works in terms of sustainability after applying LCA on C&DW analysis or management:

- LCA demonstrates the reduction of this new material on the global warming potential of concrete. Reductions from 66 to 70% are possible for high strength concrete with low clinker content and 7–35% with a higher clinker content [135]. Similarly, [132] developed a comparative analysis of recycled and conventional concrete. Results demonstrated a reduction of (−30%) environmental impacts for Eco-Indicator 99 and ecological scarcity.

- Transportation stage plays a critical role in the recycling process of C&DW [69,78,129,137]. Depending on the distance to the destination (in the case of production exportation), it can be one of the most predominant stages from the environmental perspective [143].
- It is necessary to analyze the environmental performance of a system from different perspectives before decision-making. Recycling waste is not always the best alternative for C&DW; it depends on regional differences in operations and waste composition [43].
- Benefits from substituting primary raw materials can be overset by the increased impacts due to additional energy requirements of the selective demolition compared to the traditional one. Consequently, the environmental sustainability of selective demolition should be addressed on a case-by-case basis [124].
- Preventative models can support the preparation of national waste programs and could serve as an instrumental tool to simulate the environmental impacts of construction waste management scenarios that include waste prevention [34].
- LCA and GIS (geographical information systems) provide beneficial results to analyze EOL scenarios by considering the number, size, type, and location of recycling plants [141].

4. Discussion

After performing the literature search, identifying the selected works, and developing the bibliometric and content-based analysis, several issues can be highlighted concerning the concept of LCA and C&DW.

Firstly, LCA approaches and C&DW are a growing and trending topic that will be gaining more relevance in waste management and the design and planning of new industrial and residential projects worldwide. Furthermore, there is a generalized pressure for finding more environmentally friendly solutions and the almost mandatory requirement of measuring and reducing impacts during the whole life cycle of buildings. Therefore, more research efforts are required worldwide to contribute more specialized knowledge to aid designers, architects, and engineers during the conception of new building projects and the EOL management of existing ones. LCA is still challenging due to the data required, the availability of indicators, and characterization factors related to each case study. However, it is a helpful tool to analyze and make decisions related to environmental issues for using and managing C&DW.

Following the previous idea, it is notable that only a few countries have an important research advance in evaluating C&DW materials from a holistic life cycle perspective. For example, countries such as China, Spain, and Italy have a well-established research agenda in the field, proposing strategies and engineering approaches to manage and recover material from existing buildings and comparing technical performances between C&DW and virgin materials. Although most research is limited to a particular region, province, or city, methodological approaches based on geographic information systems, optimization techniques to analyze logistic burdens and costs, detailed environmental analysis, and statistical approaches can be extrapolated to other countries and regions. Furthermore, it is necessary to include in such an analysis the source of virgin materials and construction products from a worldwide point of view since several materials are imported depending on the demanding country or region.

In terms of objectives or purpose of selected works, most are dedicated to evaluate, assess, and quantify technical attributes (environmental, economic, and related to mechanical properties) of C&DW. Such works focus on providing helpful information to make decisions related to select materials, determine the most suitable waste management strategy (recycling onsite, offsite, and landfill) existing C&DW. Nevertheless, the practical application of such works is limited by local conditions and other factors, such as current legislation, transportation infrastructure, and technology availability. In terms of materials, management alternatives for C&DW need to be addressed individually due to the vast amount and variety of construction materials currently available. It is necessary to consolidate comprehensive databases, preferably including the origin and destination of raw materials.

Some works propose new methodologies, guidelines, or indicators to study C&DW materials compared to conventional virgin materials. However, these studies are fewer compared to those dedicated to evaluating, assessing, and quantifying, which demonstrates a lack of methodological approaches, not only for the management of C&DW, but also for the whole life cycle, starting from early design phases. Nevertheless, unfortunately, most existing buildings closer to their EOL stage were constructed without considering life cycle implications, especially those built before 2000. Lastly, in a minor proportion, several works aimed to explore or analyze previous research results. Such articles are review-type and consolidate information and data from LCA application cases on buildings, and highlight research gaps in a general perspective. As an important opportunity, it is remarkable that LCA shows improved results when it is combined with other methodological tools such as material flow analysis, life cycle cost, environmental life cycle costing, multicriteria analysis, among others. Selected works consider a wide variety of C&DW materials that include concrete, steel, brick, plaster, insulation, glass, aluminum, gypsum, board, ash, timber, wood, and rubber.

Moreover, most analytical works consider five to twelve materials, including all possible materials obtained from demolition tasks. This provides a robust characterization in terms of LCA. However, to fully assess many materials through LCA (primary production, secondary production, transportation, use, and EOL) requires a vast amount of information that can be difficult to find in the literature, especially when a local context analysis is necessary and when the chemical composition of materials can vary according to their geographical source. The complexity of this issue is increased when eight or more LCA parameters or indicators are required.

Indicators employed in LCA include various chemical, physical, and technical parameters that provide full detail of sustainability burdens in different life cycle stages (manufacturing, use, and EOL). However, there is no homogeneity in the use of LCA indicators, which does not allow a technical comparison of research works. Therefore, to avoid confusion and misunderstanding of results from the case study analysis, it is necessary to dedicate research efforts to the definition of worldwide accepted indicators or measurement parameters.

5. Barriers and Future Challenges

From the analysis of selected works, several common barriers were identified. The most relevant are described in detail below:

- LCA is a method based on data inventory; therefore, the robustness of each analysis depends on the availability of data to measure and compare environmental indicators. A significant proportion of selected works (30% approximately) employed a complete set of environmental indicators. Meanwhile, most of the works (50% approximately) employ one to four indicators. This demonstrates a bias in the implementation of the method in the group of selected works. This situation makes it difficult to make a detailed comparison among the research works with traceability and reliability.
- There are no widely extended policies related to the disposal of debris in construction sites and the threshold of recycled-content building materials [17,144]. This is evidenced in the type of approach proposed by most research works based on local analysis (municipalities, cities, or regions). Therefore, there is no formal approach to evaluate C&DW through LCA in a world unified methodology. The consequences of this situation can be seen on a larger scale in the European Union (Figure 9), where although the recovery average of C&DW is around 30%, the situation varies from country to country. For example, southern countries such as Spain, Portugal, and Greece

only recover approximately 5% of the generated C&DW, while northern countries such as Holland, Belgium, and Denmark recover more than 80% of the CD&DW.

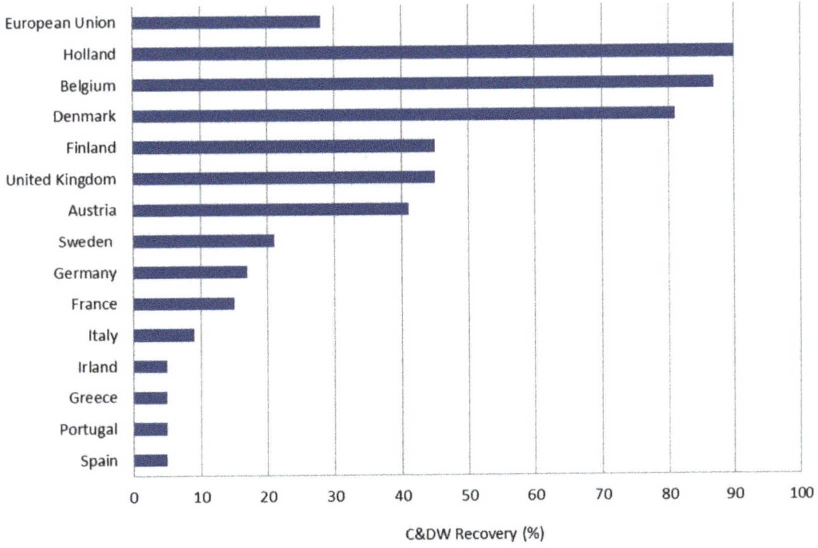

European Union
Holland
Belgium
Denmark
Finland
United Kingdom
Austria
Sweden
Germany
France
Italy
Irland
Greece
Portugal
Spain
0 10 20 30 40 50 60 70 80 90 100
C&DW Recovery (%)

Figure 9. C&DW recovery in European Union countries. Adapted from ref. [153].

- The economic viability of C&DW management is insufficient to guarantee its implementation. Other factors must be included in the analysis, such as logistic deadlines, availability of resources, transportation issues, derived environmental impacts, among others [150]. It is necessary to have an entire database of environmental and technical data from all raw materials (including geographical origin) to develop robust analysis based on a sustainability perspective that includes environmental and social aspects.
- Commonly, it is concluded that recycling is more sustainable, but from an absolute perspective, it can cause an unacceptable impact on the environment [150]. The impact of recycling must be considered in any LCA analysis using a case-by-case basis.
- According to the results previously explained and discussed, five main research challenges need to be addressed to improve the applicability and effectiveness of LCA approaches to study C&DW.
- LCA implementations found in the existing literature focus on specific locations or cities; there is no traceability of the primary source of materials that can even be imported from distant regions. A robust LCA approach should consider the entire life cycle, considering socioeconomic conditions of locations or countries where raw material is obtained and processed until it is delivered as a construction material or product. The same consideration should be included to select the most suitable EOL alternative of C&DW, involving long travel distances.
- Most research works are focused on recycling concrete or aggregates of different nature. However, it is necessary to propose new approaches that provide different circular economy strategies such as repair or repurpose to avoid reprocessing of C&DW material that can be more expensive in terms of cost and environment. Recycling is considered one of the less desirable among the circular economy strategies [154,155], since it involves using resources (energy, water, additional supplies) that can be significant compared to the extraction of virgin material. Nevertheless, due to the nature of construction materials, it is difficult to apply many strategies compared to electronics, industrial machines, and domestic appliances.

- Dedicated design methodologies that include prediction of environmental impact, maintenance tasks, and demolition processes must integrate LCA data from early design phases. Building information modeling presents vast potential as a tool for including LCA aspects and facilitating the waste management of buildings during their EOL stage.
- Industry 4.0 technology and advanced techniques such as artificial intelligence, machine learning, and digital twin are required to boost new research efforts for building project design and waste management. In addition, simulation tools and modeling software need to include material property databases to assess and select materials from early design phases. Thus, it facilitates calculations and the decision-making of construction materials and their fractions when C&DW sources are considered for replacing virgin material.
- More research is required to evaluate the role of legislation and policies in different countries related to the waste management of C&DW and their evolution during the last decade. For example, stimulating and controlling approaches should be considered to move construction to a circular economy model. In addition, some new business models associated with the servitization of buildings (co-living, working spaces, rent-based offices) must be assessed in the use stage to compensate for possible costs during the EOL phase.

6. Conclusions

This article aimed to consolidate and summarize relevant trends and insights from a systematic literature review of works related to LCA and C&DW during the last two decades. The works identified and selected were analyzed following a bibliometric and content-based analysis. Global bibliometric parameters included the evolution of works in time, geographical distribution of works, most relevant research centers, featured journals, and research clusters. The content-based analysis covered objectives and methodology, practical application, C&DW materials studied, and LCA parameters. Research gaps lie in the need for more research dedicated to design methodologies that provide helpful guidelines to consider the whole life cycle of buildings from early design stages, to include a more circular economy perspective to generate additional alternatives to recycling and recovery of C&DW, and to more broadly analyze globalized supply chains to consider the entire life cycle impact of raw materials. Additionally, there is a need to integrate Industry 4.0 and new data-driven methods to optimize design and decision-making around the management of construction materials. Furthermore, research is necessary to analyze the impact of uneven evolution of legislation and policies worldwide and evaluate their impact on long-term sustainability performance in the world construction sector.

Author Contributions: Conceptualization, J.A.M., and A.M.-R.; methodology, J.A.M.; validation, J.A.M., C.F.-R., and A.M.-R.; formal analysis, J.A.M.; investigation, J.A.M. and C.F.-R.; resources, J.A.M.; data curation, J.A.M.; writing—original draft preparation, J.A.M.; writing—review and editing, J.A.M. and C.F.-R.; visualization, J.A.M.; supervision, C.F.-R. and A.M.-R. All authors have read and agreed to the published version of the manuscript.

Funding: This research received no external funding.

Institutional Review Board Statement: Not applicable.

Informed Consent Statement: Not applicable.

Data Availability Statement: Not applicable.

Acknowledgments: The authors would like to thank the Engineering Faculties of Pontificia Universidad Javeriana and Universidad El Bosque University for their support during the development of this study.

Conflicts of Interest: The authors declare no conflict of interest.

Appendix A. Full Search Query Employed in SCOPUS Database

TITLE-ABS-KEY (("CDW" OR "Construction waste" OR "demolition waste") AND ("LCA" OR "Life cycle Assessment" OR "Life cycle Analysis" OR "Lifecycle Assessment")) AND (LIMIT-TO (SRCTYPE, "j") OR LIMIT-TO (SRCTYPE, "p")) AND (LIMIT-TO (DOCTYPE, "ar") OR LIMIT-TO (DOCTYPE, "cp") OR LIMIT-TO (DOCTYPE, "re")) AND (LIMIT-TO (LANGUAGE, "English")) AND (LIMIT-TO (EXACTKEYWORD, "Life Cycle") OR LIMIT-TO (EXACTKEYWORD, "Demolition") OR LIMIT-TO (EXACTKEYWORD, "Construction And Demolition Waste") OR LIMIT-TO (EXACTKEYWORD, "Life Cycle Assessment") OR LIMIT-TO (EXACTKEYWORD, "Life Cycle Assessment (LCA)") OR LIMIT-TO (EXACTKEYWORD, "Life Cycle Analysis") OR LIMIT-TO (EXACTKEYWORD, "Demolition Wastes")) AND (EXCLUDE (OA, "all")).

Results: 201 records (21 April 2021).

References

1. Nasab, T.J.; Monavari, S.M.; Jozi, S.A.; Majedi, H. Assessment of carbon footprint in the construction phase of high-rise constructions in Tehran. *Int. J. Environ. Sci. Technol.* **2020**, *17*, 3153–3164. [CrossRef]
2. Ruuska, A.; Häkkinen, T. Material efficiency of building construction. *Buildings* **2014**, *4*, 266–294. [CrossRef]
3. Ortiz, O.; Pasqualino, J.C.; Castells, F. Environmental performance of construction waste: Comparing three scenarios from a case study in Catalonia, Spain. *Waste Manag.* **2010**, *30*, 646–654. [CrossRef]
4. Hu, M.; Kleijn, R.; Bozhilova-Kisheva, K.P.; di Maio, F. An approach to LCSA: The case of concrete recycling. *Int. J. Life Cycle Assess.* **2013**, *18*, 1793–1803. [CrossRef]
5. Guignot, S.; Touzé, S.; von der Weid, F.; Ménard, Y.; Villeneuve, J. Recycling Construction and Demolition Wastes as Building Materials: A Life Cycle Assessment. *J. Ind. Ecol.* **2015**, *19*, 1030–1043. [CrossRef]
6. Junior, A.C.P.; Jacinto, C.; Oliveira, T.M.; Polisseni, A.E.; Brum, F.M.; Teixeira, E.R.; Mateus, R. Characterisation and life cycle assessment of pervious concrete with recycled concrete aggregates. *Crystals* **2021**, *11*, 209. [CrossRef]
7. Zhang, C.; Hu, M.; Dong, L.; Xiang, P.; Zhang, Q.; Wu, J.; Li, B.; Shi, S. Co-benefits of urban concrete recycling on the mitigation of greenhouse gas emissions and land use change: A case in Chongqing metropolis, China. *J. Clean. Prod.* **2018**, *201*, 481–498. [CrossRef]
8. Zhang, C.; Hu, M.; Dong, L.; Gebremariam, A.; Miranda-Xicotencatl, B.; Di Maio, F.; Tukker, A. Eco-efficiency assessment of technological innovations in high-grade concrete recycling. *Resour. Conserv. Recycl.* **2019**, *149*, 649–663. [CrossRef]
9. Jalaei, F.; Zoghi, M.; Khoshand, A. Life cycle environmental impact assessment to manage and optimize construction waste using Building Information Modeling (BIM). *Int. J. Constr. Manag.* **2019**, 1–18. [CrossRef]
10. Zoghi, M.; Kim, S. Dynamic modeling for life cycle cost analysis of BIM-based construction waste management. *Sustainability* **2020**, *12*, 2483. [CrossRef]
11. Ram, V.G.; Kishore, K.C.; Kalidindi, S.N. Environmental benefits of construction and demolition debris recycling: Evidence from an Indian case study using life cycle assessment. *J. Clean. Prod.* **2020**, *255*, 120258. [CrossRef]
12. Narayanaswamy, A.H.; Walker, P.; Reddy, B.V.V.; Heath, A.; Maskell, D. Mechanical and thermal properties, and comparative life-cycle impacts, of stabilised earth building products. *Constr. Build. Mater.* **2020**, *243*, 118096. [CrossRef]
13. Kvočka, D.; Lešek, A.; Knez, F.; Ducman, V.; Panizza, M.; Tsoutis, C.; Bernardi, A. Life cycle assessment of prefabricated geopolymeric façade cladding panels made from large fractions of recycled construction and demolition waste. *Materials* **2020**, *13*, 3931. [CrossRef] [PubMed]
14. Di Maria, A.; Eyckmans, J.; van Acker, K. Downcycling versus recycling of construction and demolition waste: Combining LCA and LCC to support sustainable policy making. *Waste Manag.* **2018**, *75*, 3–21. [CrossRef]
15. Yeheyis, M.; Hewage, K.; Alam, M.S.; Eskicioglu, C.; Sadiq, R. An overview of construction and demolition waste management in Canada: A life-cycle analysis approach to sustainability. *Clean Technol. Environ. Policy* **2013**, *15*, 81–91. [CrossRef]
16. Di Maria, A.; Eyckmans, J.; van Acker, K. Use of LCA and LCC to help decision-making between downcycling versus recycling of construction and demolition waste. In *Advances in Construction and Demolition Waste Recycling*; Woodhead Publishing: Sawston, UK, 2020; pp. 537–558.
17. Blengini, G.A. Life cycle of buildings, demolition and recycling potential: A case study in Turin, Italy. *Build. Environ.* **2009**, *44*, 319–330. [CrossRef]
18. Laurent, A.; Bakas, I.; Clavreul, J.; Bernstad, A.; Niero, M.; Gentil, E.; Hauschild, M.Z.; Christensen, T.H. Review of LCA studies of solid waste management systems-Part I: Lessons learned and perspectives. *Waste Manag.* **2014**, *34*, 573–588. [CrossRef]
19. Bovea, M.D.; Powell, J.C. Developments in life cycle assessment applied to evaluate the environmental performance of construction and demolition wastes. *Waste Manag.* **2016**, *50*, 151–172. [CrossRef]
20. Vieira, D.R.; Calmon, J.L.; Coelho, F.Z. Life cycle assessment (LCA) applied to the manufacturing of common and ecological concrete: A review. *Constr. Build. Mater.* **2016**, *124*, 656–666. [CrossRef]
21. Wu, H.; Zuo, J.; Yuan, H.; Zillante, G.; Wang, J. A review of performance assessment methods for construction and demolition waste management. *Resour. Conserv. Recycl.* **2019**, *150*, 104407. [CrossRef]

22. Rosado, L.P.; Penteado, C.S.G. Municipal management of construction and demolition waste: Influence of disposal fees. *Ambient. Soc.* **2020**, *23*, 1–21. [CrossRef]
23. Salgado, R.A.; Apul, D.; Guner, S. Life cycle assessment of seismic retrofit alternatives for reinforced concrete frame buildings. *J. Build. Eng.* **2020**, *28*, 101064. [CrossRef]
24. Sözer, H.; Sözen, H. Waste capacity and its environmental impact of a residential district during its life cycle. *Energy Rep.* **2020**, *6*, 286–296. [CrossRef]
25. Sözer, H.; Sözen, H.; Utkucu, D. Waste potential of a building through gate-to-grave approach based on Life Cycle Assessment (LCA). *Int. J. Sustain. Dev. Plan.* **2020**, *15*, 165–171. [CrossRef]
26. Umadevi, P.; Pradeep, T. A review on environmental, economic and social facets of construction & demolition waste management. *Int. J. Adv. Sci. Technol.* **2020**, *29*, 4692–4697.
27. Zhang, J.; Ding, L.; Li, F.; Peng, J. Recycled aggregates from construction and demolition wastes as alternative filling materials for highway subgrades in China. *J. Clean. Prod.* **2020**, *255*, 120223. [CrossRef]
28. Zhao, Z.; Courard, L.; Groslambert, S.; Jehin, T.; Léonard, A.; Xiao, J. Use of recycled concrete aggregates from precast block for the production of new building blocks: An industrial scale study. *Resour. Conserv. Recycl.* **2020**, *157*, 104786. [CrossRef]
29. Agrela, F.; Díaz-López, J.L.; Rosales, J.; Cuenca-Moyano, G.M.; Cano, H.; Cabrera, M. Environmental assessment, mechanical behavior and new leaching impact proposal of mixed recycled aggregates to be used in road construction. *J. Clean. Prod.* **2021**, *280*. [CrossRef]
30. Bogoviku, L.; Waldmann, D. Modelling of mineral construction and demolition waste dynamics through a combination of geospatial and image analysis. *J. Environ. Manag.* **2021**, *282*. [CrossRef]
31. Bonoli, A.; Zanni, S.; Serrano-Bernardo, F. Sustainability in building and construction within the framework of circular cities and european new green deal. The contribution of concrete recycling. *Sustainability* **2021**, *13*, 2139. [CrossRef]
32. Caneda-martínez, L.; Monasterio, M.; Moreno-juez, J.; Martínez-ramírez, S.; García, R.; Frías, M. Behaviour and properties of eco-cement pastes elaborated with recycled concrete powder from construction and demolition wastes. *Materials* **2021**, *14*, 1299. [CrossRef] [PubMed]
33. Li, J.; Zhang, J.; Ni, S.; Liu, L.; Walubita, L.F. Mechanical performance and environmental impacts of self-compacting concrete with recycled demolished concrete blocks. *J. Clean. Prod.* **2021**, *293*, 126129. [CrossRef]
34. Llatas, C.; Bizcocho, N.; Soust-Verdaguer, B.; Montes, M.V.; Quiñones, R. An LCA-based model for assessing prevention versus non-prevention of construction waste in buildings. *Waste Manag.* **2021**, *126*, 608–622. [CrossRef] [PubMed]
35. Nguyen, W.; Martinez, D.M.; Jen, G.; Duncan, J.F.; Ostertag, C.P. Interaction between global warming potential, durability, and structural properties of fiber-reinforced concrete with high waste materials inclusion. *Resour. Conserv. Recycl.* **2021**, *169*, 105453. [CrossRef]
36. Resch, E.; Andresen, I.; Cherubini, F.; Brattebø, H. Estimating dynamic climate change effects of material use in buildings—Timing, uncertainty, and emission sources. *Build. Environ.* **2021**, *187*, 107399. [CrossRef]
37. Sormunen, P.; Deviatkin, I.; Horttanainen, M.; Kärki, T. An evaluation of thermoplastic composite fillers derived from construction and demolition waste based on their economic and environmental characteristics. *J. Clean. Prod.* **2021**, *280*. [CrossRef]
38. Zhou, A.; Zhang, W.; Wei, H.; Liu, T.; Zou, D.; Guo, H. A novel approach for recycling engineering sediment waste as sustainable supplementary cementitious materials. *Resour. Conserv. Recycl.* **2021**, *167*, 105435. [CrossRef]
39. Wang, J.; Wu, H.; Duan, H.; Zillante, G.; Zuo, J.; Yuan, H. Combining life cycle assessment and Building Information Modelling to account for carbon emission of building demolition waste: A case study. *J. Clean. Prod.* **2016**, *172*, 3154–3166. [CrossRef]
40. Wu, Z.; Yu, A.T.W.; Poon, C.S. An off-site snapshot methodology for estimating building construction waste composition—A case study of Hong Kong. *Environ. Impact Assess. Rev.* **2019**, *77*, 128–135. [CrossRef]
41. Butera, S.; Christensen, T.H.; Astrup, T.F. Life cycle assessment of construction and demolition waste management. *Waste Manag.* **2015**, *44*, 196–205. [CrossRef]
42. Cao, X.; Li, X.; Zhu, Y.; Zhang, Z. A comparative study of environmental performance between prefabricated and traditional residential buildings in China. *J. Clean. Prod.* **2015**, *109*, 131–143. [CrossRef]
43. Dahlbo, H.; Bachér, J.; Lähtinen, K.; Jouttijärvi, T.; Suoheimo, P.; Mattila, T.; Sironen, S.; Myllymaa, T.; Saramäki, K. Construction and demolition waste management—A holistic evaluation of environmental performance. *J. Clean. Prod.* **2015**, *107*, 333–341. [CrossRef]
44. Fu, F.; Sun, J.; Pasquire, C. Carbon Emission Assessment for Steel Structure Based on Lean Construction Process. *J. Intell. Robot. Syst. Theory Appl.* **2015**, *79*, 401–416. [CrossRef]
45. Hol, G.H.P. Building a green swimming pool by using concrete with aggregates from demolition waste. *Procedia Eng.* **2015**, *125*, 613–616. [CrossRef]
46. Ng, W.Y.; Chau, C.K. New Life of the Building Materials-Recycle, Reuse and Recovery. *Energy Procedia* **2015**, *75*, 2884–2891. [CrossRef]
47. Tošić, N.; Marinković, S.; Dašić, T.; Stanić, M. Multicriteria optimization of natural and recycled aggregate concrete for structural use. *J. Clean. Prod.* **2015**, *87*, 766–776. [CrossRef]
48. Udawatta, N.; Zuo, J.; Chiveralls, K.; Zillante, G. Improving waste management in construction projects: An Australian study. *Resour. Conserv. Recycl.* **2015**, *101*, 73–83. [CrossRef]

49. Estanqueiro, B.; Silvestre, J.D.; de Brito, J.; Pinheiro, M.D. Environmental life cycle assessment of coarse natural and recycled aggregates for concrete. *Eur. J. Environ. Civ. Eng.* **2018**, *22*, 429–449. [CrossRef]
50. Faleschini, F.; Zanini, M.A.; Pellegrino, C.; Pasinato, S. Sustainable management and supply of natural and recycled aggregates in a medium-size integrated plant. *Waste Manag.* **2016**, *49*, 146–155. [CrossRef]
51. Kucukvar, M.; Egilmez, G.; Tatari, O. Life cycle assessment and optimization-based decision analysis of construction waste recycling for a LEED-certified university building. *Sustainability* **2016**, *8*, 89. [CrossRef]
52. Gayarre, F.L.; Pérez, J.G.; Pérez, C.L.; López, M.S.; Martínez, A.L. Life cycle assessment for concrete kerbs manufactured with recycled aggregates. *J. Clean. Prod.* **2016**, *113*, 41–53. [CrossRef]
53. Mortaheb, M.M.; Mahpour, A. Integrated construction waste management, a holistic approach. *Sci. Iran.* **2016**, *23*, 2044–2056. [CrossRef]
54. Sou, W.I.; Chu, A.; Chiueh, P.T. Sustainability assessment and prioritisation of bottom ash management in Macao. *Waste Manag. Res.* **2016**, *34*, 1275–1282. [CrossRef] [PubMed]
55. Suárez, S.; Roca, X.; Gasso, S. Product-specific life cycle assessment of recycled gypsum as a replacement for natural gypsum in ordinary Portland cement: Application to the Spanish context. *J. Clean. Prod.* **2016**, *117*, 150–159. [CrossRef]
56. Hossain, M.U.; Poon, C.S.; Lo, I.M.C.; Cheng, J.C.P. Comparative environmental evaluation of aggregate production from recycled waste materials and virgin sources by LCA. *Resour. Conserv. Recycl.* **2016**, *109*, 67–77. [CrossRef]
57. Hossain, M.U.; Poon, C.S.; Lo, I.M.C.; Cheng, J.C.P. Evaluation of environmental friendliness of concrete paving eco-blocks using LCA approach. *Int. J. Life Cycle Assess.* **2016**, *21*, 70–84. [CrossRef]
58. Tang, X.; Li, C.; Hu, S.; Liu, Y.; Geng, H. Evaluating extended land consumption in building life cycle to improve land conservation: A case study in Shenyang, China. *Resour. Conserv. Recycl.* **2016**, *109*, 78–89. [CrossRef]
59. Zambrana-Vasquez, D.; Zabalza-Bribián, I.; Jáñez, A.; Aranda-Usón, A. Analysis of the environmental performance of life-cycle building waste management strategies in tertiary buildings. *J. Clean. Prod.* **2016**, *130*, 143–154. [CrossRef]
60. Penteado, C.S.G.; Rosado, L.P. Comparison of scenarios for the integrated management of construction and demolition waste by life cycle assessment: A case study in Brazil. *Waste Manag. Res.* **2016**, *34*, 1026–1035. [CrossRef] [PubMed]
61. Arm, M.; Wik, O.; Engelsen, C.J.; Erlandsson, M.; Hjelmar, O.; Wahlström, M. How Does the European Recovery Target for Construction & Demolition Waste Affect Resource Management? *Waste Biomass Valoriz.* **2017**, *8*, 1491–1504. [CrossRef]
62. Cuenca-Moyano, G.M.; Zanni, S.; Bonoli, A.; Valverde-Palacios, I. Development of the life cycle inventory of masonry mortar made of natural and recycled aggregates. *J. Clean. Prod.* **2017**, *140*, 1272–1286. [CrossRef]
63. Ghanbari, M.; Abbasi, A.M.; Ravanshadnia, M. Economic and Environmental Evaluation and Optimal Ratio of Natural and Recycled Aggregate Production. *Adv. Mater. Sci. Eng.* **2017**, *2017*. [CrossRef]
64. Ghanbari, M.; Abbasi, A.M.; Ravanshadnia, M. Environmental Life Cycle Assessment and Cost Analysis of Aggregate Production Industries. *Appl. Ecol. Environ. Res.* **2017**, *15*, 1577–1593. [CrossRef]
65. Ghose, A.; Pizzol, M.; McLaren, S.J. Consequential LCA modelling of building refurbishment in New Zealand- an evaluation of resource and waste management scenarios. *J. Clean. Prod.* **2017**, *165*, 119–133. [CrossRef]
66. Mah, C.M.; Fujiwara, T.; Ho, C.S. Concrete waste management decision analysis based on life cycle assessment. *Chem. Eng. Trans.* **2017**, *56*, 25–30. [CrossRef]
67. Mastrucci, A.; Marvuglia, A.; Popovici, E.; Leopold, U.; Benetto, E. Geospatial characterization of building material stocks for the life cycle assessment of end-of-life scenarios at the urban scale. *Resour. Conserv. Recycl.* **2017**, *123*, 54–66. [CrossRef]
68. Rosado, L.P.; Vitale, P.; Penteado, C.S.G.; Arena, U. Life cycle assessment of natural and mixed recycled aggregate production in Brazil. *J. Clean. Prod.* **2017**, *151*, 634–642. [CrossRef]
69. Shan, X.; Zhou, J.; Chang, V.W.C.; Yang, E.H. Life cycle assessment of adoption of local recycled aggregates and green concrete in Singapore perspective. *J. Clean. Prod.* **2017**, *164*, 918–926. [CrossRef]
70. Hossain, M.U.; Wu, Z.; Poon, C.S. Comparative environmental evaluation of construction waste management through different waste sorting systems in Hong Kong. *Waste Manag.* **2017**, *69*, 325–335. [CrossRef]
71. Vitale, P.; Arena, U. An attributional life cycle assessment for an Italian residential multifamily building. *Environ. Technol.* **2017**, *39*, 3033–3045. [CrossRef]
72. Vitale, P.; Arena, N.; di Gregorio, F.; Arena, U. Life cycle assessment of the end-of-life phase of a residential building. *Waste Manag.* **2017**, *60*, 311–321. [CrossRef] [PubMed]
73. Bizcocho, N.; Llatas, C. Inclusion of prevention scenarios in LCA of construction waste management. *Int. J. Life Cycle Assess.* **2019**, *24*, 468–484. [CrossRef]
74. Borghi, G.; Pantini, S.; Rigamonti, L. Life cycle assessment of non-hazardous Construction and Demolition Waste (CDW) management in Lombardy Region (Italy). *J. Clean. Prod.* **2018**, *184*, 815–825. [CrossRef]
75. Mah, C.M.; Fujiwara, T.; Ho, C.S. Environmental impacts of construction and demolition waste management alternatives. *Chem. Eng. Trans.* **2018**, *63*, 343–348. [CrossRef]
76. Cuenca-Moyano, G.M.; Martín-Morales, M.; Bonoli, A.; Valverde-Palacios, I. Environmental assessment of masonry mortars made with natural and recycled aggregates. *Int. J. Life Cycle Assess.* **2019**, *24*, 191–210. [CrossRef]
77. Ghanbari, M.; Abbasi, A.M.; Ravanshadnia, M. Production of natural and recycled aggregates: The environmental impacts of energy consumption and CO_2 emissions. *J. Mater. Cycles Waste Manag.* **2018**, *20*, 810–822. [CrossRef]

78. Göswein, V.; Gonçalves, A.B.; Silvestre, J.D.; Freire, F.; Habert, G.; Kurda, R. Transportation matters—Does it? GIS-based comparative environmental assessment of concrete mixes with cement, fly ash, natural and recycled aggregates. *Resour. Conserv. Recycl.* **2018**, *137*, 1–10. [CrossRef]
79. Heeren, N.; Hellweg, S. Tracking Construction Material over Space and Time: Prospective and Geo-referenced Modeling of Building Stocks and Construction Material Flows. *J. Ind. Ecol.* **2019**, *23*, 253–267. [CrossRef]
80. Lockrey, S.; Verghese, K.; Crossin, E.; Nguyen, H. Concrete recycling life cycle flows and performance from construction and demolition waste in Hanoi. *J. Clean. Prod.* **2018**, *179*, 593–604. [CrossRef]
81. Palacios-Munoz, B.; Gracia-Villa, L.; Zabalza-Bribián, I.; López-Mesa, B. Simplified structural design and LCA of reinforced concrete beams strengthening techniques. *Eng. Struct.* **2018**, *174*, 418–432. [CrossRef]
82. Pantini, S.; Borghi, G.; Rigamonti, L. Towards resource-efficient management of asphalt waste in Lombardy region (Italy): Identification of effective strategies based on the LCA methodology. *Waste Manag.* **2018**, *80*, 423–434. [CrossRef] [PubMed]
83. Seco, A.; Omer, J.; Marcelino, S.; Espuelas, S.; Prieto, E. Sustainable unfired bricks manufacturing from construction and demolition wastes. *Constr. Build. Mater.* **2018**, *167*, 154–165. [CrossRef]
84. Silgado, S.S.; Valdiviezo, L.C.; Domingo, S.G.; Roca, X. Multi-criteria decision analysis to assess the environmental and economic performance of using recycled gypsum cement and recycled aggregate to produce concrete: The case of Catalonia (Spain). *Resour. Conserv. Recycl.* **2018**, *133*, 120–131. [CrossRef]
85. Hossain, M.U.; Wang, L.; Yu, I.K.M.; Tsang, D.C.W.; Poon, C.S. Environmental and technical feasibility study of upcycling wood waste into cement-bonded particleboard. *Constr. Build. Mater.* **2018**, *173*, 474–480. [CrossRef]
86. Vitale, P.; Spagnuolo, A.; Lubritto, C.; Arena, U. Environmental performances of residential buildings with a structure in cold formed steel or reinforced concrete. *J. Clean. Prod.* **2018**, *189*, 839–852. [CrossRef]
87. Wang, T.; Wang, J.; Wu, P.; Wang, J.; He, Q.; Wang, X. Estimating the environmental costs and benefits of demolition waste using life cycle assessment and willingness-to-pay: A case study in Shenzhen. *J. Clean. Prod.* **2018**, *172*, 14–26. [CrossRef]
88. Wang, J.; Wu, H.; Tam, V.W.Y.; Zuo, J. Considering life-cycle environmental impacts and society's willingness for optimizing construction and demolition waste management fee: An empirical study of China. *J. Clean. Prod.* **2019**, *206*, 1004–1014. [CrossRef]
89. Yazdanbakhsh, A.; Bank, L.C.; Baez, T.; Wernick, I. Comparative LCA of concrete with natural and recycled coarse aggregate in the New York City area. *Int. J. Life Cycle Assess.* **2018**, *23*, 1163–1173. [CrossRef]
90. Yazdanbakhsh, A. A bi-level environmental impact assessment framework for comparing construction and demolition waste management strategies. *Waste Manag.* **2018**, *77*, 401–412. [CrossRef] [PubMed]
91. Zanni, S.; Simion, I.M.; Gavrilescu, M.; Bonoli, A. Life Cycle Assessment Applied to Circular Designed Construction Materials. *Procedia CIRP* **2018**, *69*, 154–159. [CrossRef]
92. Dastjerdi, B.; Strezov, V.; Kumar, R.; Behnia, M. An evaluation of the potential of waste to energy technologies for residual solid waste in New South Wales, Australia. *Renew. Sustain. Energy Rev.* **2019**, *115*, 109398. [CrossRef]
93. Hertwich, E.G.; Ali, S.; Ciacci, L.; Fishman, T.; Heeren, N.; Masanet, E.; Asghari, F.N.; Olivetti, E.; Pauliuk, S.; Tu, Q.; et al. Material efficiency strategies to reducing greenhouse gas emissions associated with buildings, vehicles, and electronics—A review. *Environ. Res. Lett.* **2019**, *14*. [CrossRef]
94. Liikanen, M.; Grönman, K.; Deviatkin, I.; Havukainen, J.; Hyvärinen, M.; Kärki, T.; Varis, J.; Soukka, R.; Horttanainen, M. Construction and demolition waste as a raw material for wood polymer composites—Assessment of environmental impacts. *J. Clean. Prod.* **2019**, *225*, 716–727. [CrossRef]
95. Nußholz, J.L.K.; Rasmussen, F.N.; Milios, L. Circular building materials: Carbon saving potential and the role of business model innovation and public policy. *Resour. Conserv. Recycl.* **2018**, *141*, 308–316. [CrossRef]
96. Pantini, S.; Giurato, M.; Rigamonti, L. A LCA study to investigate resource-efficient strategies for managing post-consumer gypsum waste in Lombardy region (Italy). *Resour. Conserv. Recycl.* **2019**, *147*, 157–168. [CrossRef]
97. Park, W.J.; Kim, T.; Roh, S.; Kim, R. Analysis of life cycle environmental impact of recycled aggregate. *Appl. Sci.* **2019**, *9*, 1021. [CrossRef]
98. Pavlu, T.; Kocí, V.; Hájek, P. Environmental assessment of two use cycles of recycled aggregate concrete. *Sustainability* **2019**, *11*, 6185. [CrossRef]
99. Rosado, L.P.; Vitale, P.; Penteado, C.S.G.; Arena, U. Life cycle assessment of construction and demolition waste management in a large area of São Paulo State, Brazil. *Waste Manag.* **2019**, *85*, 477–489. [CrossRef]
100. Santos, A.C.; Mendes, P.; Teixeira, M.R. Social life cycle analysis as a tool for sustainable management of illegal waste dumping in municipal services. *J. Clean. Prod.* **2019**, *210*, 1141–1149. [CrossRef]
101. Sözer, H.; Sözen, H. Evaluating the capacity of a building's waste and the potential for savings using the life cycle assessment methodology. *WIT Trans. Ecol. Environ.* **2019**, *231*, 159–170. [CrossRef]
102. Erduran, D.Ü.; Elias-Ozkan, S.T.; Ulybin, A. Assessing potential environmental impact and construction cost of reclaimed masonry walls. *Int. J. Life Cycle Assess.* **2020**, *25*. [CrossRef]
103. Vares, S.; Hradil, P.; Sansom, M.; Ungureanu, V. Economic potential and environmental impacts of reused steel structures. *Struct. Infrastruct. Eng.* **2020**, *16*, 750–761. [CrossRef]
104. Vidal, R.; Sánchez-Pantoja, N.; Martínez, G. Análisis del ciclo de vida de un edificio con estructura de madera contralaminada en Granada-España. *Inf. Constr.* **2019**, *71*, 289. [CrossRef]

105. Yazdanbakhsh, A.; Lagouin, M. The effect of geographic boundaries on the results of a regional life cycle assessment of using recycled aggregate in concrete. *Resour. Conserv. Recycl.* **2019**, *143*, 201–209. [CrossRef]
106. Akbarieh, A.; Jayasinghe, L.B.; Waldmann, D.; Teferle, F.N. BIM-based end-of-lifecycle decision making and digital deconstruction: Literature review. *Sustainability* **2020**, *12*, 2670. [CrossRef]
107. Colangelo, F.; Navarro, T.G.; Farina, I.; Petrillo, A. Comparative LCA of concrete with recycled aggregates: A circular economy mindset in Europe. *Int. J. Life Cycle Assess.* **2020**, *25*, 1790–1804. [CrossRef]
108. Foster, G.; Kreinin, H. A review of environmental impact indicators of cultural heritage buildings: A circular economy perspective. *Environ. Res. Lett.* **2020**, *15*. [CrossRef]
109. Guzdek, S.; Malinowski, M.; Religa, A.; Liszka, D.; Petryk, A. Economic and ecological assessment of transport of various types of waste. *J. Ecol. Eng.* **2020**, *21*, 19–26. [CrossRef]
110. Hafez, H.; Kurda, R.; Kurda, R.; Al-Hadad, B.; Mustafa, R.; Ali, B. A critical review on the influence of fine recycled aggregates on technical performance, environmental impact and cost of concrete. *Appl. Sci.* **2020**, *10*, 1018. [CrossRef]
111. Hasan, U.; Whyte, A.; al Jassmi, H. Life cycle assessment of roadworks in United Arab Emirates: Recycled construction waste, reclaimed asphalt pavement, warm-mix asphalt and blast furnace slag use against traditional approach. *J. Clean. Prod.* **2020**, *257*, 120531. [CrossRef]
112. Huang, B.; Gao, X.; Xu, X.; Song, J.; Geng, Y.; Sarkis, J.; Fishman, T.; Kua, H.; Nakatani, J. A Life Cycle Thinking Framework to Mitigate the Environmental Impact of Building Materials. *One Earth* **2020**, *3*, 564–573. [CrossRef]
113. Illankoon, I.M.C.S.; Lu, W. Cost implications of obtaining construction waste management-related credits in green building. *Waste Manag.* **2020**, *102*, 722–731. [CrossRef]
114. Ismaeel, W.S.E.; Ali, A.A.M.M. Assessment of eco-rehabilitation plans: Case study 'Richordi Berchet' palace. *J. Clean. Prod.* **2020**, *259*, 120857. [CrossRef]
115. Jain, S.; Singhal, S.; Pandey, S. Environmental life cycle assessment of construction and demolition waste recycling: A case of urban India. *Resour. Conserv. Recycl.* **2019**, *155*, 104642. [CrossRef]
116. Khan, S.M.; Qaiser, N.; Shaikh, S.F.; Hussain, M.M. Design Analysis and Human Tests of Foil-Based Wheezing Monitoring System for Asthma Detection. *IEEE Trans. Electron. Devices* **2020**, *67*, 249–257. [CrossRef]
117. Kirthika, S.K.; Singh, S.K.; Chourasia, A. Alternative fine aggregates in production of sustainable concrete- A review. *J. Clean. Prod.* **2020**, *268*, 122089. [CrossRef]
118. Li, J.; Liang, J.; Zuo, J.; Guo, H. Environmental impact assessment of mobile recycling of demolition waste in Shenzhen, China. *J. Clean. Prod.* **2020**, *263*, 121371. [CrossRef]
119. Liu, J.; Huang, Z.; Wang, X. Economic and environmental assessment of carbon emissions from demolition waste based on LCA and LCC. *Sustainability* **2020**, *12*, 6683. [CrossRef]
120. Iodice, S.; de Toro, P. Waste and wasted landscapes: Focus on abandoned industrial areas. *Detritus* **2020**, *11*, 103–120. [CrossRef]
121. Marrero, M.; Rivero-Camacho, C.; Alba-Rodríguez, M.D. What are we discarding during the life cycle of a building? Case studies of social housing in Andalusia, Spain. *Waste Manag.* **2020**, *102*, 391–403. [CrossRef]
122. Grabois, T.M.; Caldas, L.R.; Julião, N.R.; Filho, R.D.T. An experimental and environmental evaluation of mortars with recycled demolition waste from a hospital implosion in Rio de Janeiro. *Sustainability* **2020**, *12*, 8945. [CrossRef]
123. Mousavi, M.; Ventura, A.; Antheaume, N. Decision-based territorial life cycle assessment for the management of cement concrete demolition waste. *Waste Manag. Res.* **2020**, *38*, 1405–1419. [CrossRef]
124. Pantini, S.; Rigamonti, L. Is selective demolition always a sustainable choice? *Waste Manag.* **2020**, *103*, 169–176. [CrossRef]
125. Park, W.J.; Kim, R.; Roh, S.; Ban, H. Identifying the major construction wastes in the building construction phase based on life cycle assessments. *Sustainability* **2020**, *12*, 8096. [CrossRef]
126. Pešta, J.; Pavlů, T.; Fortová, K.; Kočí, V. Sustainable masonry made from recycled aggregates: LCA case study. *Sustainability* **2020**, *12*, 1581. [CrossRef]
127. Bianchini, F.; Hewage, K. How 'green' are the green roofs? Life-cycle analysis of green roof materials. *Build. Environ.* **2012**, *48*, 57–65. [CrossRef]
128. Chau, C.K.; Hui, W.K.; Ng, W.Y.; Powell, G. Assessment of CO_2 emissions reduction in high-rise concrete office buildings using different material use options. *Resour. Conserv. Recycl.* **2012**, *61*, 22–34. [CrossRef]
129. Mercante, I.T.; Bovea, M.D.; Ibáñez-Forés, V.; Arena, A.P. Life cycle assessment of construction and demolition waste management systems: A Spanish case study. *Int. J. Life Cycle Assess.* **2012**, *17*, 232–241. [CrossRef]
130. Coelho, A.; de Brito, J. Environmental analysis of a construction and demolition waste recycling plant in Portugal—Part I: Energy consumption and CO_2 emissions. *Waste Manag.* **2013**, *33*, 1258–1267. [CrossRef] [PubMed]
131. Coelho, A.; de Brito, J. Environmental analysis of a construction and demolition waste recycling plant in Portugal—Part II: Environmental sensitivity analysis. *Waste Manag.* **2013**, *33*, 147–161. [CrossRef]
132. Knoeri, C.; Sanyé-Mengual, E.; Althaus, H.J. Comparative LCA of recycled and conventional concrete for structural applications. *Int. J. Life Cycle Assess.* **2013**, *18*, 909–918. [CrossRef]
133. Li, D.Z.; Chen, H.X.; Hui, E.C.M.; Zhang, J.B.; Li, Q.M. A methodology for estimating the life-cycle carbon efficiency of a residential building. *Build. Environ.* **2013**, *59*, 448–455. [CrossRef]
134. Biswas, W.K. Carbon footprint and embodied energy assessment of a civil works program in a residential estate of Western Australia. *Int. J. Life Cycle Assess.* **2014**, *19*, 732–744. [CrossRef]

135. de Schepper, M.; van den Heede, P.; van Driessche, I.; de Belie, N. Life Cycle Assessment of Completely Recyclable Concrete. *Materials* **2014**, *7*, 6010–6027. [CrossRef] [PubMed]
136. Kucukvar, M.; Egilmez, G.; Tatari, O. Evaluating environmental impacts of alternative construction waste management approaches using supply-chain-linked life-cycle analysis. *Waste Manag. Res.* **2014**, *32*, 500–508. [CrossRef]
137. Vossberg, C.; Mason-Jones, K.; Cohen, B. An energetic life cycle assessment of C&D waste and container glass recycling in Cape Town, South Africa. *Resour. Conserv. Recycl.* **2014**, *88*, 39–49. [CrossRef]
138. Zeng, L.; Zhu, H.; Ma, Y.; Huang, J.; Li, G. Greenhouse gases emissions from solid waste: An analysis of Expo 2010 Shanghai, China. *J. Mater. Cycles Waste Manag.* **2014**, *16*, 616–622. [CrossRef]
139. Martínez, E.; Nuñez, Y.; Sobaberas, E. End of life of buildings: Three alternatives, two scenarios. A case study. *Int. J. Life Cycle Assess.* **2013**, *18*, 1082–1088. [CrossRef]
140. Bianchini, F.; Hewage, K. Probabilistic social cost-benefit analysis for green roofs: A life-cycle approach. *Build. Environ.* **2012**, *58*, 152–162. [CrossRef]
141. Blengini, G.A.; Garbarino, E. Resources and waste management in Turin (Italy): The role of recycled aggregates in the sustainable supply mix. *J. Clean. Prod.* **2010**, *18*, 1021–1030. [CrossRef]
142. Blengini, G.A.; Garbarino, E. Integrated life cycle management of aggregates quarrying, processing and recycling: Definition of a common LCA methodology in the SARMa project. *Int. J. Sustain. Soc.* **2011**, *3*, 327–344. [CrossRef]
143. Ibáñez-Forés, V.; Bovea, M.D.; Simó, A. Life cycle assessment of ceramic tiles. Environmental and statistical analysis. *Int. J. Life Cycle Assess.* **2011**, *16*, 916–928. [CrossRef]
144. Yu, D.; Tan, H.; Ruan, Y. A future bamboo-structure residential building prototype in China: Life cycle assessment of energy use and carbon emissi. *Energy Build.* **2011**, *43*, 2638–2646. [CrossRef]
145. Coelho, A.; de Brito, J. Influence of construction and demolition waste management on the environmental impact of buildings. *Waste Manag.* **2012**, *32*, 532–541. [CrossRef]
146. Barbudo, A.; Galvín, A.P.; Agrela, F.; Ayuso, J.; Jiménez, J.R. Correlation analysis between sulphate content and leaching of sulphates in recycled aggregates from construction and demolition wastes. *Waste Manag.* **2012**, *32*, 1229–1235. [CrossRef]
147. Weil, M.; Jeske, U.; Schebek, L. Closed-loop recycling of construction and demolition waste in Germany in view of stricter environmental threshold values. *Waste Manag. Res.* **2006**, *24*, 197–206. [CrossRef]
148. Dosho, Y. Development of a sustainable concrete waste recycling system: Application of recycled aggregate concrete produced by aggregate replacing method. *J. Adv. Concr. Technol.* **2007**, *5*, 27–42. [CrossRef]
149. Tiruta-Barna, L.; Benetto, E.; Perrodin, Y. Environmental impact and risk assessment of mineral wastes reuse strategies: Review and critical analysis of approaches and applications. *Resour. Conserv. Recycl.* **2007**, *50*, 351–379. [CrossRef]
150. Klang, A.; Vikman, P.Å.; Brattebø, H. Sustainable management of demolition waste—An integrated model for the evaluation of environmental, economic and social aspects. *Resour. Conserv. Recycl.* **2003**, *38*, 317–334. [CrossRef]
151. Simion, I.M.; Fortuna, M.E.; Bonoli, A.; Gavrilescu, M. Comparing environmental impacts of natural inert and recycled construction and demolition waste processing using LCA. *J. Environ. Eng. Landsc. Manag.* **2013**, *21*, 273–287. [CrossRef]
152. Khan, M.W.; Ali, Y. Sustainable construction: Lessons learned from life cycle assessment (LCA) and life cycle cost analysis (LCCA). *Constr. Innov.* **2020**, *20*, 191–207. [CrossRef]
153. Knauf GmbH España. Reciclaje y cierre del ciclo de vida de las placas de yeso laminado. *Anales Sectoriales-Reciclaje y Gestión de Residuos*. June 2013. Available online: https://www.interempresas.net/Reciclaje/Articulos/109556-Reciclaje-y-cierre-del-ciclo-de-vida-de-las-placas-de-yeso-laminado.html (accessed on 9 June 2021).
154. Cramer, J. *Moving Towards a Circular Economy in the Netheralnds: Challenges and Directions*; Utrecht University: Utrecht, The Netherlands, 2014; pp. 1–9.
155. Mesa, J.; González-Quiroga, A.; Maury, H. Developing an indicator for material selection based on durability and environmental footprint: A Circular Economy perspective. *Resour. Conserv. Recycl.* **2020**, *160*, 104887. [CrossRef]

 *sustainability*

Review

Some Remarks towards a Better Understanding of the Use of Concrete Recycled Aggregate: A Review

Anna M. Grabiec [1], Jeonghyun Kim [2,*], Andrzej Ubysz [2] and Pilar Bilbao [3]

1 Faculty of Environmental Engineering and Mechanical Engineering, Poznań University of Life Sciences, Piątkowska 94, 60-649 Poznań, Poland; agra@up.poznan.pl
2 Faculty of Civil Engineering, Wrocław University of Science and Technology, Wybrzeże Wyspiańskiego 27, 50-370 Wrocław, Poland; andrzej.ubysz@pwr.edu.pl
3 Faculty of Architecture and Urbanism, National University of La Plata, 7 Avenue 776, La Plata B1900, Argentina; pilibilbao@gmail.com
* Correspondence: jeonghyun.kim@pwr.edu.pl

Abstract: Research on recycled concrete aggregates (RCAs) has been progressively advanced. Beyond replacing natural aggregates with RCA, discussions have been held on the effect of the parent concrete and repeatedly recycled aggregate concrete. Although it has been reported that RCA can be technically used for structural concrete, due to several other factors, RCA is mainly used for sub-bases. Therefore, identifying these factors is the key to promoting the use of RCA. Therefore, this review study first briefly summarizes the physical and chemical characteristics of RCA compared to natural aggregate, and reviews the effects of parent concrete and repeatedly recycled aggregate on next generation concrete. This study also briefly discusses the RCA standards of various countries and the factors that hinder the widespread use of RCA. The results show that there is a correlation in properties between parent concrete and the next generation concrete, and the properties of concrete decrease when RCA is used repeatedly. In addition, on the basis of the literature review, factors hindering the use of RCA were found to be unstable supply and demand, economic feasibility, and negative perceptions.

Keywords: recycled concrete aggregate; recycled aggregate concrete; residual mortar; reusing; sustainability

Citation: Grabiec, A.M.; Kim, J.; Ubysz, A.; Bilbao, P. Some Remarks towards a Better Understanding of the Use of Concrete Recycled Aggregate: A Review. *Sustainability* **2021**, *13*, 13336. https://doi.org/10.3390/su132313336

Academic Editor: Jorge de Brito

Received: 18 October 2021
Accepted: 29 November 2021
Published: 2 December 2021

1. Introduction

According to the annual Mineral Commodity Summaries issued by the US Geological Survey [1], 88% of about 1.5 billion tons of crushed aggregate consumed in the United States in 2020 was used in construction and cement manufacturing. Moreover, 960 million tons of construction sand and gravel were produced in the United States in 2020, of which 46% was used for concrete.

In Europe, it is also reported that 25% and 15% of aggregates are used in ready-mixed concrete and precast concrete, respectively [2]. The global construction aggregate and cement market, including developing countries such as China and India, is expected to grow steadily by 2030 [3]. However, the dark side of this growth in construction is related to environmental issues. The cement industry is responsible for about 7% of global anthropogenic carbon dioxide (CO_2) emission, and 36.4% of the total waste generated in the EU-28 in 2016 came from the construction sector [4]. The annual construction and demolition (C&D) waste in the United States has increased by about 3.5 times from 170 million tons in 2005 to 600 million tons in 2018 [5]. Shi and Xu [6] have predicted that the mass of C&D waste in China will soon reach 650 million tons. According to the literature of Sobotka et al. [7] published in 2016, C&D waste generated in Poland in 2022 was predicted to be 5.5 million tons, but in 2018 it already exceeded 6.7 million tons [8]. That confirms the C&D waste is expected to increase due to deteriorated infrastructures [9,10]. Therefore, the effective recycling of C&D waste is an immediate challenge.

The time when the use of recycled concrete aggregate (RCA) to make concrete began was in the 40s of the 20th century, when on the one hand, there was a need to cope with the huge amount of building debris resulting from the destruction of the war, and on the other hand, to find a low-cost and easily accessible material for the construction of new buildings. According to Levy and Helene 2002, via Aragão [11], it is for this reason that 1946 can be considered as the starting date of the period in which the new concept of recycled aggregate concrete (RAC) was developed. The focus was on the construction of new buildings using concrete and ceramic debris, mainly for foundations. The ecological aspects, if one can speak of them at all, were certainly of very little importance. While after 1945, the increase in interest in RCA was due to the social and economic crisis (shortage of financial resources and building materials), today's increase in interest is the outcome of the environmental and social crisis. The subsequent reports of the Intergovernmental Panel of Climate Changes (IPPC) and the UN report Ecosystems and Human Well-being: Assessment [12] are significant in this regard. Quite free, and in many cases uncontrolled, use of natural resources for the purposes of civilization, runs the risk of disrupting many natural systems [13]. The current crisis is therefore forcing an intensification of efforts to protect the environment in all areas of human industrial activity, including the building industry.

Concrete waste, one of the major compositions of C&D waste, is produced as RCA through the crushing, separating, and screening process, and it is used as a substitute for natural aggregate (NA). However, since the RCA consists of original aggregate and residual mortar, RCA and NA can be considered as similar materials, but the properties of the two materials are different. Therefore, the properties of concrete using RCA are different from those of using NA. In general, concrete made with RCA not only decreases the density and workability in the fresh state but also decreases the mechanical properties and durability-related performance of hardened state concrete compared to natural aggregate concrete (NAC) [13–16].

The undisputed leader in concrete recycling is Japan, where the recycling of materials from demolition and deconstructing already accounted for 92% in 2005 [17]. Moreover, about 88% of the construction waste generated in the EU is reused [18], but due to the poor properties of RAC mentioned above, the RCA is mainly used in the most primary way, such as road base and road leveling [19]. Therefore, the value-added of RCA was one of the challenging tasks, and a variety of studies were conducted to improve the properties of RAC for using it in various areas. As research data were accumulated, it was reported that RCA could be used for the production of structural concrete [20,21]. However, apart from the performance of RAC, there are other major factors that hinder the use of RCA and RAC which have not been discussed sufficiently.

Basically, the untreated RAC does not act as well as NAC in terms of mechanical properties and durability performance [22–26]. However, the applicability of the RCA to concrete is clear [27,28], and RCA is practically being used for concrete production [10,29]. In this context, research on RCA and RAC is progressing. Beyond the study on the possibility of replacing NA with RCA, the effect of parent concrete on the properties of next-generation concrete and the effect of repeated use of RCA on the properties of RCA and RAC were investigated. The former case is important in terms of better utilization of RCA and prediction of the property development of RAC, but the latter is particularly crucial in the long term as it relates to the reuse of RCA. Although extensive review papers on RCA and RAC have been published by some researchers (e.g., [30–35]), many of these deal with the correlation between RCA replacement and the performance of RAC, there has been little discussion on the effect of the properties of the parent concrete on the properties of the next-generation RAC, and in particular the properties of RAC that is repeatedly recycled. This gap may be because there are not enough data to discuss, as studies on the reuse of RCA [36–42] have been conducted by a few researchers in recent years. Nevertheless, identifying the factors hindering the widespread use of RCA and evaluating the effects of multiple recycling of RCA on the properties of concrete need to be addressed as they are key points for achieving sustainability in the construction industry.

This paper is structured as follows. Section 1 describes the significance of this research. Section 2 briefly addresses the properties and production of RCA. Section 3 deals with the effect of parent concrete and multiple uses of RCA on the properties of the next-generation RCA and RAC. This section includes a discussion of RCA standards for concrete in selected countries and several factors that hinder the use of RCA. The last section presents the conclusions.

2. Recycled Concrete Aggregate

2.1. Characteristics

RCA produced from C&D waste is made up of old residual mortar and NA (i.e., original aggregate) (Figure 1). This residual mortar sticks to the original aggregate and is considered a major factor that distinguishes RCA from NA [43]. The presence of the residual mortar lowers densities of the RCA and raises its water absorption capacity, and it contributes to the reduction in the mechanical strength and durability performance of RAC [44–47].

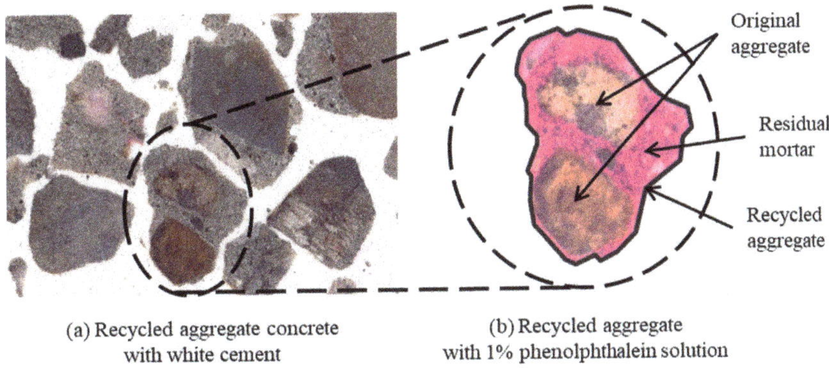

Figure 1. A sectional view of RAC (**a**) and RCA (**b**).

Figure 2 presents the relationship between residual mortar content, water absorption and specific gravity of NA and RCA studied by various research groups. The specific gravity of NA is mostly distributed in the range of 2.6 to 2.8 and the water absorption ratio is less than 2%, whereas, for the RCA, the water absorption ranges between 2% and 8%, and the specific gravity is distributed between about 2.3 and 2.5. A direct comparison of water absorption and density by residual mortar content may not be meaningful, as the original aggregate of the RCA used in each literature is not the same, but it seems clear that the content of residual mortar is inversely proportional to the density of RCA and directly proportional to the water absorption capacity.

The content of residual mortar may also be closely related to the size of RCA. Juan and Gutiérrez [48] investigated the effect of the size of RCA on the residual mortar content and noted that the residual mortar content of 4–8 mm fraction of RCA ranges from 33% to 55%, while it has 23% to 44% for 8–16 mm fraction. Similarly, in the study by Suryawanshi et al. [49], it was observed that the residual mortar content in the 4.75–10 mm fraction was 27% higher than that in the 10–20 mm fraction. On the contrary, of the total three groups of RCA used in the study [44], two had around 3% higher residual mortar content in the 5–10 mm fraction, while the other RCA had about 10% lower residual mortar content in the 5–10 mm fraction. Abbas et al. [50] sorted two different RCAs into four groups by size, respectively, and tested the residual mortar content. One of the two RCAs had the highest residual mortar content in the order of 19 mm, 4.75 mm, 12.7 mm, and

9.5 mm, while for another RCA, the residual mortar content gradually increased from the smallest to the largest aggregates.

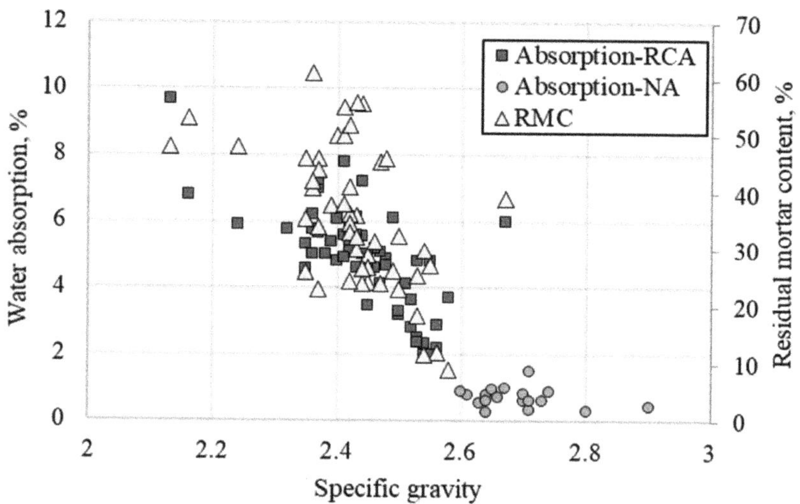

Figure 2. Physical properties of NA and RCA [44,48,51–72].

In addition to physical characteristics, RCA differs in chemical composition from NA. Figure 3 shows the relationship of chemical composition between calcium oxide (CaO) and silicon dioxide (SiO_2) in ordinary Portland cement, NA and RCA. Since the cement is produced to meet the composition required by standards [73,74], CaO and SiO_2 were concentrated at about 62% and 22%, respectively [71,75–77]. For NA used in various studies, the amount of SiO_2 was in the range of about 50–63%, and CaO was mainly less than 10% [58,75,76,78–80]. For RCA, which is a mixture of original aggregate and residual mortar, the SiO_2 content was lower and the CaO was higher than those of NA, distributed between ordinary Portland cement and NA [58,75,76,78–83].

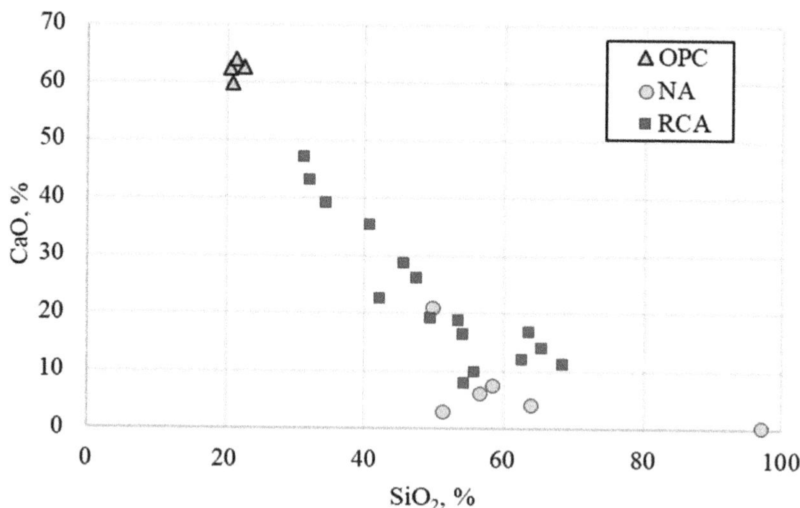

Figure 3. Chemical composition of ordinary Portland cement (OPC), NAs and RCAs [58,71,75–83].

Although it may not be appropriate to compare the properties of RCAs used in different studies due to geological differences, there appears to be a correlation between the chemical composition and physical properties of RCA. According to a study by Kim et al. [58], when RCAs are obtained from the same source of parent concrete, the CaO content decreases and the SiO_2 content increases with a lower residual mortar content.

2.2. Production of Recycled Concrete Aggregate

In order to produce RCA with the suitable size as a construction material (0–40 mm) from massive C&D waste, concrete debris must go through a series of essential processes in which mechanical crushing is mainly employed. This basically consists of crushing to obtain size fractions suitable for use; magnetic separation to remove ferrous metals; a separation by air blowers to get rid of the nonmagnetic substance, such as paper, wood, plastic and dust; and screenings to sort the produced RCA by size (Figure 4). The quality of the RCA produced is affected by the type and specification of the recycling facility [84,85].

Regarding the effect of the type of crushers, Matias et al. [86] reported that RCA subjected to primary crushing by a jaw crusher had a lower shape index than RCA subjected to secondary crushing using an impact crusher. In other words, the aggregate that has only undergone the primary crushing has a more angular shape, and the one that has gone through the secondary crushing was more rounded. Öztürk et al. [87] fabricated RAC mixtures with RCA produced by impact crusher and vertical shaft crusher, respectively, and reported that RAC with RCA from the impact crusher showed a higher compressive strength than that of RAC from the vertical shaft crusher. On the contrary, Cepuritis et al. [88] noted that sand produced by the vertical shaft crusher showed a consistent shape and is more favorable for concrete than sand produced by milling or a cone crusher. Moreover, Ulsen et al. [89] reported that the residual mortar content, density, and particle size distribution of RCA produced by the jaw crusher and impact crusher were similar. However, the influence of the crusher type on the aggregate may vary due to factors such as rock types [90,91].

A meaningful trend is observed between the number of crushing and the physical characteristics of RCA. Nagataki et al. [92] crushed RCA three times by a combination of the impact crusher and the mechanical grinder (i.e., impact crusher—mechanical grinder—mechanical grinder), and compared its density, water absorption capacity, and residual mortar content. Compared to the primary crushed RCA, the tertiary crushed RCA increased in density by about 4.5%, and water absorption and residual mortar content decreased by 1.7–2.5% and 20%, respectively. Moreover, cracks in the RCA were observed after mechanical grinding, but the authors noted that it was negligible. Similarly, Pedro et al. [93] first crushed RCA with a jaw crusher and secondarily crushed the RCA with a hammer mill, and compared the density and water absorption for each RCA. As a result, a significant change was not observed in the density, but the water absorption of all six aggregates used in the test decreased by 0.2–1.5%. Regarding the number of crushing, the Korea Resource Recycling Association divided the production of RCA into four stages based on the purpose of use: primary crushing for the use of filling and leveling; secondary crushing for sub-base of the road pavement; tertiary crushing for the production of high-quality RCA; quaternary crushing for the production of sand for concrete [94].

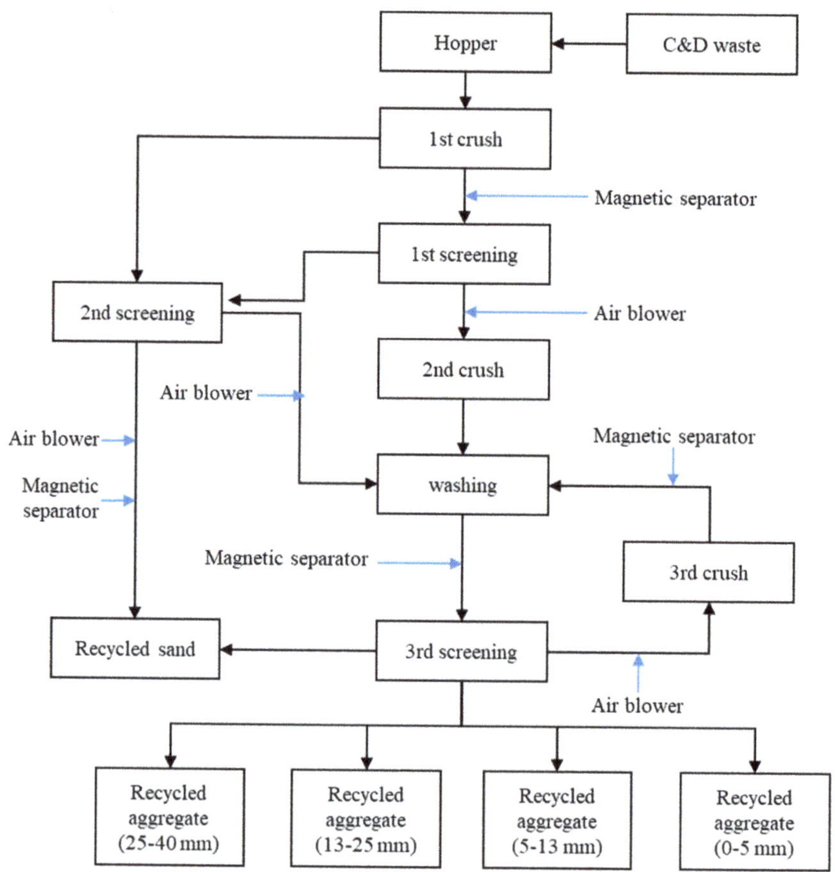

Figure 4. Scheme of RCA production (adapted from [94]).

RCA can be produced in mobile or stationary recycling plants. Regarding the recycling of C&D waste, the place where RCA is produced is one of the crucial factors. In general, mobile plants that produce RCA at construction and demolition sites do not need to transport C&D waste to stationary recycling plants, but the ability to remove impurities and residual mortar is relatively low. On the other hand, stationary recycling plants can produce high-quality RCA through several stages of crushing and screening, but transportation of C&D waste from the construction site to the recycling plant is involved [33]. Tam and Tam [95] stated that collecting concrete waste from different sources and managing it in one plant can cause an average effect, which can hinder the use of high-quality RCA and low-quality RCA for each purpose. On the contrary, Hiete [96] argued that the down-cycle of using RCA as a sub-base for road construction and reclamation remains dominant. Thus, for up-cycling, RCA needs to be processed in stationary plants capable of producing high-quality RCA.

The ultimate purpose of the concrete waste recycling process is to produce high-quality RCA that can minimize performance degradation of concrete when replacing NA with RCA by removing the residual mortar as much as possible. Thus, the following methods have been proposed to remove residual mortar: the hydrochloric acid dissolution method [97,98], the freeze-and-thaw method [50], the thermal shock method [48], the microwave method [99], and the high-performance sonic impulse method [100]. Moreover, technologies utilizing CO_2 [101] and bacteria [102] have been reported to show the potential to improve the performance of RCA and RAC. However, in a study on the energy consumption of RCA processing techniques conducted by Quattrone et al. [103], the fuel-fed thermo-mechanical processes are found to consume about 36–62 times more energy in producing RCA than traditional recycling processes consisting of crushing-separating-screening.

3. Recycled Aggregate Concrete

3.1. The Effect of Parent Concrete on the Properties of Next Generation Concrete

Research on RAC has further progressed from the effect of the RCA replacement ratio to the effect of the parent concrete on the next generation concrete, and this replacement, as well as the compressive strength of parent concrete are crucial in influencing the quality of next generation concrete.

Ahmad Bhat [104] produced low-, medium- and high-strength RCAs by crushing parent concretes with different compressive strengths of 20 MPa, 40 MPa, and 60 MPa, respectively. Using the three RCAs produced, a total of 18 series of RACs were manufactured: 6 series of RAC with target strengths of 20, 40 and 60 MPa at 50% and 100% replacement ratios using RCA obtained from the 20 MPa parent concrete; 6 series of RAC with target strengths of 20, 40 and 60 MPa at 50% and 100% replacement ratios using RCA obtained from 40 MPa parent concrete; another 6 series of RAC with target strengths of 20, 40 and 60 MPa at 50% and 100% replacement ratios using RCA obtained from 60 MPa parent concrete (Figure 5). The compressive strength of the low strength concrete with 50% RCA replacement obtained from high-strength concrete was 9% higher than that of parent concrete, and the tensile strength was the same as that of parent concrete with NA (Figures 6 and 7). A similar result was noted that RCA obtained from high strength parent concrete can achieve a similar compressive and tensile strength to normal strength concrete with NA [105]. In addition, as frequently observed in other studies [106,107], RAC with 100% RCA replacement showed lower mechanical strength than RAC with 50% RCA replacement. Considering the lower physical performance of RCA compared to NA, it is an expected result, and, the relationship between the strength of the parent concrete and the strength of the next generation concrete is also observed.

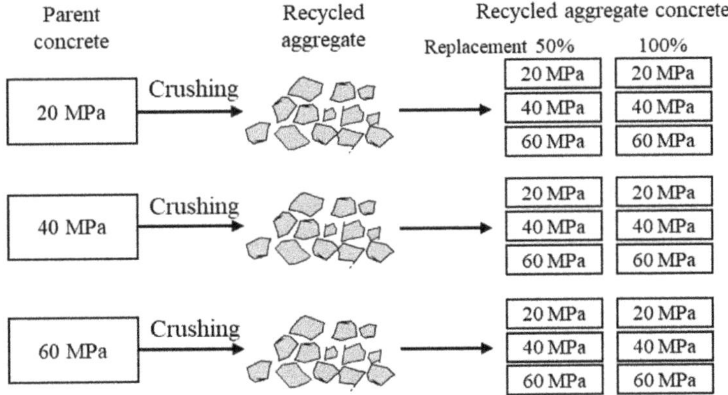

Figure 5. Experimental scheme of the effect of the strength of the parent concrete on the next generation concrete [104].

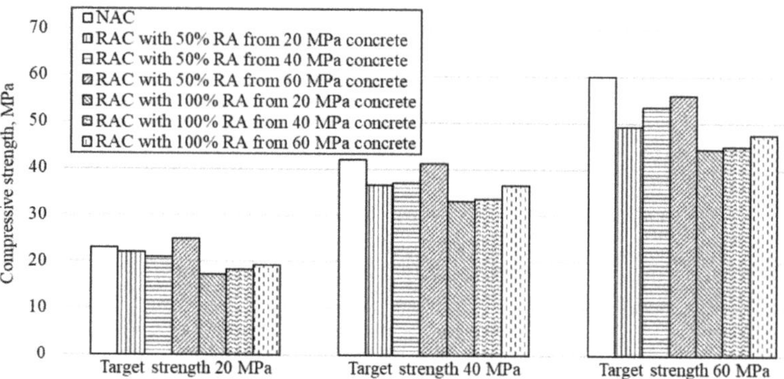

Figure 6. Compressive strength of RAC made from RCA produced from concrete with different strengths [104].

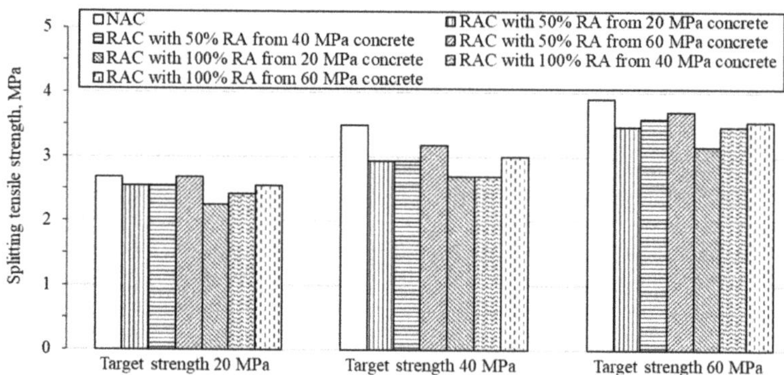

Figure 7. Splitting the tensile strength of RAC made from RCA produced from concrete with different strengths [104].

Kou and Poon [108] assessed the basic characteristics of five RCAs obtained from concrete with different compressive strengths of 30 MPa, 45 MPa, 60 MPa, 80 MPa, and 100 MPa. As the strength of the parent concrete increased, the water absorption of RCAs decreased, resulting in better quality. Afterwards, a total of 5 series of RACs with a 100% replacement rate were manufactured using each RCA, and mechanical properties and durability performances were analyzed. As a result of the mechanical properties test of compressive, tensile strength, and elastic modulus, RAC made from RCA obtained from high-strength concrete was found to be superior to RACs with RCA from low-strength concrete in all tests except for the slump value. Similar trends were observed in drying shrinkage and chloride resistance tests. The authors interpreted that the low water absorption ratio of RCA from high-strength concrete contributes to lower drying shrinkage and chloride penetration.

Gholampour and Ozbakkaloglu [71] made concrete mixtures with target strengths of 40 MPa and 80 MPa using RCAs produced from concretes with a compressive strength of 20 MPa, 40 MPa, and 110 MPa, and evaluated the mechanical properties. As shown in Table 1, the performance of concrete made of RCA from high-strength concrete was superior to those of medium- and low-strength concrete with all the selected properties. Unlike the test results in [108], the slump values of RAC made of RCA from high-strength concrete were the same as that of concrete made of RCA produced from medium-strength concrete, and were more workable than that made from low-strength concrete. In particular, for the creep deformation, RAC with low-strength RA was 60% higher than that of high-strength

RCA. On the contrary, Padmini et al. [61] reported that low-strength parent concrete, which has a relatively weaker binding force between the residual mortar and the original aggregates than that of high-strength concrete, separates most residual mortar from the RCA surface during the crushing process; thus, a specific gravity of RCA decreases and the water absorption increases as the strength of parent concrete increases. It was observed that the residual mortar content in RCA obtained from the low-strength concrete was less, and for this reason, the RCA of a small size (i.e., 10 mm) may have more residual mortar rather than aggregates, which could lead to a greater reduction in mechanical properties. Wang et al. [109] noted that the level of residual mortar content may not be an absolute factor in determining the properties of concrete. In the study, it was reported that the RCA from high-strength parent concrete reduced the shrinkage of RAC at a water-to-cement ratio of 0.30, even though the RCA had a higher residual mortar content compared to that of RAC with RCA from low-strength parent concrete. However, at the water-to-cement ratios of 0.45 and 0.60, the shrinkage of RAC increased in the order of RCA from low-strength, medium-strength, and high-strength concrete. The authors explained that the lower porosity and uniformly hardened residual mortar of RCA from high-strength concrete could contribute to an increased shrinkage deformation of RAC.

Table 1. Fresh and hardened properties of RAC with RCA from low-, medium-, and high strength parent concrete [71].

Test		RAC1	RAC2	RAC3	RAC4
Compressive strength of parent concrete, MPa		40	110	20	110
Target strength, MPa		40		80	
Slump, mm		115	115	140	150
Density, kg/m^3	Fresh	2351	2364	2356	2407
	Hardened	2330	2345	2341	2381
Compressive strength, MPa		32.0	36.8	64.3	80.1
Elastic modulus, GPa		26.7	29.1	31.9	41.5
Tensile strength, MPa		3.22	3.36	3.90	5.11

Geng et al. [110] evaluated the effect of the service time of parent concrete (i.e., 1 year, 18 years, and 40 years) on strength development. The compressive strength development of NAC at 1 day was 33% compared to that at 28 days, while for the RAC with the service life of 1 year, 18 years and 40 years, the strength development decreased to 30%, 24%, 11%, respectively. However, after longer periods of hardening after 28 days to 90 days, the strength of NAC increased by 17%, whereas RAC gained 21–29%. Therefore, it was emphasized that the existing prediction model for strength development for NAC may not be suitable for RAC.

Something which seems to be of particular importance in the context of RCA use for production of RAC is the possible reactivity of RA as a consequence of NA's reactivity. It is well known that alkali-silica reaction (ASR) is one of the major causes of concrete degradation. In the case of RAC, this problem can be significant depending on the percentage of RCA as a substitute of NA and because of the obvious crushing process [22].

3.2. The Effect of Repeated Use of Recycled Concrete Aggregate

Generally, research on RAC has been focused on concrete made from RCA obtained from NAC. However, some researchers studied the reuse of RCA by investigating the effect of repeatedly reused RCA on the properties of RAC. The scheme of repeated use of RCA is shown in Figure 8.

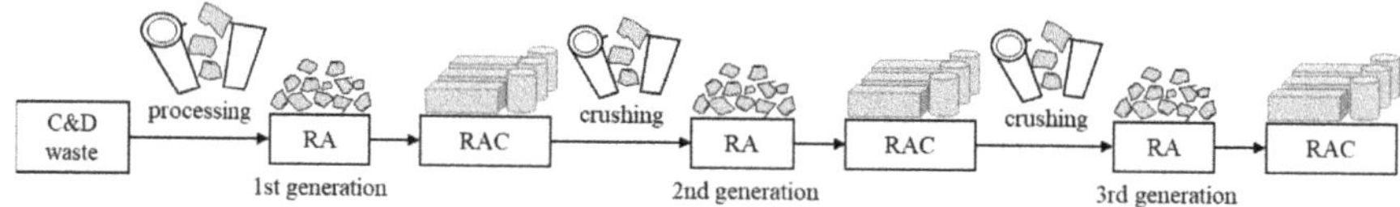

Figure 8. Scheme of the repeated use of RCA and RAC.

Huda and Alam [37] evaluated the effect of RCA reuse on the properties of RCA and RAC by repeating the crushing and casting of RAC a total of three times. For the physical characteristics of RCA, compared to the first generation RCA, the density of the second and third generation RCA decreased by 10% and 14%, respectively, and the water absorption rate increased by 1.9% and 4.2%. For the fresh state properties of the RAC, as the number of reuse of RCA increased from the first generation to the third generation, the slump decreased by 6% and 15%, and air content increased by 0.3% and 0.5%, respectively. For the mechanical properties, compared with the first generation RAC, the compressive strength and tensile strength of the third generation RAC decreased by 21% and 36%, respectively, and the elastic modulus decreased by 5%. However, the strengths of the second generation RAC increased by 1% to 2%. Similar observations were found in several studies. In the study performed by Salesa et al. [41], the compressive strength and elastic modulus of the first and second generation RACs were evaluated. Except for the slump of the second generation RAC, which decreased by 9% compared to that of the first generation RAC, significant changes were not observed in the compressive strength and elastic modulus. Zhu et al. [40] observed that the residual mortar content of RCA increased to 38%, 55% and 62% in proportion to the number of recycling. As expected, the slump value, compressive strength and elastic modulus decreased. According to Zhu et al. [36], decreases in the durability performance of RAC, such as resistance against the freeze–thaw action, chloride penetration and carbonation were observed as RCA was repeatedly reused.

Comprehensively, it seems clear that the multiple uses of RCA result in the low quality of RAC [38]. Marie and Quiasrawi [42] noted that the second generation RCA can replace NA up to 20% to make the acceptable quality of RAC, while Abed et al. [111] concluded that up to 50% of the second generation RCA can optimally replace NA. However, the studies referenced in this section appear to address changes in the properties of RCA and RAC that occur when RCA is reused without pretreatment. Therefore, further study can be recommended to evaluate the physical properties of concrete after improving the quality by pre-treating RCA such as removing residual mortar. In addition, a detailed investigation into the durability characteristics of repeatedly recycled RAC is required [112]. Current research into repeatedly recycled RAC evaluates the characteristics of concrete manufactured and crushed in a laboratory, and the consideration of carbonation exposed to the natural environment is insufficient. Carbonation is one of the factors to be considered as it reduces the pH of concrete and contributes to the corrosion of reinforcement [113,114].

The authors of this review paper believe that it is purposeful, although it is not the only possibility in RAC technology, to consider the repeated use of RCA. The main argument seems to be the depletion of NA deposits, as well as the ecological aspect related to the life cycle of concrete. Concrete can become more environmentally friendly when CO_2 absorption of not only concrete elements but also the RCA is more accurately recognized. Concrete during its life cycle absorbs CO_2 emitted during cement production and other anthropogenic activities. Different values are given in the literature, from 7.6 to 24%. The disagreement in evaluation is due to different factors influencing concrete carbonation (among others, compressive strength, cement amount, age of concrete structure). This amount can be increased if concrete is treated as a source of material with a greater CO_2 absorption capacity as a result of the larger specific surface area of the aggregate [115]. Repeated use of RCA would provide an even greater opportunity for this. Undoubtedly, it leads to a polemic as to what is more important, whether these include technological or ecological aspects. Therefore, it is even more legitimate to carry out research and analysis on both of them.

3.3. Standards of Recycled Aggregate for Concrete

The first approaches to the standardization of established technological procedures concerning RCA appeared in the late 1970s and early 1980s, first in Japan in 1977 and then in the USA (ASTM C 32-82 and C 125-79) [116] and in some European countries (Germany—DIN 4226-100, Denmark, Great Britain—BS 8500-2) and Brazil (following [11]).

The lack of specific legislation regulating the requirements for the properties of RCA and concretes in some countries is due to the currently still insufficient natural reserves in these countries and the general perception that concrete with RCA has significantly worse properties. In Poland, for example, issues concerning RCA and RAC are only reflected in standards PN-EN 12620:2013 [117] and PN-EN—206:2014 [118]. The Polish Committee for Standardization has adopted draft standards (FprEN 12620:2017, FprEN 13139:2017 and FprEN 13242:2017 [119–121]), which include requirements for RCA, but so far they have not been incorporated by the European Parliament in the harmonized standards package.

The standards of RCA for concrete established in selected countries and several requirements are shown in Table 2.

Table 2. Physical characteristics of RCA specified in some standards.

Standard	Classification	Minimal Density, kg/m^3	Maximal Water Absorption, %	Maximal Chloride, Content, %	Maximal Sulphate Content, %
JIS A5021 [124] (Japan)	High	2500	3	0.04	-
JIS A5022 [123] (Japan)	Medium	2300	5	0.04	-
JIS A5023 [122] (Japan)	Low	-	7	0.04	-
GB/T 25177 [125] (China)	Class 1	2450	3	0.06	2
	Class 2	2350	5	0.06	2
	Class 3	2250	8	0.06	2
WBTC-No.12 [126] (Hong Kong)	-	2000	10	0.05	-
KS F2527 [127] (Korea)	-	2500	3	-	-
RILEM [128]	Type I	1500	20	-	1
	Type II	2000	10	-	1
	Type III	2400	3	-	1
HB-155 [131] (Australia)	-	2100	6	0.05	0.5
NBR-15116 [129] (Brazil)	-	-	7	1	1
LNEC-E471 [130] (Portugal)	-	2200	7	-	0.8
EHE-08 [132] (Spain)	-	-	7	-	1

Japanese standards classify RCAs for concrete into three levels: low quality [122], medium quality [123] and high quality [124] based on the purpose of use. The low-quality RCA is produced by crushing concrete waste and is used for members where do not require high strength and high durability, and the medium-quality RCA can mainly be used where they are not affected by drying shrinkage and freezing and thawing action except for specific cases. The high-quality RCA is obtained by advanced treatments, and there seems to be no restriction on its use. Chinese standard [125] also divides the quality of RCA into three classes, and the density of RCA is based on the saturated surface-dry condition instead of an oven-dry state. Differently, the Hong Kong [126] and Korean standards [127] do not divide RCA into several groups. The Korean standard [127] requires a minimum oven-dry density of 2500 kg/m^3 and a maximum water absorption of 3%, etc. Compared to the other Asian counties mentioned above, Hong Kong specifications [126] seem to have more relaxed requirements (i.e., a minimum oven-dry density of 2000 kg/m^3 and a maximum water absorption of 10%); however, the use of RAC with 100% RCA replacement is limited to a minor concrete structure, such as concrete benches and planter walls. RILEM [128] classified RCA into three categories: RCA obtained from masonry waste (type 1), RCA obtained from concrete waste (type 2), and a combination of RCA and NA (type 3). NBR-15116 [129], LNEC-E471 [130], HB-155 [131] EHE-08 [132] referred to the study by Gonçalves and Brito [133]. Compared to the Asian standards mentioned, the water absorption of these standards is rather high, in the range of 6–7%, but the sulphate content is tightly controlled. Spanish standards EHE-08 [132] recommend the use of a superplasticizer when replacing more than 20% of RCA due to the high absorption capacity of RCA. The Polish standard PN-EN 12620:2013 [117] classifies aggregates into natural, recycled and artificial and allows their use in concrete technology provided they meet criteria for impurities that may affect the quality of concrete, including water soluble sulphate content and alkaline reactivity for RCAs.

Although the standards may have been established according to the individual circumstances of each country, the density and water absorption ratio required by Japanese, Korean, and Chinese standards are higher than those of other standards. Taking into account the distribution of the absorption rate of RCAs shown in Figure 2, obtaining the high-quality RCA with a water absorption ratio of less than 3% is considered to need more advanced facilities than the typical RCA production process.

3.4. Barriers to the Use of Recycled Aggregate Concrete in Practice

Various studies have been conducted on the feasibility of improving the mechanical properties and durability performance of RAC [134–137]. Hossain et al. [138] conducted a life cycle assessment of RCA and NA production, and concluded that RCA produced from C&D waste reduced non-renewable energy consumption and greenhouse gas emission by 58% and 65%, respectively. Similarly, the better environmental benefits of replacing NA with RCA have been reported [139–142], but there are some obstacles to using RCA and RAC in practice.

3.4.1. Imbalance in Supply and Demand

According to a case study carried out by Böhmer et al. [143], concrete waste accounts for about 25–40% of C&D waste. Assuming that the proportion of the aggregates are 50–60%, due to the losses incurred in the various treatment processes, the actual amount that can be obtained will be less. With the exception of studies to maximize the use of powder and dust generated during the waste concrete process [46,144,145], imbalance in supply and demand of RCA can be a barrier to the recycling of C&D waste [146,147]. In Korea, the mandatory use ratio of RCA is specified for concrete work in certain construction projects, but if the supply of RCA is difficult due to the distance between the construction site and the RCA supplier, the mandatory use is exempted [148,149].

3.4.2. Economic Viability

Economic viability is an important factor influencing C&D waste management. Manowong [150] mentioned that metal-oriented recycling, which is easily perceived as having economic value, is prioritized in some countries. In the interview study conducted by Wu et al. [151], one interviewee noted that if a regulatory infraction is the most profitable option, it is more likely to be chosen. In fact, due to the limit of monitoring by governments and easy access to illegal reclamation, cases of illegal waste dumping and reclamation are frequently reported [152–154].

Martínez-Lage et al. [155] analyzed the economic feasibility of recycling C&D waste and stated that distance is one of the major factors affecting cost. A case study in a region in Spain showed that the insufficient number of recycling plants and unfavorable locations would increase transportation distances and, as a result, the use of RAC could increase costs further. Similar analyses were found by [156,157]. Hiete [96] argued that recycling of C&D waste is most attractive in high-density areas where demand and supply are matched due to the short transport distance, while areas with a low population density are not economically effective. Nunes et al. [158] analyzed two recycling plants with a production capacity of 20 tons and 100 tons per hour, respectively. The authors concluded that C&D

waste recycling plants operated by private companies may not be financially suitable under current market conditions in Brazil.

3.4.3. Negative Perceptions

Several studies mentioned the negative perception of stakeholders as one of the obstacles to the use of C&D waste [159–161]. It has been mentioned above that the quality of RCA varies significantly depending on the processing process, and that various equipment and technologies are required to obtain the high-quality RCA. Moreover, the quality may vary depending on the environment exposed during the service life [162].

According to Jong et al. and Taylor-Lange et al. [163,164], concrete with fly ash slightly increased the possibility of indoor radioactive pollution compared to conventional concrete. A large amount of uranium is detected in the waste concrete generated by the demolition of nuclear facilities [165]. In 2011, in Seoul, Korea, radioactivity was detected on asphalt roads mixed with C&D waste, resulting in a negative perception of the Korean public about C&D waste materials. These are, for now, individual suggestions in the literature, which in the opinion of the authors of this paper, should be analyzed and emphasized more distinctly.

4. Conclusions

The analysis of the literature cited in this paper has allowed the authors to select scientific content, some of which is worthy of research development, particularly in the ecological aspect. The most important threads are presented below.

It is obvious that concrete with RCA is a technologically demanding material of worse quality. The amount of the residual mortar in RCA is considered to be one of the factors that directly reduce properties of RAC. Due to the presence of the residual mortar, RCA has higher water absorption and a lower density in comparison to natural aggregate. However, the rational approach to using RCA for producing next generation concrete, first of all by considering an optimal RCA replacement ratio and the selection of parent concrete types of higher compressive strength seems to be an opportunity to mitigate quality loss of concrete. These two factors are crucial in influencing the quality of next generation concrete with RCA.

Moreover, it is advisable to look for advanced technologies, giving a chance to produce RCA of higher quality with a significant amount of residual mortar removed, and, as a result, RAC of comparable or even higher quality in comparison to concrete types with natural aggregate. This is particularly justified in the case of repeated use of RCA when the residual mortar content attached to a single RCA increases with the increasing number of RCA reuse cycles. It is the opinion of the authors of this review paper that the repeated use of RCA is prosperous from the ecological viewpoint, reducing the depletion of natural aggregate deposits and enhancing the life cycle of concrete. Clearly, more research data including carbonation degree evaluation of RAC are needed to identify a clear trend.

Undoubtedly, harmonization and unification of standards for RCA in individual countries would optimize the rational management not only on an individual scale, but also globally. The latter seems particularly important in the context of a widely perceived and environmentally aware protection. It seems to be crucial because barriers such as economic viability and negative perception of stakeholders in many countries act as factors hindering the practical use of RCA and RAC.

Author Contributions: Conceptualization, J.K. and P.B.; methodology, J.K. and P.B.; formal analysis, J.K. and A.M.G.; data curation, J.K. and A.M.G.; writing—original draft preparation, J.K.; writing—review and editing, A.M.G.; visualization, J.K. and P.B.; supervision, A.M.G. and A.U.; project administration, A.M.G. and A.U. All authors have read and agreed to the published version of the manuscript.

Funding: This research received no external funding.

Institutional Review Board Statement: Not applicable.

Informed Consent Statement: Not applicable.

Data Availability Statement: Not applicable.

Conflicts of Interest: The authors declare no conflict of interest.

References

1. USGS. *Mineral Commodity Summaries 2021*; USGS: Reston, VA, USA, 2021.
2. UEPG. *A Sustainable Industry for a Sustainable Europe*; Annual Report; UEPG: Brussels, Belgium, 2020.
3. PMR. *Global Construction Aggregates Market*; PMR: New York, NY, USA, 2020.
4. Eurostat. Available online: https://ec.europa.eu/eurostat/tgm/refreshTableAction.do?tab=table&plugin=1&pcode=ten00106&language=en (accessed on 2 March 2021).
5. EPA. Construction and Demolition Debris: Material-Specific Data. Available online: https://www.epa.gov/facts-and-figures-about-materials-waste-and-recycling/construction-and-demolition-debris-material (accessed on 7 July 2021).
6. Shi, J.; Xu, Y. Estimation and forecasting of concrete debris amount in China. *Resour. Conserv. Recycl.* **2006**, *49*. [CrossRef]
7. Sobotka, A.; Sagan, J.; Sikora, A. Logistyka odzysku w remontach obiektów budowlanych. *Mater. Bud.* **2016**, 134–136. [CrossRef]
8. Eurostat. Generation of Waste by Waste Category, Hazardousness and NACE Rev. 2 Activity. Available online: https://ec.europa.eu/eurostat/databrowser/view/env_wasgen/default/table?lang=en (accessed on 14 November 2021).
9. Jin, J.; Wang, Z.; Ran, S. Solid waste management in Macao: Practices and challenges. *Waste Manag.* **2006**, *26*, 1045–1051. [CrossRef] [PubMed]
10. Kim, J. Construction and demolition waste management in Korea: Recycled aggregate and its application. *Clean Technol. Environ. Policy* **2021**, *23*, 2223–2234. [CrossRef]
11. Aragão, H.G. *Análise Estrutural de Lajes Pré-Moldadas Produzidas com Concreto Reciclado de Construção e Demolição*; Universidade Federal de Alagoas: Maceió, Brazil, 2007.
12. Assessment, M.E. *Ecosystems and Human Well-Being*; Island Press: Washington, DC, USA, 2005; Volume 5.
13. Grabiec, A.M.; Zawal, D.; Rasaq, W.A. The effect of curing conditions on selected properties of recycled aggregate concrete. *Appl. Sci.* **2020**, *10*, 4441. [CrossRef]
14. Rahal, K. Mechanical properties of concrete with recycled coarse aggregate. *Build. Environ.* **2007**, *42*, 407–415. [CrossRef]
15. Tavakoli, M.; Soroushian, P. Strengths of recycled aggregate concrete made using field-demolished concrete as aggregate. *Mater. J.* **1996**, *93*, 178–181.
16. Ajdukiewicz, A.B.; Kliszczewicz, A.T. Comparative tests of beams and columns made of recycled aggregate concrete and natural aggregate concrete. *J. Adv. Concr. Technol.* **2007**, *5*, 259–273. [CrossRef]
17. Storey, J.B. *Construction Materials Stewardship: The Status Quo in Selected Countries: CIB W115*; Centre for Building Performance Research, Victoria University of Wellington: Wellington, New Zealand, 2008; ISBN 906363059X.
18. Eurostat. Record Recycling Rates and Use of Recycled Materials in the EU. Available online: https://ec.europa.eu/eurostat/documents/2995521/9629294/8-04032019-BP-EN.pdf/295c2302-4ed1-45b9-af86-96d1bbb7acb1 (accessed on 7 July 2021).
19. Tam, V.W.Y.; Soomro, M.; Evangelista, A.C.J. A review of recycled aggregate in concrete applications (2000–2017). *Constr. Build. Mater.* **2018**, *172*, 272–292. [CrossRef]
20. Malešev, M.; Radonjanin, V.; Marinković, S. Recycled concrete as aggregate for structural concrete production. *Sustainability* **2010**, *2*, 1204–1225. [CrossRef]
21. Fathifazl, G.; Razaqpur, A.G.; Burkan Isgor, O.; Abbas, A.; Fournier, B.; Foo, S. Shear capacity evaluation of steel reinforced recycled concrete (RRC) beams. *Eng. Struct.* **2011**, *33*, 1025–1033. [CrossRef]
22. Barreto Santos, M.; de Brito, J.; Santos Silva, A. A review on alkali-silica reaction evolution in recycled aggregate concrete. *Materials* **2020**, *13*, 2625. [CrossRef]
23. Bravo, M.; Duarte, A.P.C.; de Brito, J.; Evangelista, L. Tests and simulation of the bond-slip between steel and concrete with recycled aggregates from CDW. *Buildings* **2021**, *11*, 40. [CrossRef]
24. Bravo, M.; de Brito, J.; Pontes, J.; Evangelista, L. Shrinkage and creep performance of concrete with recycled aggregates from CDW plants. *Mag. Concr. Res.* **2017**, *69*, 974–995. [CrossRef]
25. Bravo, M.; de Brito, J.; Pontes, J.; Evangelista, L. Durability performance of concrete with recycled aggregates from construction and demolition waste plants. *Constr. Build. Mater.* **2015**, *77*, 357–369. [CrossRef]
26. Junak, J.; Sicakova, A. Effect of surface modifications of recycled concrete aggregate on concrete properties. *Buildings* **2017**, *8*, 2. [CrossRef]
27. Silva, R.V.; de Brito, J.; Dhir, R.K. Fresh-state performance of recycled aggregate concrete: A review. *Constr. Build. Mater.* **2018**, *178*, 19–31. [CrossRef]
28. Mi, R.; Pan, G.; Liew, K.M.; Kuang, T. Utilizing recycled aggregate concrete in sustainable construction for a required compressive strength ratio. *J. Clean. Prod.* **2020**, *276*, 124249. [CrossRef]
29. Silva, R.V.; de Brito, J.; Dhir, R.K. Use of recycled aggregates arising from construction and demolition waste in new construction applications. *J. Clean. Prod.* **2019**, *236*. [CrossRef]
30. Silva, R.V.; de Brito, J.; Dhir, R.K. Tensile strength behaviour of recycled aggregate concrete. *Constr. Build. Mater.* **2015**, *83*, 108–118. [CrossRef]
31. Silva, R.V.; de Brito, J.; Dhir, R.K. Comparative analysis of existing prediction models on the creep behaviour of recycled aggregate concrete. *Eng. Struct.* **2015**, *100*, 31–42. [CrossRef]

32. Papatzani, S.; Paine, K. Construction, demolition and excavation waste management in EU/Greece and its potential use in concrete. *Fresenius Environ. Bull* **2017**, *26*, 5572–5580.
33. Silva, R.V.; de Brito, J.; Dhir, R.K. Availability and processing of recycled aggregates within the construction and demolition supply chain: A review. *J. Clean. Prod.* **2017**, *143*, 598–614. [CrossRef]
34. Behera, M.; Bhattacharyya, S.K.; Minocha, A.K.; Deoliya, R.; Maiti, S. Recycled aggregate from C&D waste & its use in concrete—A breakthrough towards sustainability in construction sector: A review. *Constr. Build. Mater.* **2014**, *68*, 501–516. [CrossRef]
35. Silva, R.V.; de Brito, J.; Dhir, R.K. Properties and composition of recycled aggregates from construction and demolition waste suitable for concrete production. *Constr. Build. Mater.* **2014**, *65*, 202–217. [CrossRef]
36. Zhu, P.; Hao, Y.; Liu, H.; Wei, D.; Liu, S.; Gu, L. Durability evaluation of three generations of 100% repeatedly recycled coarse aggregate concrete. *Constr. Build. Mater.* **2019**, *210*, 442–450. [CrossRef]
37. Huda, S.B.; Alam, M.S. Mechanical behavior of three generations of 100% repeated recycled coarse aggregate concrete. *Constr. Build. Mater.* **2014**, *65*, 574–582. [CrossRef]
38. Abreu, V.; Evangelista, L.; de Brito, J. The effect of multi-recycling on the mechanical performance of coarse recycled aggregates concrete. *Constr. Build. Mater.* **2018**, *188*, 480–489. [CrossRef]
39. Thomas, C.; de Brito, J.; Gil, V.; Sainz-Aja, J.A.; Cimentada, A. Multiple recycled aggregate properties analysed by X-ray microtomography. *Constr. Build. Mater.* **2018**, *166*, 171–180. [CrossRef]
40. Zhu, P.; Zhang, X.; Wu, J.; Wang, X. Performance degradation of the repeated recycled aggregate concrete with 70% replacement of three-generation recycled coarse aggregate. *J. Wuhan Univ. Technol. Sci. Ed.* **2016**, *31*, 989–995. [CrossRef]
41. Salesa, Á.; Pérez-Benedicto, J.A.; Colorado-Aranguren, D.; López-Julián, P.L.; Esteban, L.M.; Sanz-Baldúz, L.J.; Sáez-Hostaled, J.L.; Ramis, J.; Olivares, D. Physico-mechanical properties of multi-recycled concrete from precast concrete industry. *J. Clean. Prod.* **2017**, *141*, 248–255. [CrossRef]
42. Marie, I.; Quiasrawi, H. Closed-loop recycling of recycled concrete aggregates. *J. Clean. Prod.* **2012**, *37*, 243–248. [CrossRef]
43. Mi, R.; Pan, G.; Li, Y.; Kuang, T.; Lu, X. Distinguishing between new and old mortars in recycled aggregate concrete under carbonation using iron oxide red. *Constr. Build. Mater.* **2019**, *222*, 601–609. [CrossRef]
44. Duan, Z.H.; Poon, C.S. Properties of recycled aggregate concrete made with recycled aggregates with different amounts of old adhered mortars. *Mater. Des.* **2014**, *58*, 19–29. [CrossRef]
45. Thomas, C.; Setién, J.; Polanco, J.A.; Alaejos, P.; Sánchez De Juan, M. Durability of recycled aggregate concrete. *Constr. Build. Mater.* **2013**, *40*, 1054–1065. [CrossRef]
46. Shi, M.; Ling, T.C.; Gan, B.; Guo, M.Z. Turning concrete waste powder into carbonated artificial aggregates. *Constr. Build. Mater.* **2019**, *199*, 178–184. [CrossRef]
47. Etxeberria, M.; Vázquez, E.; Marí, A.; Barra, M. Influence of amount of recycled coarse aggregates and production process on properties of recycled aggregate concrete. *Cem. Concr. Res.* **2007**, *37*, 735–742. [CrossRef]
48. De Juan, M.S.; Gutiérrez, P.A. Study on the influence of attached mortar content on the properties of recycled concrete aggregate. *Constr. Build. Mater.* **2009**, *23*, 872–877. [CrossRef]
49. Suryawanshi, S.R.; Singh, B.; Bhargava, P. Characterization of recycled aggregate concrete. In *Advances in Structural Engineering*; Springer: Berlin/Heidelberg, Germany, 2015; pp. 1813–1822.
50. Abbas, A.; Fathifazl, G.; Burkan Isgor, O.; Razaqpur, A.G.; Fournier, B.; Foo, S. Proposed method for determining the residual mortar content of recycled concrete aggregates. *J. ASTM Int.* **2008**, *5*, 1–12. [CrossRef]
51. Arezoumandi, M.; Smith, A.; Volz, J.S.; Khayat, K.H. An experimental study on flexural strength of reinforced concrete beams with 100% recycled concrete aggregate. *Eng. Struct.* **2015**, *88*, 154–162. [CrossRef]
52. Mi, R.; Liew, K.M.; Pan, G.; Kuang, T. Carbonation resistance study and inhomogeneity evolution of recycled aggregate concretes under loading effects. *Cem. Concr. Compos.* **2021**, *118*, 103916. [CrossRef]
53. Sidorova, A.; Vazquez-Ramonich, E.; Barra-Bizinotto, M.; Roa-Rovira, J.J.; Jimenez-Pique, E. Study of the recycled aggregates nature's influence on the aggregate-cement paste interface and ITZ. *Constr. Build. Mater.* **2014**, *68*, 677–684. [CrossRef]
54. Afroughsabet, V.; Biolzi, L.; Ozbakkaloglu, T. Influence of double hooked-end steel fibers and slag on mechanical and durability properties of high performance recycled aggregate concrete. *Compos. Struct.* **2017**, *181*, 273–284. [CrossRef]
55. Butler, L.J.; West, J.S.; Tighe, S.L. Towards the classification of recycled concrete aggregates: Influence of fundamental aggregate properties on recycled concrete performance. *J. Sustain. Cem. Mater.* **2014**, *3*, 140–163. [CrossRef]
56. Nuaklong, P.; Sata, V.; Chindaprasirt, P. Influence of recycled aggregate on fly ash geopolymer concrete properties. *J. Clean. Prod.* **2016**, *112*, 2300–2307. [CrossRef]
57. Gokce, A.; Nagataki, S.; Saeki, T.; Hisada, M. Freezing and thawing resistance of air-entrained concrete incorporating recycled coarse aggregate: The role of air content in demolished concrete. *Cem. Concr. Res.* **2004**, *34*, 799–806. [CrossRef]
58. Kim, N.; Kim, J.; Yang, S. Mechanical strength properties of RCA concrete made by a modified EMV method. *Sustainability* **2016**, *8*, 924. [CrossRef]
59. Yang, S.; Lee, H. Mechanical properties of recycled aggregate concrete proportioned with modified equivalent mortar volume method for paving applications. *Constr. Build. Mater.* **2017**, *136*, 9–17. [CrossRef]
60. Fan, Y.; Xiao, J.; Tam, V.W.Y. Effect of old attached mortar on the creep of recycled aggregate concrete. *Struct. Concr.* **2014**, *15*, 169–178. [CrossRef]

61. Padmini, A.K.; Ramamurthy, K.; Mathews, M.S. Influence of parent concrete on the properties of recycled aggregate concrete. *Constr. Build. Mater.* **2009**, *23*, 829–836. [CrossRef]
62. Dimitriou, G.; Savva, P.; Petrou, M.F. Enhancing mechanical and durability properties of recycled aggregate concrete. *Constr. Build. Mater.* **2018**, *158*, 228–235. [CrossRef]
63. Rajhans, P.; Chand, G.; Kisku, N.; Panda, S.K.; Nayak, S. Proposed mix design method for producing sustainable self compacting heat cured recycled aggregate concrete and its microstructural investigation. *Constr. Build. Mater.* **2019**, *218*, 568–581. [CrossRef]
64. Pandurangan, K.; Dayanithy, A.; Om Prakash, S. Influence of treatment methods on the bond strength of recycled aggregate concrete. *Constr. Build. Mater.* **2016**, *120*, 212–221. [CrossRef]
65. Domingo-Cabo, A.; Lázaro, C.; López-Gayarre, F.; Serrano-López, M.A.; Serna, P.; Castaño-Tabares, J.O. Creep and shrinkage of recycled aggregate concrete. *Constr. Build. Mater.* **2009**, *23*, 2545–2553. [CrossRef]
66. Sasanipour, H.; Aslani, F. Durability properties evaluation of self-compacting concrete prepared with waste fine and coarse recycled concrete aggregates. *Constr. Build. Mater.* **2020**, *236*, 117540. [CrossRef]
67. Fan, Y.; Niu, H.; Zhang, X. Impact of the properties of old mortar on creep prediction model of recycled aggregate concrete. *Constr. Build. Mater.* **2020**, *239*, 117772. [CrossRef]
68. Abbas, A.; Fathifazl, G.; Isgor, O.B.; Razaqpur, A.G.; Fournier, B.; Foo, S. Durability of recycled aggregate concrete designed with equivalent mortar volume method. *Cem. Concr. Compos.* **2009**, *31*, 555–563. [CrossRef]
69. Anike, E.E.; Saidani, M.; Ganjian, E.; Tyrer, M.; Olubanwo, A.O. Evaluation of conventional and equivalent mortar volume mix design methods for recycled aggregate concrete. *Mater. Struct.* **2020**, *53*, 1–15. [CrossRef]
70. McGinnis, M.J.; Davis, M.; de la Rosa, A.; Weldon, B.D.; Kurama, Y.C. Strength and stiffness of concrete with recycled concrete aggregates. *Constr. Build. Mater.* **2017**, *154*, 258–269. [CrossRef]
71. Gholampour, A.; Ozbakkaloglu, T. Time-dependent and long-term mechanical properties of concretes incorporating different grades of coarse recycled concrete aggregates. *Eng. Struct.* **2018**, *157*, 224–234. [CrossRef]
72. Kim, Y.; Hanif, A.; Usman, M.; Park, W. Influence of bonded mortar of recycled concrete aggregates on interfacial characteristics—Porosity assessment based on pore segmentation from backscattered electron image analysis. *Constr. Build. Mater.* **2019**, *212*, 149–163. [CrossRef]
73. KATS. *KS L5201 Portland Cement*; Korean Agency Technology Standards: Eumseong-gun, Korea, 2016.
74. ASTM. *ASTM C150/C150M-20 Standard Specification for Portland Cement*; ASTM: West Conshohocken, PA, USA, 2020.
75. Limbachiya, M.C.; Marrocchino, E.; Koulouris, A. Chemical-mineralogical characterisation of coarse recycled concrete aggregate. *Waste Manag.* **2007**, *27*, 201–208. [CrossRef]
76. Bui, N.K.; Satomi, T.; Takahashi, H. Recycling woven plastic sack waste and PET bottle waste as fiber in recycled aggregate concrete: An experimental study. *Waste Manag.* **2018**, *78*, 79–93. [CrossRef]
77. Kim, J.; Kim, N.; Yang, S. The effect of the residual mortar of recycled concrete aggregate on alkali silica reaction. *Int. J. Highw. Eng.* **2015**, *17*, 19–24. [CrossRef]
78. Medina, C.; Zhu, W.; Howind, T.; Frías, M.; de Sánchez Rojas, M.I. Effect of the constituents (asphalt, clay materials, floating particles and fines) of construction and demolition waste on the properties of recycled concretes. *Constr. Build. Mater.* **2015**, *79*, 22–33. [CrossRef]
79. Sánchez-Cotte, E.H.; Pacheco-Bustos, C.A.; Fonseca, A.; Triana, Y.P.; Mercado, R.; Yepes-Martínez, J.; Lagares Espinoza, R.G. The chemical-mineralogical characterization of recycled concrete aggregates from different sources and their potential reactions in asphalt mixtures. *Materials* **2020**, *13*, 5592. [CrossRef] [PubMed]
80. Saravanakumar, P.; Abhiram, K.; Manoj, B. Properties of treated recycled aggregates and its influence on concrete strength characteristics. *Constr. Build. Mater.* **2016**, *111*, 611–617. [CrossRef]
81. Martínez, I.; Etxeberria, M.; Pavón, E.; Díaz, N. Influence of demolition waste fine particles on the properties of recycled aggregate masonry mortar. *Int. J. Civ. Eng.* **2018**, *16*, 1213–1226. [CrossRef]
82. Moreno-Pérez, E.; Hernández-Ávila, J.; Rangel-Martínez, Y.; Cerecedo-Sáenz, E.; Arenas-Flores, A.; Reyes-Valderrama, M.; Salinas-Rodríguez, E. Chemical and mineralogical characterization of recycled aggregates from construction and demolition waste from Mexico City. *Minerals* **2018**, *8*, 237. [CrossRef]
83. Krour, H.; Trauchessec, R.; Lecomte, A.; Diliberto, C.; Barnes-Davin, L.; Bolze, B.; Delhay, A. Incorporation rate of recycled aggregates in cement raw meals. *Constr. Build. Mater.* **2020**, *248*, 118217. [CrossRef]
84. Park, S.S.; Kim, S.J.; Chen, K.Q.; Lee, Y.J.; Lee, S.B. Crushing characteristics of a recycled aggregate from waste concrete. *Constr. Build. Mater.* **2018**, *160*, 100–105. [CrossRef]
85. Rajan, B.; Singh, D. Investigation on effects of different crushing stages on morphology of coarse and fine aggregates. *Int. J. Pavement Eng.* **2020**, *21*, 177–195. [CrossRef]
86. Matias, D.; de Brito, J.; Rosa, A.; Pedro, D. Mechanical properties of concrete produced with recycled coarse aggregates—Influence of the use of superplasticizers. *Constr. Build. Mater.* **2013**, *44*, 101–109. [CrossRef]
87. Öztürk, A.U.; Erdem, R.T.; Kozanoglu, C. Investigation of crushing type of concrete aggregates on mechanical properties of concrete. *Int. J. Mater. Eng.* **2012**, *2*, 6–9. [CrossRef]
88. Cepuritis, R.; Garboczi, E.J.; Jacobsen, S. Three dimensional shape analysis of concrete aggregate fines produced by VSI crushing. *Powder Technol.* **2017**, *308*, 410–421. [CrossRef]

89. Ulsen, C.; Tseng, E.; Angulo, S.C.; Landmann, M.; Contessotto, R.; Balbo, J.T.; Kahn, H. Concrete aggregates properties crushed by jaw and impact secondary crushing. *J. Mater. Res. Technol.* **2019**, *8*, 494–502. [CrossRef]
90. Kamani, M.; Ajalloeian, R. The effect of rock crusher and rock type on the aggregate shape. *Constr. Build. Mater.* **2020**, *230*, 117016. [CrossRef]
91. Liu, Y.; Sun, W.; Nair, H.; Lane, D.S.; Wang, L. Quantification of aggregate morphologic characteristics with the correlation to uncompacted void content of coarse aggregates in Virginia. *Constr. Build. Mater.* **2016**, *124*, 645–655. [CrossRef]
92. Nagataki, S.; Gokce, A.; Saeki, T.; Hisada, M. Assessment of recycling process induced damage sensitivity of recycled concrete aggregates. *Cem. Concr. Res.* **2004**, *34*, 965–971. [CrossRef]
93. Pedro, D.; de Brito, J.; Evangelista, L. Performance of concrete made with aggregates recycled from precasting industry waste: Influence of the crushing process. *Mater. Struct.* **2015**, *48*, 3965–3978. [CrossRef]
94. KORAS. *Casebook of Recycled Aggregate and Recycled Aggregate Product*; KORAS: Seoul, Korea, 2020. Available online: https://www.me.go.kr/home/file/readDownloadFile.do?fileId=158651&fileSeq=1 (accessed on 2 March 2021).
95. Tam, V.W.Y.; Tam, C.M. Crushed aggregate production from centralized combined and individual waste sources in Hong Kong. *Constr. Build. Mater.* **2007**, *21*, 879–886. [CrossRef]
96. Hiete, M. Waste management plants and technology for recycling construction and demolition (C&D) waste: State-of-the-art and future challenges. In *Handbook of Recycled Concrete and Demolition Waste*; Elsevier Inc.: Amsterdam, The Netherlands, 2013; pp. 53–75, ISBN 9780857096906.
97. Tam, V.W.Y.; Tam, C.M.; Le, K.N. Removal of cement mortar remains from recycled aggregate using pre-soaking approaches. *Resour. Conserv. Recycl.* **2007**, *50*, 82–101. [CrossRef]
98. Kim, Y.; Hanif, A.; Kazmi, S.M.S.; Munir, M.J.; Park, C. Properties enhancement of recycled aggregate concrete through pretreatment of coarse aggregates—Comparative assessment of assorted techniques. *J. Clean. Prod.* **2018**, *191*, 339–349. [CrossRef]
99. Akbarnezhad, A.; Ong, K.C.G.; Zhang, M.H.; Tam, C.T.; Foo, T.W.J. Microwave-assisted beneficiation of recycled concrete aggregates. *Constr. Build. Mater.* **2011**, *25*, 3469–3479. [CrossRef]
100. Linß, E.; Mueller, A. High-performance sonic impulses—An alternative method for processing of concrete. *Proc. Int. J. Miner. Process.* **2004**, *74*, S199–S208. [CrossRef]
101. Tam, V.W.Y.; Butera, A.; Le, K.N.; Li, W. Utilising CO_2 technologies for recycled aggregate concrete: A critical review. *Constr. Build. Mater.* **2020**, *250*, 118903. [CrossRef]
102. Grabiec, A.M.; Klama, J.; Zawal, D.; Krupa, D. Modification of recycled concrete aggregate by calcium carbonate biodeposition. *Constr. Build. Mater.* **2012**, *34*, 145–150. [CrossRef]
103. Quattrone, M.; Angulo, S.C.; John, V.M. Energy and CO_2 from high performance recycled aggregate production. *Resour. Conserv. Recycl.* **2014**, *90*, 21–33. [CrossRef]
104. Ahmad Bhat, J. Effect of strength of parent concrete on the mechanical properties of recycled aggregate concrete. *Mater. Today Proc.* **2021**, *42*, 1462–1469. [CrossRef]
105. Tabsh, S.W.; Abdelfatah, A.S. Influence of recycled concrete aggregates on strength properties of concrete. *Constr. Build. Mater.* **2009**, *23*, 1163–1167. [CrossRef]
106. Xiao, J.; Li, J.; Zhang, C. Mechanical properties of recycled aggregate concrete under uniaxial loading. *Cem. Concr. Res.* **2005**, *35*, 1187–1194. [CrossRef]
107. Vieira, G.L.; Schiavon, J.Z.; Borges, P.M.; da Silva, S.R.; de Oliveira Andrade, J.J. Influence of recycled aggregate replacement and fly ash content in performance of pervious concrete mixtures. *J. Clean. Prod.* **2020**, *271*, 122665. [CrossRef]
108. Kou, S.C.; Poon, C.S. Effect of the quality of parent concrete on the properties of high performance recycled aggregate concrete. *Constr. Build. Mater.* **2015**, *77*, 501–508. [CrossRef]
109. Wang, Q.; Geng, Y.; Wang, Y.; Zhang, H. Drying shrinkage model for recycled aggregate concrete accounting for the influence of parent concrete. *Eng. Struct.* **2020**, *202*, 109888. [CrossRef]
110. Geng, Y.; Wang, Q.; Wang, Y.; Zhang, H. Influence of service time of recycled coarse aggregate on the mechanical properties of recycled aggregate concrete. *Mater. Struct.* **2019**, *52*, 97. [CrossRef]
111. Abed, M.; Nemes, R.; Tayeh, B.A. Properties of self-compacting high-strength concrete containing multiple use of recycled aggregate. *J. King Saud Univ. Eng. Sci.* **2020**, *32*, 108–114. [CrossRef]
112. Thomas, C.; de Brito, J.; Cimentada, A.; Sainz-Aja, J.A. Macro- and micro- properties of multi-recycled aggregate concrete. *J. Clean. Prod.* **2020**, *245*, 118843. [CrossRef]
113. Mi, R.; Pan, G.; Kuang, T. Reducing the carbonation zone and steel corrosion zone widths of recycled aggregate concrete by optimizing its mixing process. *J. Mater. Civ. Eng.* **2021**, *33*, 04021061. [CrossRef]
114. Mi, R.; Pan, G.; Li, Y.; Kuang, T. Carbonation degree evaluation of recycled aggregate concrete using carbonation zone widths. *J. CO_2 Util.* **2021**, *43*, 101366. [CrossRef]
115. Possan, E.; Felix, E.F.; Thomaz, W.A. CO_2 uptake by carbonation of concrete during life cycle of building structures. *J. Build. Pathol. Rehabil.* **2016**, *1*, 1–9. [CrossRef]
116. Hansen, T.C.; Hedegkd, S.E. Properties of recycled aggregate concretes as affected by admixtures in original concretes. *Proc. J. Proc.* **1984**, *81*, 21–26.
117. PKN. *PN-EN 12620:2013 Aggregates for Concrete*; Polish Committee for Standardization: Warsaw, Poland, 2013.

118. PKN. *PN-EN—206:2014 Concrete—Requirements, Properties and Conformity*; Polish Committee for Standardization: Warsaw, Poland, 2014.
119. CENELEC. *FprEN 12620:2017 Aggregates for Concrete*; CENELEC: Brussels, Belgium, 2017.
120. CENELEC. *FprEN 13139:2017 Aggregates for Mortar*; CENELEC: Brussels, Belgium, 2017.
121. CENELEC. *FprEN 13242:2017 Aggregates for Unbound and Hydraulically Bound Materials for Use in Civil Engineering Work and Road Construction*; CENELEC: Brussels, Belgium, 2017.
122. JSA. *JIS A 5023 Recycled Aggregate for Concrete-Class L*; Japanese Standards Association: Tokyo, Japan, 2018.
123. JSA. *JIS A 5022 Recycled Aggregate for Concrete-Class M*; Japanese Standards Association: Tokyo, Japan, 2018.
124. JSA. *JIS A 5021 Recycled Aggregate for Concrete-Class H*; Japanese Standards Association: Tokyo, Japan, 2018.
125. CNS. *GB/T 25177-2010 Standard for Recycled Coarse Aggregate for Concrete*; China National Standard: Beijing, China, 2010.
126. WBTC. *WBTC No.12/2002 Specifications Facilitating the Use of Recycled Aggregates*; Works Bureau Technical Circular: Hong Kong, China, 2004.
127. KATS. *KS F2527 Aggregates for Concrete*; Korean Agency for Technology Standards: Eumseong-gun, Korea, 2020.
128. Rilem, T.C. 121-DRG. Specifications for concrete with recycled aggregates. *Mater. Struct.* **1994**, *27*, 557–559.
129. ABNT. *NBR 15116 Recycled Aggregates from Construction and Demolition Waste: Non-Structural Concrete—Requirements*; Assoiacao Brasileira de Normas Tecnicas: São Paulo, Brazil, 2005.
130. LNEC. *LNEC E 471 Guideline for the Use of Recycled Coarse Aggregates in Hydraulic Binders' Concrete*; LNEC: Lisboa, Portugal, 2006.
131. CRISO. *HB-155, 2002 Guide to the Use of Recycled Concrete and Masonry Materials*; Commonwealth Scientific and Industrial Research Organisation: Canberra, Austraila, 2002.
132. Spanish Ministry of Public Works. *EHE-08 Instrucción Del Hormigón Estructural EHE-08 (Spanish Structural Concrete Code)*; Spanish Ministry of Public Works: Madrid, Spain, 2008.
133. Gonçalves, P.; de Brito, J. Recycled aggregate concrete (RAC)–comparative analysis of existing specifications. *Mag. Concr. Res.* **2010**, *62*, 339–346. [CrossRef]
134. Pani, L.; Francesconi, L.; Rombi, J.; Mistretta, F.; Sassu, M.; Stochino, F. Effect of parent concrete on the performance of recycled aggregate concrete. *Sustainability* **2020**, *12*, 9399. [CrossRef]
135. Katz, A. Treatments for the improvement of recycled aggregate. *J. Mater. Civ. Eng.* **2004**, *16*, 597–603. [CrossRef]
136. Kim, J. Properties of recycled aggregate concrete designed with equivalent mortar volume mix design. *Constr. Build. Mater.* **2021**, *301*, 124091. [CrossRef]
137. Bravo-German, A.M.; Bravo-Gómez, I.D.; Mesa, J.A.; Maury-Ramírez, A. Mechanical properties of concrete using recycled aggregates obtained from old paving stones. *Sustainability* **2021**, *13*, 44. [CrossRef]
138. Hossain, M.U.; Poon, C.S.; Lo, I.M.C.; Cheng, J.C.P. Comparative environmental evaluation of aggregate production from recycled waste materials and virgin sources by LCA. *Resour. Conserv. Recycl.* **2016**, *109*, 67–77. [CrossRef]
139. Ding, T.; Xiao, J.; Tam, V.W.Y. A closed-loop life cycle assessment of recycled aggregate concrete utilization in China. *Waste Manag.* **2016**, *56*, 367–375. [CrossRef]
140. Zhang, C.; Hu, M.; Dong, L.; Xiang, P.; Zhang, Q.; Wu, J.; Li, B.; Shi, S. Co-benefits of urban concrete recycling on the mitigation of greenhouse gas emissions and land use change: A case in Chongqing metropolis, China. *J. Clean. Prod.* **2018**, *201*, 481–498. [CrossRef]
141. Estanqueiro, B.; Dinis Silvestre, J.; de Brito, J.; Duarte Pinheiro, M. Environmental life cycle assessment of coarse natural and recycled aggregates for concrete. *Eur. J. Environ. Civ. Eng.* **2018**, *22*, 429–449. [CrossRef]
142. Mesa, J.A.; Fúquene-Retamoso, C.; Maury-Ramírez, A. Life cycle assessment on construction and demolition waste: A systematic literature review. *Sustainability* **2021**, *13*, 7676. [CrossRef]
143. Böhmer, S.; Moser, G.; Neubauer, C.; Peltoniemi, M.; Schachermayer, E.; Tesar, M.; Walter, B.; Winter, B. *Aggregates Case Study*; Contract No 150787-2007 F1SC-AT; ISO: Geneva, Switzerland, 2008.
144. Xiao, J.; Ma, Z.; Sui, T.; Akbarnezhad, A.; Duan, Z. Mechanical properties of concrete mixed with recycled powder produced from construction and demolition waste. *J. Clean. Prod.* **2018**, *188*, 720–731. [CrossRef]
145. Ahmari, S.; Ren, X.; Toufigh, V.; Zhang, L. Production of geopolymeric binder from blended waste concrete powder and fly ash. *Constr. Build. Mater.* **2012**, *35*, 718–729. [CrossRef]
146. Earle, J.; Ergun, D.; Gorgolewski, M. *Barriers for Deconstruction and Reuse/Recycling of Construction Materials in Canada*; CIB: Ottawa, ON, Canada, 2014; Volume 20.
147. Tam, V.W.Y. Economic comparison of concrete recycling: A case study approach. *Resour. Conserv. Recycl.* **2008**, *52*, 821–828. [CrossRef]
148. MoE. *Construction Waste Recycling Promotion Act, No. 16317*; Ministry of Environment: Seoul, Korea, 2019.
149. MOLIT. *Regulations on the Mandatory Use of Recycled Aggregates for Construction Projects*; Ministry of Land, Infrastructure and Transport: Seoul, Korea, 2009.
150. Manowong, E. Investigating factors influencing construction waste management efforts in developing countries: An experience from Thailand. *Waste Manag. Res.* **2012**, *30*, 56–71. [CrossRef] [PubMed]
151. Wu, Z.; Yu, A.T.W.; Shen, L. Investigating the determinants of contractor's construction and demolition waste management behavior in Mainland China. *Waste Manag.* **2017**, *60*, 290–300. [CrossRef]

152. Nagapan, S.; Rahman, I.A.; Asmi, A.; Memon, A.H.; Latif, I. Issues on construction waste: The need for sustainable waste management. In Proceedings of the 2012 IEEE Colloquium on Humanities, Science and Engineering (CHUSER), Kota Kinabalu, Malaysia, 3–4 December 2012.
153. Lu, W. Big data analytics to identify illegal construction waste dumping: A Hong Kong study. *Resour. Conserv. Recycl.* **2019**, *141*, 264–272. [CrossRef]
154. Dladla, I.; Machete, F.; Shale, K. A review of factors associated with indiscriminate dumping of waste in eleven African countries. *Afr. J. Sci. Technol. Innov. Dev.* **2016**, *8*, 475–481. [CrossRef]
155. Martínez-Lage, I.; Vázquez-Burgo, P.; Velay-Lizancos, M. Sustainability evaluation of concretes with mixed recycled aggregate based on holistic approach: Technical, economic and environmental analysis. *Waste Manag.* **2020**, *104*, 9–19. [CrossRef]
156. Marinković, S.; Radonjanin, V.; Malešev, M.; Ignjatović, I. Comparative environmental assessment of natural and recycled aggregate concrete. *Waste Manag.* **2010**, *30*, 2255–2264. [CrossRef] [PubMed]
157. Faleschini, F.; Zanini, M.A.; Pellegrino, C.; Pasinato, S. Sustainable management and supply of natural and recycled aggregates in a medium-size integrated plant. *Waste Manag.* **2016**, *49*, 146–155. [CrossRef]
158. Nunes, K.R.A.; Mahler, C.F.; Valle, R.; Neves, C. Evaluation of investments in recycling centres for construction and demolition wastes in Brazilian municipalities. *Waste Manag.* **2007**, *27*, 1531–1540. [CrossRef]
159. Jin, R.; Li, B.; Zhou, T.; Wanatowski, D.; Piroozfar, P. An empirical study of perceptions towards construction and demolition waste recycling and reuse in China. *Resour. Conserv. Recycl.* **2017**, *126*, 86–98. [CrossRef]
160. Shooshtarian, S.; Caldera, S.; Maqsood, T.; Ryley, T. Using Recycled Construction and demolition waste products: A review of stakeholders' perceptions, decisions, and motivations. *Recycling* **2020**, *5*, 31. [CrossRef]
161. Rao, A.; Jha, K.N.; Misra, S. Use of aggregates from recycled construction and demolition waste in concrete. *Resour. Conserv. Recycl.* **2007**, *50*, 71–81. [CrossRef]
162. Min, B.; Park, J.-W.; Choi, W.-K.; Lee, K.-W. Separation of radionuclide from dismantled concrete waste. *J. Nucl. Fuel Cycle Waste Technol.* **2009**, *7*, 79–86.
163. Taylor-Lange, S.C.; Stewart, J.G.; Juenger, M.C.G.; Siegel, J.A. The contribution of fly ash toward indoor radon pollution from concrete. *Build. Environ.* **2012**, *56*, 276–282. [CrossRef]
164. De Jong, P.; Van Dijk, W.; Van Hulst, J.G.A.; Van Heijningen, R.J.J. The effect of the composition and production process of concrete on the 222Rn Exhalation rate. In Proceedings of the 1995 6th International Symposium on the Natural Radiation Environment, Montreal, QC, Canada, 5–9 June 1995; Elsevier Science Ltd.: Amsterdam, The Netherlands, 1997; Volume 22, pp. 287–293.
165. Lee, Y.J.; Hwang, D.S.; Lee, K.W.; Jeong, G.H.; Moon, J.K. Characterization of cement waste form for final disposal of decommissioned concrete waste. *J. Nucl. Fuel Cycle Waste Technol.* **2013**, *11*, 271–280. [CrossRef]

MDPI
St. Alban-Anlage 66
4052 Basel
Switzerland
Tel. +41 61 683 77 34
Fax +41 61 302 89 18
www.mdpi.com

Sustainability Editorial Office
E-mail: sustainability@mdpi.com
www.mdpi.com/journal/sustainability

www.ingramcontent.com/pod-product-compliance
Lightning Source LLC
LaVergne TN
LVHW071001170726
843515LV00006B/1050

* 9 7 8 3 0 3 6 5 5 0 5 0 3 *